Current Advances on the Assessment and Mitigation of Fire Risk in Buildings and Urban Areas — 1st Edition

Current Advances on the Assessment and Mitigation of Fire Risk in Buildings and Urban Areas — 1st Edition

Editor

Tiago Miguel Ferreira

MDPI

Basel • Beijing • Wuhan • Barcelona • Belgrade • Novi Sad • Cluj • Manchester

Editor
Tiago Miguel Ferreira
College of Arts, Technology
and Environment (CATE)
University of the West of
England (UWE Bristol)
Bristol
United Kingdom

Editorial Office
MDPI
St. Alban-Anlage 66
4052 Basel, Switzerland

This is a reprint of articles from the Special Issue published online in the open access journal *Fire* (ISSN 2571-6255) (available at: www.mdpi.com/journal/fire/special_issues/2I58372153).

For citation purposes, cite each article independently as indicated on the article page online and as indicated below:

Lastname, A.A.; Lastname, B.B. Article Title. *Journal Name* **Year**, *Volume Number*, Page Range.

ISBN 978-3-7258-0314-9 (Hbk)
ISBN 978-3-7258-0313-2 (PDF)
doi.org/10.3390/books978-3-7258-0313-2

Cover image courtesy of Tiago Miguel Ferreira

Contents

About the Editor vii

Preface ix

Ruben Dobler Strand and Torgrim Log
A Cold Climate Wooden Home and Conflagration Danger Index: Justification and Practicability for Norwegian Conditions
Reprinted from: *Fire* **2023**, *6*, 377, doi:10.3390/fire6100377 1

Jie Wang, Meilin Yang, Tianming Li, Xuepeng Jiang and Kaihua Lu
Research and Application of Improved Multiple Imputation Based on R Language in Fire Prediction
Reprinted from: *Fire* **2023**, *6*, 235, doi:10.3390/fire6060235 27

Sepideh Noori, Alireza Mohammadi, Tiago Miguel Ferreira, Ata Ghaffari Gilandeh and Seyed Jamal Mirahmadzadeh Ardabili
Modelling and Mapping Urban Vulnerability Index against Potential Structural Fire-Related Risks: An Integrated GIS-MCDM Approach
Reprinted from: *Fire* **2023**, *6*, 107, doi:10.3390/fire6030107 40

Md Kamrul Hassan, Md Rayhan Hasnat, Kai Png Loh, Md Delwar Hossain, Payam Rahnamayiezekavat and Grahame Douglas et al.
Effect of Interlayer Materials on Fire Performance of Laminated Glass Used in High-Rise Building: Cone Calorimeter Testing
Reprinted from: *Fire* **2023**, *6*, 84, doi:10.3390/fire6030084 58

Yang Yang, Hongbo Du and Gang Yao
A Scientometric Research on Applications and Advances of Fire Safety Evacuation in Buildings
Reprinted from: *Fire* **2023**, *6*, 83, doi:10.3390/fire6030083 80

Musab Omar, Abdelgadir Mahmoud and Sa'ardin Bin Abdul Aziz
Critical Factors Affecting Fire Safety in High-Rise Buildings in the Emirate of Sharjah, UAE
Reprinted from: *Fire* **2023**, *6*, 68, doi:10.3390/fire6020068 102

Razieh Khaksari Haddad and Zambri Harun
Development of a Novel Quantitative Risk Assessment Tool for UK Road Tunnels
Reprinted from: *Fire* **2023**, *6*, 65, doi:10.3390/fire6020065 122

Samson Tan, Khalid Moinuddin and Paul Joseph
The Ignition Frequency of Structural Fires in Australia from 2012 to 2019
Reprinted from: *Fire* **2023**, *6*, 35, doi:10.3390/fire6010035 140

André Samora-Arvela, José Aranha, Fernando Correia, Diogo M. Pinto, Cláudia Magalhães and Fantina Tedim
Understanding Building Resistance to Wildfires: A Multi-Factor Approach
Reprinted from: *Fire* **2023**, *6*, 32, doi:10.3390/fire6010032 158

Vigneshwaran Narayanan, Anene Oguaka and Richard Shaun Walls
Reduced Scale Experiments on Fire Spread Involving Multiple Informal Settlement Dwellings
Reprinted from: *Fire* **2022**, *5*, 199, doi:10.3390/fire5060199 173

Maria Mahamed (Polinova), Lea Wittenberg, Haim Kutiel and Anna Brook
Fire Risk Assessment on Wildland–Urban Interface and Adjoined Urban Areas: Estimation Vegetation Ignitability by Artificial Neural Network
Reprinted from: *Fire* **2022**, *5*, 184, doi:10.3390/fire5060184 . . . . **205**

Dener Silva, Hugo Rodrigues and Tiago Miguel Ferreira
Assessment and Mitigation of the Fire Vulnerability and Risk in the Historic City Centre of Aveiro, Portugal
Reprinted from: *Fire* **2022**, *5*, 173, doi:10.3390/fire5050173 . . . . **218**

Charlie Hopkin, Michael Spearpoint, Yusuf Muhammad and William Makant
Estimating the Suppression Performance of an Electronically Controlled Residential Water Mist System from BS 8458:2015 Fire Test Data
Reprinted from: *Fire* **2022**, *5*, 144, doi:10.3390/fire5050144 . . . . **232**

About the Editor

Tiago Miguel Ferreira

Tiago Miguel Ferreira is a Lecturer and a Deputy Programme Leader for the MSc in Civil Engineering at the University of the West of England (UWE Bristol), in the UK, and an Invited Associate Professor at the University of Coimbra, Portugal. Before joining UWE Bristol, he was a researcher at the Institute for Sustainability and Innovation in Structural Engineering (ISISE) at the University of Minho, Portugal, where he was also part of the teaching team for the Advanced Masters in Structural Analysis of Monuments and Historical Constructions. His academic focus revolves around the study of the structural vulnerability of historical buildings and urban areas in the face of natural and anthropogenic hazards, including fires, earthquakes, floods, and landslides. More recently, his focus has expanded to explore the interactions between these hazards and their impact on both physical and social vulnerability, encompassing both single and compound, as well as cascading hazards. Recognised as one of the top 2% more influential scientists globally by Elsevier BV and Stanford University in 2021, 2022, and 2023, his academic contributions extend to 200+ publications, including 80+ articles in leading international journals. Over the past decade, he has actively participated in 13 R&D projects, securing a total funding of £10.9+ million, and supervised and co-supervised dozens of MSc students and PhD students in various countries.

Preface

Fire safety within residential buildings and urban environments is a pressing global concern, demanding dynamic and comprehensive strategies for effective risk assessment and mitigation. The Special Issue that is being reprinted in this collection is the first of a series of Special Issues edited by the *Fire* journal to provide researchers with a platform to showcase cutting-edge developments in this critical area.

Authored by researchers and practitioners from Europe, Asia, and Australia, the thirteen contributions composing this first volume collectively offer a comprehensive perspective on fire safety, illustrating the global relevance and the imperative need for innovative risk assessment and mitigation strategies. This first volume sets the stage for subsequent Special Issues, showcasing cutting-edge developments and fostering collaborative efforts to address the evolving challenges in fire safety.

As the Editor responsible for this project, I would like to express my sincere gratitude to the authors for generously sharing their valuable knowledge and insights through their contributions and to the many peer reviewers whose rigorous evaluations significantly enhanced the quality of this Special Issue. Thank you all for your invaluable contributions.

Tiago Miguel Ferreira
Editor

fire

MDPI

Article

A Cold Climate Wooden Home and Conflagration Danger Index: Justification and Practicability for Norwegian Conditions

Ruben Dobler Strand * and Torgrim Log

Fire Disasters Research Group, Department of Safety, Chemistry and Biomedical Laboratory Sciences, Western Norway University of Applied Sciences, 5533 Haugesund, Norway; torgrim.log@hvl.no

* Correspondence: rds@hvl.no

Abstract: The vast majority of fire-related deaths occur in residential buildings. Until recently, the fire risk for these buildings was only considered through static risk assessments or period-based assessments applying to certain periods of the year, e.g., Christmas holidays. However, for homes with indoor wooden panelling, especially in the ceiling, a dynamic fire danger indicator can be predicted for cold climate regions. Recognising the effect of fuel moisture content (FMC) of indoor wooden panelling on the enclosure fire development allows for the prediction of a wooden home fire danger indicator. In the present study, dry wood fire dynamics are analysed and experimental observations are reported to support in-home wooden panel FMC as a suitable wooden home fire danger indicator. Then, from previous work, the main equation for modelling in-home FMC is considered and a generic enclosure for FMC modelling is justified based on literature data and supported through a sensitivity study for Norwegian wooden homes. Further, ten years of weather data for three selected locations in Norway, i.e., a coastal town, an inland fjord town and a mountain town, were analysed using a three-dimensional risk matrix to assess the usability of the fire risk modelling results. Finally, a cold climate wooden home fire danger index was introduced to demonstrate how the risk concept can be communicated in an intuitive way using similar gradings as the existing national forest fire index. Based on the generic enclosure, the findings support FMC as a fire risk indicator for homes with interior wooden panelling (walls and ceiling). Large differences in the number of days with arid in-home conditions were identified for the selected towns. The number of days with combined strong wind and dry wooden homes appears to depend more on the number of days with strong wind than days of in-home drought. Thus, the coastal town was more susceptible to conflagrations than the drier inland towns. This aligns well with the most significant fire disasters in Norway since 1900. In addition, it was demonstrated that the number of high-risk periods is manageable and can be addressed by local fire departments through proactive measures. In turn, the fire risk modelling and associated index respond well to the recent changes in Norwegian regulations, requiring the fire departments to have systems for detecting increased risk levels. Testing the modelling for a severe winter fire in the USA indicates that the presented approach may be of value elsewhere as well.

Keywords: risk modelling; risk analysis; wooden home fire risk indicator; cold-climate fires

Citation: Strand, R.D.; Log, T. A Cold Climate Wooden Home and Conflagration Danger Index: Justification and Practicability for Norwegian Conditions. *Fire* **2023**, *6*, 377. https://doi.org/10.3390/fire6100377

Academic Editor: Tiago Miguel Ferreira

Received: 30 August 2023
Revised: 24 September 2023
Accepted: 28 September 2023
Published: 30 September 2023

1. Introduction

1.1. Fire Risk in Cold Climates

Fire results in more than 300,000 deaths annually, making it one of the most significant causes of accidental injury and a worldwide threat [1]. In many areas of society and industry, attempts are made to dynamically model the imminent risk from fires or fire-related events, enabling situational awareness and sound risk management. Typically, dynamic fire risk modelling is applied to industries and areas where potential consequences of an unwanted event could involve major economic losses and potential fire phenomena receiving public attention, such as boiling liquid expanding vapour explosions (BLEVEs), vapour cloud explosions, jet fires, forest fires, etc. A common feature for industrial dynamic risk modelling is safety barriers and the associated condition monitoring, where system status or plant risk is considered through indicators based on the status of specific safety barriers [2]. Such an approach is only possible for systems where safe operation relies on safety systems. For systems such as forests or built environments, dynamic risk modelling cannot rely on safety barriers in the same way, and needs to be based on the status of the system itself. For forest fires, this is a well-known principle. When modelling vegetation fire risk, a generic approach is chosen for areas of similar vegetation, where typically soil and fuel moisture content (FMC) are correlated with the observed and predicted weather. This enables indications of vegetation fire risk based on how the modelled moisture content affects experienced flammability and fire spread rates. However, while statistics suggest that residential fires cause 83% of fire-related deaths [3], dynamic fire risk modelling depicting the imminent fire risk for the built environment has been lacking. The reasons for this relate to the extent of structural fires, usually affecting a single structure such as a single-family home, i.e., limited consequences and attention. More importantly, the lack of dynamic risk modelling for single-structure homes is thought to relate to the random nature of the problem, not recognizing the possibilities for a generic approach for certain types of homes, as used in forest fires.

In cold climate regions, it is well known that the fire frequency in homes increases during the winter months [4]. For Norway, from where examples are drawn in the present study, the fire frequency is highest during December and January [5]. However, a calendar-based approach to house fire risk may be insufficient to support proper risk and resource management. This approach may be too simple in an ever-changing environment where climate and cities continue to develop, posing new challenges for sufficient and safe operations and continuously stretching the limited resources of local fire departments. Hence, knowing days of high single-home fire risk could allow for proactive measures to be taken and for better risk management within a changing environment. According to the recently updated Norwegian fire safety regulations [6], the fire departments shall, from 2022, have systems to detect significant changes in the risk picture and develop plans to deal with these high-risk situations. If necessary, the fire department shall increase its emergency preparedness accordingly. Hence, tools for identifying high-risk events have become a new requirement from the Norwegian authorities as of 2022.

The cause of the increased winter fire frequency involves a combination of factors, e.g., (1) the extended use of candles during the dark season and holidays (a source of ignition), (2) the extended use of heat sources such as a fireplace or electrical ovens (ignition by flame or increased electrical loads, including chimney fires), (3) more time spent indoors, (4) alcohol consumption (Christmas holidays), and (5) dry indoor conditions favouring fire establishment and development. This latter climatic condition was identified in 1956, when the ambient dew point was suggested to explain the increased fire frequencies during the winter in cold climate regions in the USA [7]. The dry indoor conditions are especially evident in wooden homes with internal wooden surfaces, which are widespread and characteristic for wooden houses in Norway, as well as in other cold climate regions [8]. Generally, a wooden home refers to a structure with a load-bearing system and possibly external cladding made of wood. Nonetheless, in the current context, it is always assumed that wooden homes have some fraction of in-home wooden surfaces/panelling, such as a wooden ceiling and some or maybe all of the interior walls.

In recent years, increased attention from researchers towards cold climate fires has resulted in the rediscovery of the wooden home cold climate fire risk [9]. Severe conflagrations in wooden towns have been found to relate to very dry indoor and/or outdoor conditions in combination with strong wind. When indoor wooden surfaces in heated homes dry during the winter, they may highly affect fire development and the time to flashover (TTF) in the case of an enclosure fire event [10]. In turn, this leaves less time for safe escape and for the fire brigades to intervene, as well as more intense fire development post flashover and increased probability of involving new homes in the blaze.

It is evident from fire dynamics [9], statistics [5], and the literature [11–13] that wooden home fire risk (WHFR) in cold climate regions is not a static quantity. On the contrary, the WHFR is highly dynamic, changing both with the seasons and within the high-risk months of the year [11,12]. Just as heathland fire risk can become high during spring and forest fire risk during summer, the WHFR reaches a peak period during winter. It is generally known that indoor conditions become dry during the winter, as evident from gaps in parquet flooring, wooden doors sliding differently etc.; however, until recently, it was not known when this in-home drought occurs and to what extent.

Following the rediscovering of the cold climate fire risk, efforts have led to the DYNAMIC research project [14,15], which among other things emphasises the development of a WHFR and possible conflagration warning system. Herein, a model predicting indoor relative humidity (RH) and fuel moisture content (FMC) of indoor wooden surfaces was developed and validated [11] based on the findings and concepts developed in [4,16].

The present study emphasises WHFR based on indoor conditions and the associated wooden home conflagration risk. While the previously developed model for predicting indoor RH and FMC has undergone essential validation, it lacks a justified generic approach to the many Norwegian wooden homes. In addition, the WHFR indicator has not yet been justified, and the practicability of output from existing models has not been assessed. The present paper takes a step back from previous work to underpin the WHFR indicator in order to justify a generic approach and to show that explicit modelling of indoor wooden home fire risk is possible and identifies a manageable number of high fire risk events for the considered locations. In addition, the wrapping of the existing national forest fire index, in combination with the WHFR, makes an intuitive risk communication concept. The present study may serve as a guideline with respect to parameters and aspects that need consideration if attempts are made to implement the model in other cold climate regions.

1.2. Related Work

Despite little effort in dynamic fire risk modelling for homes, increased efforts have been undertaken in the field of building fire safety through fire risk quantification and assessment for structures. Rahikainen and Keski-Rahkonen [13] considered a statistical determination of ignition frequency of structural fires in Finland. As part of performance-based fire safety design, such frequencies are important variables in probabilistic quantitative fire risk assessments. The ignition frequency was considered as a function of floor area for different class structures. Periodic distribution of fire alarms was considered for the months, weeks, days, and the time of the day for 1994–1995. The results indicated that December, and particularly weeks 51 and 52, had increased alarm frequency. Another general observation was the occurrence of fire events in the afternoon, from 1500–2000, when people are at home.

Samson et al. [17] considered the ignition frequency of structural fires in Australia for different class structures, as in the previously mentioned study. The authors suggested a new coefficient for the generalised Barrois model based on statistics from Australia. A somewhat lower fire frequency was obtained than the similar study in Finland [18]. A possible explanation promoted by the authors is the widespread use of wood as a construction material in Finland, compared to wood and concrete in Australia, with the widespread use of gypsum (plasterboard) as interior surfaces. It should be mentioned that Finland and Norway are neighbouring countries in a cold climate region, both of which are known for a high number of wooden homes which internally become very dry in the wintertime due to the cold climate.

Many structural fires originate outdoors, either due to activities within the property or due to conflagration events such as wildland–urban interface (WUI) fires. A framework and implementation of a spatial incident-level fire risk model for wildfire to residential structures at the WUI was proposed by Abo et al. [19]. They argued that understanding the wildfire impact on structures in WUI areas is necessary for planning emergency and mitigating measures. Their impact assessment on structures included evaluating hazard, inventory, exposure, and impact. The implementation was presented through a case study. The framework requires knowledge on structural vulnerability, such as the wildfire vulnerability index for buildings proposed by Papathoma-Köhle et al. [20]. They considered building surroundings and characteristics, such as ground inclination, surrounding vegetation and ground covering, type of structure, roofing, and building material and shape when developing a physical vulnerability index (PVI) for buildings subjected to wildfires. These parameters were formed into an index using a random forest based automated feature selection to weight relevant indicators based on data from the documented Mati fire in Greece in 2018, which killed more than 100 people.

These studies have in common that they consider static single-structure fire risk through specific external building characteristics, as well as the ground and the vegetation around the structures. The present study emphasises dynamic modelling of in-home fuel conditions, which change with outdoor conditions.

The remainder of this paper is organised as follows. In Section 2, depth is added to the problem domain, before considering dry wood enclosure fire dynamics from the theoretical and experimental perspectives. Then, the model is briefly explained and key model parameters are qualitatively considered. Section 3 presents the sensitivity analysis of the key parameters to support the generic approach. Ten years of weather data are analysed for three selected locations in Norway in order to analyse whether the risk modelling provides a possible basis for evaluating proactive measures. The wooden home fire danger index is then presented by combining the wrapping of the existing national forest fire index with the WHFR. The findings are discussed in Section 4, while Section 5 sums up the main conclusions.

2. Materials and Methods

2.1. Wooden Homes

A characteristic of Norway is the widespread use of wood as external and partly internal panelling [8]. Nowadays, only Canada and the USA have an equal fraction of wooden homes compared to the remaining building mass [8]. While still widely used, internal wooden cladding is especially evident in older wooden houses. Norway has over 200 densely built wooden home heritage sites [21], which manifest a long tradition and several hundred years of constructional engineering development. By definition, these sites contain a minimum of 20 wooden single-structure homes with less than eight meters of separation. In addition, the buildings should primarily be constructed before the 1900s [22]. Norway does not have a tradition of high-rise buildings. Thus, low-rise single-structure wooden homes dominate how people live. Even today, wood is the primary construction material for low-rise homes [23]. While older wooden homes used wood for all internal surfaces, nowadays it is common to combine materials, and gypsum boards have become quite popular in recent years. Recently, the use of decorative wooden wall laths has gained popularity and reintroduced wood as a modern internal cladding material. In the coming years, it is not unlikely that wood as an interior building material will be increasingly used for new and refurbished homes as increased attention is brought toward its potential health effects as an indoor humidity buffer [24] and as a sustainable material. Figure 1 presents parts of a wooden home heritage site in Bergen, Norway (top), as well as a typical modern Norwegian neighbourhood in Haugesund (bottom). While both pictures present wooden homes, the old homes in Bergen are situated closer together and with can be expected to have widespread use of wood as interior panelling. The newer homes in Haugesund, while being wooden homes, are more likely to vary with respect to the indoor materials. In general, such wooden heritage sites are far more susceptible to conflagration events due to the short separation distance and use of wood as both an exterior and interior material. Additionally, many heritage sites and other wooden towns and villages are in areas without permanently manned fire stations, resulting in quite long deployment times.

Figure 1. Parts of a wooden home heritage site in Bergen, Norway (**top**, photo by Silje Marie Hatlestad, reproduced with permission) and a typical Norwegian neighbourhood in Haugesund (**bottom**).

2.2. Conflagrations

Town fire conflagrations involving several homes or other structures can originate from different scenarios contributing to fire spread. Every conflagration starts with an initiating event involving ignition, which can be separated into structure-related or vegetation-related initiating events, referring to the object or area where the initial fire originated. Structure-related events might be a house on fire, typically a fire starting within the structure, spreading to the building envelope, and then spreading further to new objects. A vegetation-related initiating event could be, e.g., a grass fire initiated by a grid failure causing sparks. Despite the initiating event, the fire may spread within a combination of vegetation and structures, i.e., a WUI fire, or only within one of the two, i.e., a wildfire or a town fire, as evident from many historical fires. Two critical factors associated with the formation of a conflagration event appear to be (1) dry fuel conditions and (2) strong wind [9,25]. The latter factor is related to wind strong enough to cause flame spread by ember transport (firebrands) or flame impingement on nearby structures or objects. This discontinuous fire spread mechanism has been a dominant fire spread mechanism for many town fire conflagrations, such as the wind-driven fires in London in 1666, Grand Forks in 1997, and the more recent Lærdal fire in 2014 [25,26].

A conflagration is imminent when the fire spreads over horizontal distances faster than firefighters can respond, quickly outgrowing the available resources. As stated in [25], "Conflagrations occur generally when strong winds drive a fire to overwhelm human suppression efforts." This concerns vegetation fires and rapid fire spread in, e.g., cured grass, and relates to the spread of resources over larger areas as well. Flame spread by firebrands from structure-to-structure (wooden home to wooden home) was in the Lærdal fire recorded at above 200 m distance [9]. During the more recent Sotra fire in Bergen 2021, a fire spread about 270 m across a fjord to the vegetation on the opposite side [27]. Such large leaps have been recorded for previous conflagrations as well, such as the crossing of the Chicago River during the 1871 Chicago Fire or the San Diego Freeway during the 1961 Bel-Air Fire [25]. Such rapid fire spread over large distances causes significant challenges for the fire brigade with respect to resources and resource redistribution. Typically, a conflagration event is of such an extent that it is not stopped or controlled by human interaction. It ends when the line of fuel ends or when the dominating mechanism for fire spread reduces in potential, e.g., reduced wind strengths or precipitation. Figure 2 illustrates potential scenarios following an initiating fire event, possibly towards a conflagration.

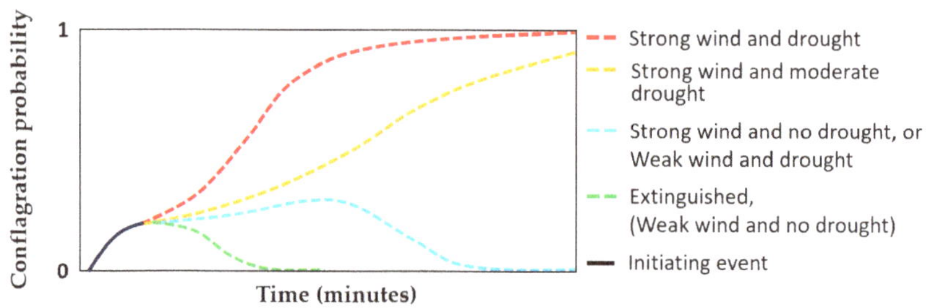

Figure 2. Illustration of possible developments following an initiating fire event.

The Lærdal fire in Norway in January 2014 is an example of a wooden town conflagration where the fire spread from structure to structure during strong wind. Over 60 buildings were damaged, of which at least 40 were lost [9]. The first home to catch fire was next to the fire station. When the part-time firefighters passed this home upon mustering at the station, 90–120 s after the emergency call, flashover had already occurred and the external wooden sidings were already burning [9]. Within an hour, two-story buildings burned to the ground. The fire risk for Lærdal at the time was reported as non-existent [28]. This reported risk was related to vegetation fire, however, and had nothing to do with the fire risk of the wooden homes in Lærdal at the time. However, in the aftermath of that fire it was found that the houses had very dry external and internal wooden panelling [9,26].

More recently, the Marshal fire in Boulder County, USA on 30 December 2021, was a wildland–urban interface (WUI) conflagration. The fire initially spread through vegetation and then secondarily from structure to structure. Estimated losses exceeded a thousand buildings, including homes and shopping centres. A moist spring and dry summer had resulted in abundant cured grass. This, combined with snow-free ground and wind gusts reaching 50 m/s on the day of the fire, resulted in large amounts of dry fuel, rapid fire spread into the town area, and loss of many wooden structures [29].

2.3. Dry Wood Fire Dynamics

It is necessary to consider the involved fire dynamics in order to understand the modelling approach for the indoor wooden home fire risk, which may be associated with the risk related to initiating a conflagration event. It is assumed that the reader is familiar with enclosure fires and how these fires may develop very differently depending on variables such as HRR, type of fuel, orientation of fuel with respect to ignition, location of fuel, ignition source, layout, volume and height of the compartment, and ventilation conditions, to mention a few. Nevertheless, the basics are provided here.

An enclosure fire typically develops through four stages [30]. The initial stage, involving ignition, lasts until the fire has been established. In this phase, average compartment temperatures are low and the heat transport is dominated by convection from hot gases and conduction within the solid materials involved. When the fire has been established, it enters the second phase, the growth phase. Heat radiation from the smoke layer becomes increasingly dominant within the growth phase, as it is proportional to the absolute temperature to the 4th power. As the fire grows in size and the temperature of the smoke layer trapped beneath the ceiling increases, a rapid, severe, and self-induced acceleration in fire development may occur, i.e., flashover. The flashover phenomenon marks the transition from the growth phase to the stage of a fully developed fire. Typically, at this stage the HRR within the enclosure becomes controlled by ventilation and all combustible surfaces become involved in the fire [31]. There are several criteria used for flashover, e.g., average smoke layer temperatures at about 550–600 °C, heat fluxes at floor level at about 20 $kW \cdot m^{-2}$, or visible flames escaping through compartment openings (indicating insufficient oxygen supply within compartment). Quite often all of these criteria are met within a short period. The last phase is the decay phase, where the fire intensity declines. This phase is not considered further in the present context.

The HRR is considered the most important parameter characterising fire behaviour [32]. It may be expressed as [30]

$$\dot{Q}_c = A_f \cdot \dot{m}_f'' \cdot \chi \cdot \Delta H_c \tag{1}$$

where A_f is the fuel surface area (m^2), $\dot{m}_f''$ is the mass flow per unit area from the fuel surface ($kg \cdot s^{-1} \cdot m^{-2}$), χ is the combustion efficiency, known to be less than unity, and ΔH_c is the heat of combustion ($J \cdot kg^{-1}$).

A heat balance for the production of volatiles for gas phase oxidation may be expressed as

$$\dot{Q}_F'' + \dot{Q}_E'' - \dot{Q}_L'' = \dot{m}_f'' \cdot L_V \tag{2}$$

where the net heat fluxes to the surface involve the heat flux from flames $\dot{Q}_F''$ ($W \cdot m^{-2}$), any external heat supply $\dot{Q}_E''$ ($W \cdot m^{-2}$), i.e., the hot smoke layer, and heat lost from water evaporation and other heat transport mechanisms is provided by $\dot{Q}_L''$ ($W \cdot m^{-2}$), and L_V ($J \cdot kg^{-1}$) is the latent heat of vaporisation and pyrolysis.

The following briefly describes an enclosure fire developing inside a living room where walls and ceiling are covered with untreated wooden panels. A potentially established fire will produce hot gases rising towards the wooden ceiling. The convective heat release is typically described by

$$\dot{Q}_{con} = \chi_{con} \cdot \dot{Q}_c \tag{3}$$

where χ_{con} is the convective fraction of the heat released by combustion. Most of the heat released interacts with the wooden ceiling, followed by the wooden walls post formation of the smoke layer. It is not unlikely that the walls are involved in a very early phase, as many fires originate close to the walls due to, e.g., electric apparatuses, socket short-circuits, candles, old fireplaces, or stove fires. The moisture content of the wooden panels interacting with the hot gases influences fire development. In principle, the wooden panels may either restrict fire development or contribute to a more rapid fire development (ignoring a possible neutral state). Restricting fire development is consistent with high FMC values. For such a scenario, the heat of combustion ΔH_c and the combustion efficiency χ presented in Equation (1) decrease as a function of increasing water content in the panels [9]. This reduces the HRR and provides less external heat flux $\dot{Q}_E''$ and preheating. Increased water levels increase the heat loss $\dot{Q}_L''$, through increased thermal conductivity as well as increasing the latent heat of vaporisation and pyrolysis L_V, as energy is consumed in the evaporation of water. Further, the evaporated water increases the heat capacity of the smoke layer and dilutes oxygen. Hence, increased FMC values reduce temperatures, and may result in prolonged time until flashover. Although this brief description emphasises wooden panels, it relates to other fuels with hygroscopic properties as well.

If the wooden panels contribute to fire development through low FMC, the wooden surfaces participate in the production of volatiles and thereby increase the net energy released within the compartment at an early stage. This would increase the smoke layer temperature and thereby the external heat flux $\dot{Q}_E$, causing increased preheating and potentially further accelerating the fire growth rate.

It should be mentioned that post-flashover combustion within the compartment is usually restricted by air access. Uncombusted hot gases exiting the ventilation openings burn on the outside of the vent openings. These external flames then represent a strong ignition source for neighbouring buildings, in addition to further accelerating fire spread on the initial structure. The drier the compartment wooden fuel is, the more combustion generally takes place outside the vent openings.

Many factors influence an enclosure fire, and any attempt to describe a possible fire scenario development would require a case-specific consideration. However, supporting the presented theory, it has been shown through experiments that for enclosure fires with indoor wooden panels the TTF is highly dependent on the FMC of the wall and ceiling panels [10]. This has been shown for different sizes of compartments and HRRs. The most recent experiments performed within the DYNAMIC research project were one half-scale ISO-room size experiments involving 12 mm wooden pine panel cladding. The one half-scale ISO-rooms were conditioned in a climate chamber three rooms at a time and burned successively. The initial HRR was kept constant, and the only parameter intentionally varied was the FMC. The FMC values ranged from 10% to approximately 18% by dry weight. The time to flashover for the wooden compartments ranged from less than 4 min to more than 18 min for the low and high FMC values, respectively. An observation from this experiments is that when the FMC is high, preheating new fuel (especially wooden ceiling panels) takes much more time (energy), as previously described, causing the currently involved wood to char to depths where the fuel becomes thermally insulated and the production of pyrolysis products is significantly reduced. This causes the fire to decrease in size, and it takes time before it increases again in intensity, resulting in the large differences in observed TTF. The intensity increases after a time when the humidity of the wood panels was removed, and the wooden surfaces were sufficiently heated to produce pyrolysis products. In the case of dry wooden compartments, very thin layers of char (less than 0.5 mm) were observed for large areas of the compartment, typically the whole ceiling and upper walls. These results indicate that when the wooden panels are dry, energy is mostly consumed by heating and pyrolysis. The fire spreads across the wooden panels faster than thermally insulating char develops, causing a rapid temperature increase and full room involvement at a much earlier stage. This experimental work is reported for the first time in the present study.

From the presented theory and experimental observations, it is evident that the FMC is expected to highly influence fire development for enclosure fires involving wooden interior surfaces (ceiling and walls). Nonetheless, despite high FMC values, the initial fuel and fire spread may be fast enough to cause rapid fire growth and flashover. Hence, large variations in fire development will be observed despite the modelled FMC indicator. Figure 3 illustrates this through skewed normal distributions. The respective distributions illustrate the variation in observed TTF for the many different wooden homes; however, depending on the FMC level, different average TTF probabilities exist. It is these expected values that are being implicitly modelled when computing the wooden home fire risk indications. In similar fire scenarios where the only difference is the FMC value of the wooden panels, lower FMC values generally result in faster fire development towards flashover and more severe fire development post-flashover.

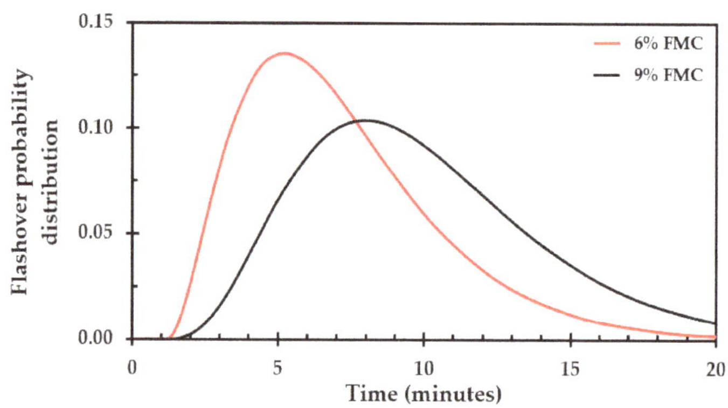

Figure 3. Skew normal distributions illustrating the probability of flashover as a function of time and fuel moisture content.

2.4. Modelling In-Home Fuel Moisture Content

Considering that the FMC is a reasonable WHFR indicator and may be used for indicating wooden home conflagration events, it needs to be modelled. From previous efforts, a model predicting indoor relative humidity (RH) and FMC of in-home wooden surfaces was developed and validated in [11]. The indoor FMC was then correlated with the TTF according to the findings in [10] and further validated in terms of producing reasonable output [12]. The model uses outdoor RH and outdoor temperature to compute the indoor RH and FMC of ceilings and walls from first principle mathematics and physics. Previous model versions have been implemented as a cloud-based microservice [12,33], as well as in a mobile application using edge computing [34]. The model does not need sophisticated tools to run and can simply be implemented and tested on a spreadsheet; however, it is the continuous (dynamic) operation of the model that requires a more advanced implementation. Using historical and predicted (forecasted) weather data, the indoor RH and FMC can be computed for the present and near future. Predictions are shown reliable for the upcoming 2–3 days [33]. The uncertainty in the predictions depends highly on the uncertainty of the weather forecast, that is, forecasted versus occurring weather. Reliable weather data for Norway are supplied by the Norwegian Meteorological Institute and made available through the application programming interfaces (APIs) of FROST [35] and MET [36].

While the mentioned model has undergone essential validation, it lacks a justified generic approach to many Norwegian wooden homes. In order to quantify the general modelling parameters, the principle of the modelling and a short discussion on the sensitivity of key parameters are presented.

The final equation to solve in order to determine the indoor water concentration, and consequently the FMC [11], is

$$\frac{dC}{dt} = \frac{\dot{m}_{surf} + \dot{m}_{ac} + \dot{m}_{supply}}{V_h} \quad (4)$$

where C ($\text{kg} \cdot \text{m}^{-3}$) is the water vapour concentration in the indoor air used to calculate the indoor RH, $\dot{m}_{surf}$ (kg/s) is the mass transfer of water vapour across wooden surfaces, $\dot{m}_{ac}$ (kg/s) is the mass transfer of water vapour through air changes induced by either a ventilation system or natural ventilation and leaks, $\dot{m}_{supply}$ (kg/s) is the water vapour production through everyday use such as cooking, showering, and respiring, and V_h (m^3) is the enclosure volume. A principal sketch of the terms in Equation (4) is shown in Figure 4. It can be seen that the water concentration of the bulk air within the enclosure changes as a function of (1) humidity exchange between wooden surfaces and bulk air, (2) air changes caused by ventilation, and (3) humidity supplied from in-home activities.

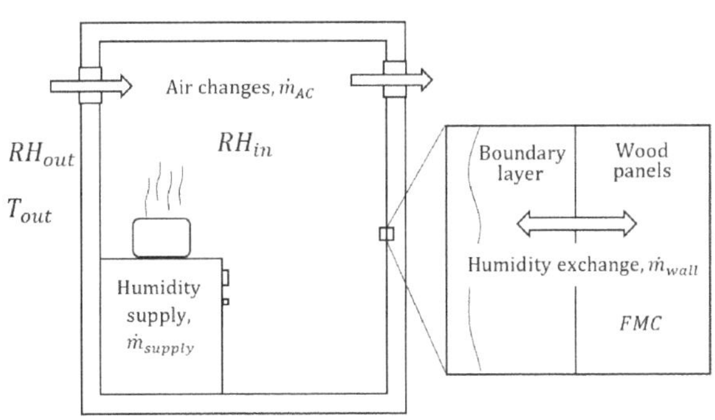

Figure 4. A principle sketch of the parameters influencing the in-home relative humidity and FMC of wooden panels.

The $\dot{m}_{surf}$ depends on the area of indoor wooden surfaces available for humidity exchange. For old wooden homes, all interior surfaces may be covered by wood; however, for newer or refurbished homes wood is more likely to be only part of the room. The fraction of wooden surface area is considered through the AV ratio, which is the ratio of the wooden surface area available for vapour exchange to the room volume ratio. The $\dot{m}_{ac}$ primarily depends on the ventilation principle, herein taken as natural ventilation. For natural ventilation, variations occur due to pressure differences caused by wind or temperature differences, i.e., the stack effect. The quantity $\dot{m}_{supply}$ depends on the intended use of the considered enclosure. In the following, these terms are briefly considered with respect to the existing literature.

2.5. Model Input

There are probably no generic wooden homes, at least not considering the modern building mass. Nonetheless, it may be argued that a generic home is more reasonable for modelling older wooden homes at heritage sites. However, for the modelling of indoor FMC the generic home is not needed per se; when modelling the risk indicator, the layout and exact content of the house are not very important, as the key influencing factors for the FMC described in Equation (4) either account for potential effects, i.e., hygroscopic properties, or are independent of these specific house-related variables. Instead of a generic home, the model uses a generic wooden home enclosure to compute a representative FMC value. Statistics from Norway show that kitchens and living rooms are the enclosures of fire origin in about 40% of all single-home fires [5]. Expanding these statistics to comprise dwellings, which might include apartments and multiple units (multiplex) houses, in which the latter may be wooden heritage townhouses, the share of fires in kitchens and living rooms increases to 50%. Compared to other compartments or areas of fire origin, kitchen and living room fires are highly representative, as can be seen from Figure 5. Hence, for modelling the FMC it is reasonable to take these enclosures as the basis. Note that fires originating from outside the building envelope are the third largest reported place of fire origin. Hence, the modelling of FMC for external wooden panels is an important topic and a part of a future conflagration warning system combining interior and exterior fire risks; however, it is not addressed in the present study.

Concerning the modelling of the FMC, the key parameters from Equation (4) and associated sub-terms are considered from statistics for a generic living room and kitchen.

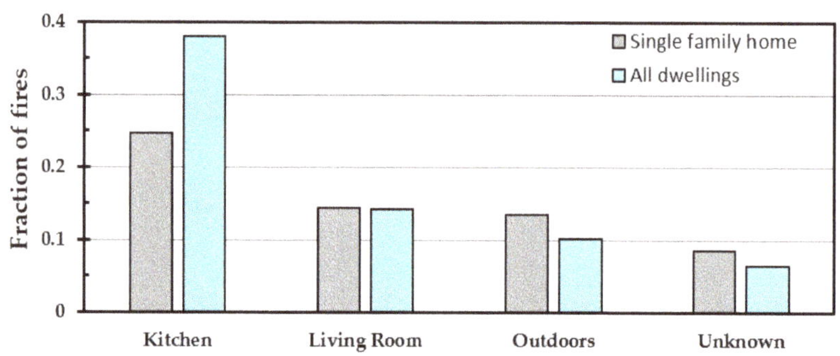

Figure 5. Fire origin in investigated fires from 2016–2023, as reported in [5].

2.5.1. Enclosure Base Area

When determining the area of a representative enclosure, it is important to consider the effects of the enclosure base area when modelling the FMC. An increased base area generally results in increased enclosure volume and increased surface area for humidity exchange. It may be assumed that the ratio between the humidity exchange area, i.e., wooden surfaces, and the enclosure volume is constant. Then, increasing the enclosure base area will result in reduced FMC values as the volume increases relative to internal humidity production. The opposite is true when decreasing the base area. Then, higher FMC values are achieved if the humidity production is kept at a constant level. The connection between these parameters is associated with a larger base area typically housing more inhabitants, causing larger humidity production. Therefore, these parameters must be seen in context.

The enclosure base area for the model is determined based on field studies and the literature. A general trend for dwellings is combining the kitchen and living room into one enclosure. This modern trend affects new as well as refurbished older homes in Norway. For old wooden houses, the kitchen is typically adjacent to the living room, though with different degrees of openness between the rooms. Hence, humidity production within the kitchen likely affects the living room, and vice versa. By taking this into account and considering the larger enclosures to generally provide a lower FMC, a general area for the kitchen and living room can be determined. According to more recent Norwegian guidelines [37], the minimum combined kitchen and living room area for a four-room dwelling is 35 m^2. However, field observations and expert opinion suggest a 40–60 m^2 range. Therefore, the present study assumes a value of 50 m^2 as quite representative. Combined with a standard room height of about 2.4 m, the compartment volume becomes 120 m^3.

2.5.2. Humidity Production

As mentioned, humidity production must be seen in context with compartment size. If considering the suggested 50 m^2 combined living room and kitchen to house a family of five, the moisture supply from respiration and sweating during five hours is in the range of 1.2 kg (Distribution; father = 70 g/h, mother = 60 g/h, Children 1 and 2 = 40 g/h and Children 3 = 30 g/h) [8]. Then, cooking and dishwashing for five persons adds another kilogram of moisture to the indoor air, adding up to at least 2 kg/day of humidity production. However, only a part of the humidity supplied to the indoor air reaches equilibrium with indoor materials, with a large portion being ventilated away before the temporarily increased humidity levels can reach a new equilibrium with the surroundings. Currently, the assumption has been that 50% of the supplied water vapour stays within the room while the remaining vapour becomes ventilated. Field measurements from old wooden homes combined with modelling efforts support a 1 kg/day humidity production distributed evenly throughout the hours of the day [11]. This even humidity distribution allows for smoothing of the FMC development, avoiding unnecessary fluctuations.

2.5.3. Air-Volume Ratio

The humidity exchange area to the enclosure volume ratio, or AV ratio, is another important aspect of the generic enclosure. Increasing the wooden exchange area for a set enclosure base area prolongs the time until equilibrium is reached, as more mass is available as a humidity buffer. In turn, a smaller wooden exchange area results in more responsive modelling, as the wooden enclosure more rapidly adapts to outdoor conditions. This latter observation might be part of a conservative modelling approach taken to ensure that dry indoor periods are identified early. The AV ratio should stay above 0.4 to ensure that wooden panels cover at least an exchange area consistent with the enclosure roof. Based on refurbishments and the identified lag (if too large wooden areas are assumed), it is suggested to keep the AV ratio in the range of 0.4–0.6, with a recommended value of 0.5. The suggested value corresponds to at least the ceiling (most importantly) and about one wall being covered with wooden panels.

2.6. Natural Ventilation

Regarding ventilation and the associated $\dot{m}_{ac}$, older homes are unlikely to have balanced ventilation systems, and, as previously mentioned, natural ventilation is assumed. When considering ventilation, the air change rate per hour (ACH) expresses the fraction of the building or enclosure volume being replaced in an hour. In the model, the ACH is a function of temperature differences (stack effect). The specific ACH then depends on the temperature difference between outdoor air, indoor air, and a ventilation factor γ used to match proper ventilation rates, as described in [11]. For the considered climatic region, the literature suggests an average ACH for naturally ventilated single homes equal to 0.32 [8]. This translates into a recommended γ in the range 350–400, with a suggested value of 380 based on an experimental best fit approach.

The ranges and recommendations of the key parameters related to Equation (4) in the modelling of the FMC are summarised in Table 1. While the model parameters can be altered to describe a particular case, the recommended values should be used for a general FMC calculation, e.g., for a town or city.

Table 1. Recommended values for key parameters used in the modelling of in-home RH and FMC.

Parameter	Recommended Value	Recommended Range	Recommended Range as % Change from Reference	Description
Enclosure Base Area	50 m^2	40–60 m^2	±20%	–
AV-ratio	0.5	0.4–0.6	±20%	Humidity exchange area to enclosure volume
Gamma, γ	380	350–400	−8–5%	Ventilation factor
Humidity production	1 kg/day	0.8–1.2 kg/day	±20%	From respiring, cooking, plants, etc.

3. Results

3.1. Sensitivity Analysis

The previously discussed key model parameters were assessed through a single-parameter sensitivity analysis and a lightweight combined-parameter analysis, i.e., changing more than one parameter at a time. In addition, the outdoor weather data of relative humidity and temperature collected from external services and used as input data for the model were analysed.

In previous work, the TTF was correlated with the FMC [10], expressing the WHFR in units of minutes. The TTF has subsequently been used as a measure for the WHFR as part of a risk communication concept. For this reason, the TTF is used in the following analysis.

The sensitivity study for the key model parameters of room size, AV ratio, ventilation factor γ, and humidity production are presented in Figure 6. The results are presented as the percentage change in the 5th percentile of the computed TTF values versus the relative change in parameter value compared to the recommended values. The chosen measure was tested and found to be suitable. It captures the resulting changes at the low TTF values, a critical model output range. The recommended values, ranges, and corresponding percentage change in range, as provided in Section 2.4, are presented in Table 1. The latter can be used to better interpret Figure 6. Most parameters were suggested within a range of ±20% of the recommended value. However, the presented analysis shows a greater range of change for insight purposes. A parameter is termed sensitive if the 5th percentile TTF change exceeds 0.5 times a single TTF class. This is further discussed in Section 4.

From Figure 6, it can be seen that the 5th percentile TTF develops nearly linearly with the relative change in parameter size for both AV ratio and humidity production. Within the suggested range of ±20% of the recommended value, both parameters appear with a relatively small slope, resulting in minor changes to the 5th percentile TTF. The most significant differences can be observed at the boundaries, at ±20%. For the AV ratio, the corresponding change in 5th percentile TTF is 0.71% to −1.22%. In terms of the computed TTF values and a reference TTF at 4 min, this translates to TTF values in the range of 3.96–4.02 min, i.e., insignificant changes. The humidity production has a steeper slope, and for the same range the changes in the 5th percentile TTF equal 2.5% to −2.5%. When increasing the range to ±40%, the largest difference in the 5th percentile TTF can be observed for the humidity production, at about ±5%. Hence, concerning model output, the two parameters are not very sensitive within the recommended range. The humidity production is further considered in Section 4.

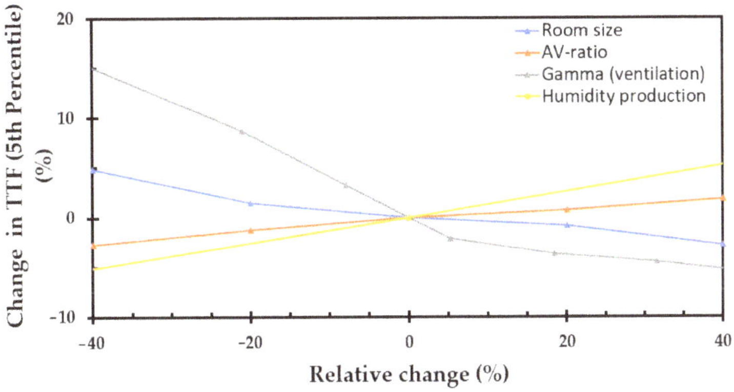

Figure 6. Sensitivity analysis of key parameters associated with modelling the in-home RH and FMC (TTF). A relative change of 0% represents modelling using the recommended values in Table 1.

When considering the room size and ventilation factor γ, less linearity is observed, i.e., they both appear to have an inverse-like dependency to the 5th percentile TTF when expanding the axis of relative change, as shown in Figure 7. This is understandable, as reducing the room size and/or ventilation would increase the in-home RH and FMC. It is, however, not reasonable to consider these parameters at points far beyond the recommended range. Within this range, the most significant change in the 5th percentile TTF is observed for the ventilation factor at −20%, an 8.7% increase. This increase equals 0.35 min for a reference TTF at 4 min, and is not considered significant. Neither the ventilation factor nor the room size are found to be sensitive parameters within the recommended range with respect to model output.

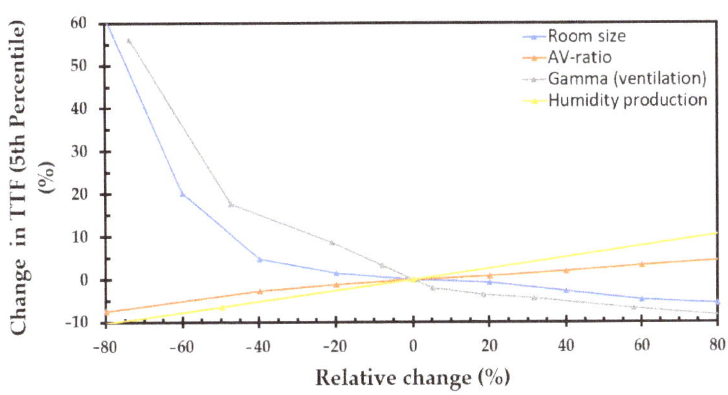

Figure 7. Sensitivity analysis of key parameters associated with modelling the in-home RH and FMC (TTF). A relative change of 0% represents modelling using the recommended values in Table 1. This figure shows an extended range of relative change compared to Figure 6.

In general, the parameters must be seen in context. When decreasing the enclosure base area, Figure 7 suggests an increase in the 5th percentile TTF. However, such a reduction in the base area needs to probably be followed by a decrease in humidity production, as smaller rooms suggest fewer people and less humidity supply.

The combined effect of all parameters was briefly assessed within the ±40% range, and it was found that the different parameters can take on different values within the range without experiencing any special combined effect on model output. While the percentage change in 5th percentile TTF cannot be summed for the different parameters, that is, adds up to the new 5th percentile TTF, it was found that the computed 5th percentile TTF from random parameter values within the ±40% range (beyond the recommended range) was within ±5% of the added values. Hence, according to Figure 6, adding the percentage change to the 5th percentile TTF could be used to estimate the combined effect of changing the parameters.

The sensitivity analysis for outdoor relative humidity and outdoor temperature is presented in Figure 8. These parameters are model input data harvested from external sources comprising historical and predicted data. The results are considered in terms of the mean (modelled) TTF at relative humidities ranging from 40–90% and temperatures in the range of −10 °C to 10 °C. The chosen ranges represent many wooden towns in Norway during December and January, including the towns of Haugesund, Lærdal, and Røros. However, the latter town may become significantly colder. For low outdoor temperatures, the modelled TTF becomes less dependent upon the outdoor RH, as presented in Figure 8. At −10 °C, the mean TTF has a spread of 0.6 min in the range of 40–90% RH. Hence, the model output is not very sensitive to the outdoor RH at low temperatures. However, the dependency on the outdoor RH increases with temperature, and at 10 °C the spread across the different RHs corresponds to a mean TTF ranging from 4.6–8.1 min, which is significant.

It is important to note that the outdoor RH and temperature are taken from input data harvested from high-end weather data sources, such as the Norwegian Meteorological Institute [35,36]. These results support the need for mathematical modelling to keep track of the changing conditions.

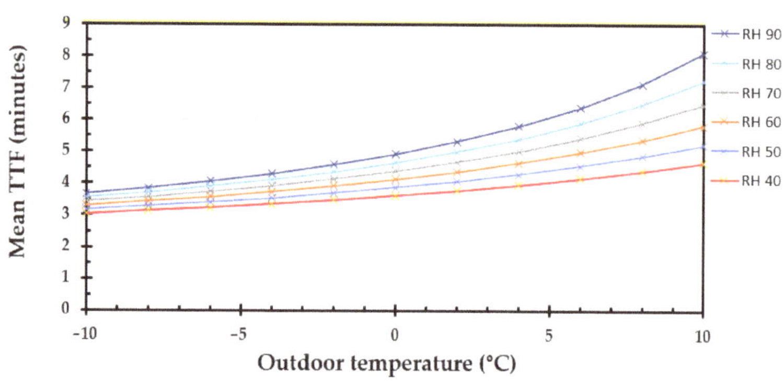

Figure 8. Sensitivity analysis of outdoor relative humidity and outdoor temperature.

3.2. *Practicability*

An important aspect of risk modelling is the practicability (usefulness) of the modelled results. Herein, paying attention to the different users of the modelled WHFR is important. If only considering the TTF (FMC), predictions can be related to the susceptibility and consequences of a potential fire developing indoors for a single home, i.e., a responsibility of the homeowner. Considering the coexistence of dry homes and strong wind, the combination may serve as a conflagration risk indicator, i.e., a more important concern for the local fire department.

This section presents an analysis of nearly ten years of registered weather data, for the period 1 November 2013 to 1 April 2023, i.e., ten winters, for the locations of Haugesund coastal town (N 59.41°, E 5.28°), Lærdal village (N 61.10°, E 7.48°), and Røros mountain village (N 62.576°, E 11.386°), a UNESCO World Heritage Monument. The historical weather data were used as input for the WHFR model to identify days of low TTF and events where low TTF coexisted with days of strong wind (conflagration indicator). In accordance with previous work, the FMC risk indicator is primarily expressed by quantifying the correlated TTF value. The wind and computed TTF values were divided into classes as presented in

Tables 2 and 3. In the present study, strong wind refers to the classes Medium to High, while in-home drought relates to the classes Medium-high and High. This is further discussed in Section 4.

Table 2. The qualitative classification of wind with respect to wind speed.

Class	Beaufort	Description	Wind Speed, U (m/s)
High	9+	Severe gale–Hurricane force	$20.8 \leq U$
Medium high	7, 8	High wind, Near–Fresh gale	$13.9 \leq U \leq 20.7$
Medium	5, 6	Fresh–Strong breeze	$8.0 \leq U \leq 13.8$
Medium low	4	Moderate breeze	$5.5 \leq U \leq 7.9$
Low	0–3	Calm–Gentle breeze	$U \leq 5.4$

Table 3. The qualitative classification of modelled TTF and the associated FMC.

Class	Indoor RH [%]	FMC [1] [wt.%]	TTF [min]
High	$RH \leq 19$	$FMC \leq 4.35$	$TTF \leq 4$
Medium high	$19 < RH \leq 27$	$4.35 < FMC \leq 5.75$	$4 < TTF \leq 5$
Medium	$27 < RH \leq 35$	$5.75 < FMC \leq 6.90$	$5 < TTF \leq 6$
Medium low	$35 < RH \leq 41$	$6.90 < FMC \leq 7.85$	$6 < TTF \leq 7$
Low	$41 < RH$	$7.85 < FMC$	$7 < TTF$

[1] Modelled value represents indoor wooden panels at 2–3 mm depth.

The historic annual frequency of the daily mean TTF for the different TTF classes and areas for the last ten winters is presented in Table 4. Taking these results as a fair estimate of future expected frequencies, it can be seen that the expected number of days with High risk TTF (low TTF value) per year is limited to 3.2 days for Haugesund (coast), 26 days for Lærdal (inner fjord), and 97 days for Røros (inland mountain). Significant differences are observed for days of Medium high risk, where Haugesund has the lowest frequency at 31 days while Lærdal and Røros experience 110 and 108 days, respectively. For the remaining TTF classes, the differences are less substantial. However, because Røros has a high number of days with High and Medium high risk, there are fewer days of lower risk, as can be seen by comparing the 165 days of Low risk in Haugesund compared to the 73 days in Røros. The differences can generally be explained by a drier and colder inland climate for Lærdal and especially for Røros. While Lærdal is not equally cold, it typically experiences adiabatic heating of descending mountain air for most wind directions during the winter, which reduces the outdoor RH [9].

Considering the distribution of the daily mean TTF throughout the year, it has previously been shown that seasonal changes apply and that low TTF values are observed during the winter months [12].

Table 4. The annual distribution of daily mean TTF for Haugesund, Lærdal, and Røros based on weather data from 1 November 2013 to 1 April 2023 (ten winters).

Class	TTF [min]	Haugesund	Lærdal	Røros
High	$TTF \leq 4$	3.2	26	97
Medium high	$4 < TTF \leq 5$	31	110	108
Medium	$5 < TTF \leq 6$	95	71	45
Medium low	$6 < TTF \leq 7$	71	31	43
Low	$7 < TTF$	165	128	73

The risk matrix for Haugesund after analysing the modelled TTF in combination with recorded wind strengths (2013–2023) is presented in Figure 9. As can be seen, the axis of the risk matrix is made up of the different classes of wind strength and modelled TTF, making up the 25 main cells. As it is important to distinguish between the number of hours with specific wind speeds, each main cell is subdivided into four sub-cells referring to the registered duration of the wind. The subdivision of the cells can be seen in Figure 10. The numbers inside the risk matrix represent the annual frequency of the specific event registered for Haugesund.

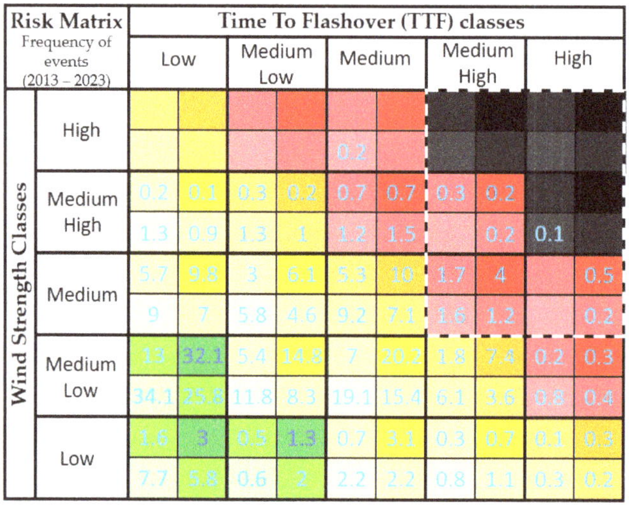

Figure 9. The computed risk matrix for Haugesund, Norway. The numbers within the matrix represent the annual frequency (days/year) of the specific TTF and wind combination and for a duration according to the sub-classes specified in Figure 10. The dotted rectangle indicates a risk level where it is recommended to implement proactive measures.

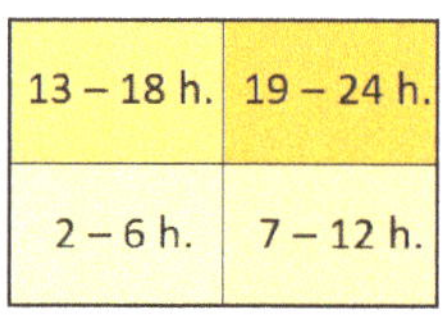

Figure 10. The sub-cells of the risk matrix refer to four different classes of wind duration (h), i.e., a single hour of registered wind above the threshold value is ignored.

It can be seen from the matrix that lower-risk events occur more often than higher-risk events, as expected. The dotted white and black squares indicate the events where, based on analysis of historic fire incidents, it is recommended to implement proactive measures. Hence, from Figure 9, events needing risk reducing measures add up to ten events per year regardless of wind duration. If only emphasising the coexistence of in-home drought and strong winds lasting more than 12 h, it adds up to 6.8 events per year. For Haugesund, the historic data shows a manageable set of high-risk events for the local fire department.

The risk matrix for Lærdal is presented in Figure 11. It can be seen that, for a specific wind speed, the frequencies are not necessarily reduced with increasing TTF classes. Hence, the frequencies have less spread for a larger portion of the matrix. For Medium-low winds, the Low and Medium-high TTF classes have quite similar frequencies in the range of 2–12 h wind duration, while for 13–24 h the Medium-high TTF has a higher frequency. This can be explained by the high number of days of both Low and Medium-high TTF for Lærdal. Considering the recommended area of proactive measures, i.e., the dotted rectangle, a frequency of 1.9 days per year is observed for Lærdal, which is less than Haugesund, despite Lærdal having a significantly higher number of days with Medium-high and High TTF. This primarily relates to Haugesund being a coastal town with many days with strong wind, resulting in more days with coexisting high wind strength and high TTF.

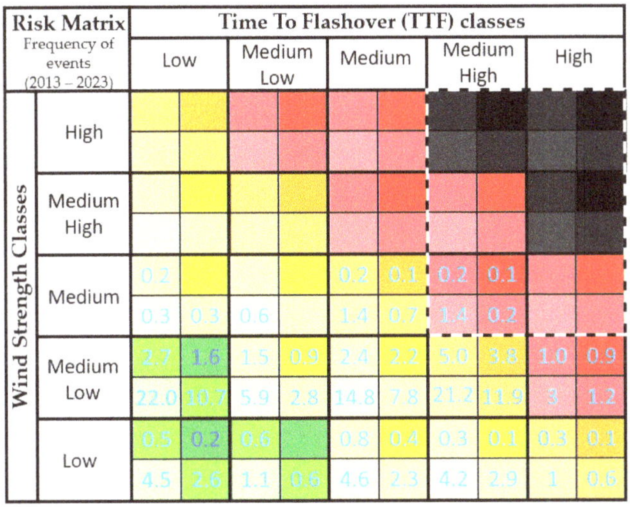

Figure 11. The computed risk matrix for Lærdal, Norway. The numbers represent the annual frequency (days/year) of the TTF and wind combination for a duration according to the sub-classes specified in Figure 10. The dotted rectangle indicates the risk level where it is recommended to implement proactive measures.

The risk matrix for Røros is presented in Figure 12. Here, similar observations as for Lærdal can be observed. For the Medium-low winds, the highest frequency can be seen for Medium-high and High TTF classes. The number of addressable events for Røros equals 10.3, regardless of wind duration. When only considering events where the registered wind speed lasts 12–24 h, 1.1 events per year are registered. As for Lærdal, the number of events recommended for proactive measures is relatively low and manageable for the local fire department. However, notice should be taken of the high number of days corresponding to the High and Medium high TTF classes. This means that Røros and Lærdal are areas where the interiors of wooden homes become very dry during the winter. In turn, this means that an occurring fire may develop rapidly and pose severe risks for both inhabitants and the buildings, many of which are of high historical value. In addition, despite the low numbers of coexisting strong winds and high TTF classes, many old wooden homes at the different heritage sites in Norway are attached. Hence, a fire may spread between structures before the arrival of the local fire department even during moderate wind speeds.

3.3. Fire Danger Communication

Recent work emphasising user-driven iterative development of a graphical user interface (GUI) for fire risk communication revealed that using the term TTF when presenting the WHFR may not be suitable [34]. While serving as a measure for quantifying the WHFR, the TTF risk communication concept needs a new approach. Developing intuitive communication concepts is challenging; thus, the focus was directed towards existing concepts, notably the Norwegian forest fire index and how that risk is communicated within Norway. The Norwegian forest fire risk is communicated as a forest fire danger, not as a fire risk, with an index of seven classes ranging from No danger to Great danger. The idea here is to use this well-established terminology in the wrapping of the FMC (TTF) indicator, thereby communicating the novel in-home wooden fire risk through an established concept. Adopting such an approach can provide a wooden home fire danger index, with the risk communicated as far as possible through a similar index with similar colours. Adjustments had to be made, as the forest fire danger index has a No danger level, which is not applicable for an in-home wooden fire risk indicator considering the year-round risk of imminent fire. The Norwegian forest fire danger index and the adopted wooden home fire danger index are presented in Figure 13.

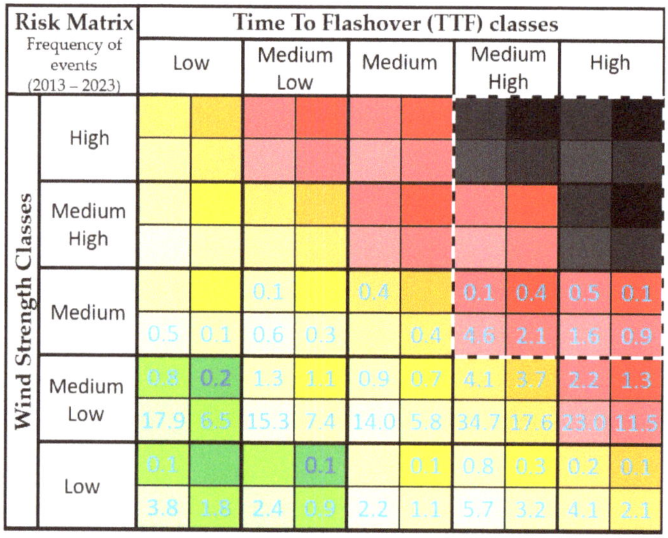

Figure 12. The computed risk matrix for Røros, Norway. The numbers within the matrix represent the annual frequency (days/year) of the TTF and wind combination for a duration according to the sub-classes specified in Figure 10. The dotted rectangle indicates the risk level where it is recommended to implement proactive measures.

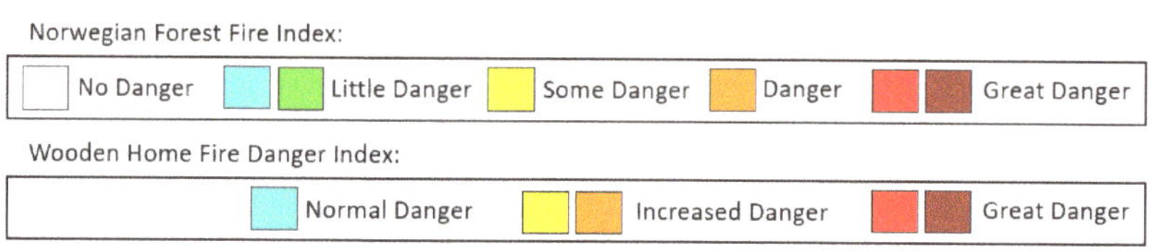

Figure 13. The different levels of danger for the Norwegian Forest Fire Index (**Top**) and the corresponding suggested levels for the Wooden Home Fire Danger Index (**Bottom**).

The indexing into the different levels of danger presented in Figure 13 follows from Table 5. The classification of the modelled TTF is based on the modelling of several historical fires. After analysing these fires and associated video materials, records, and descriptions from the fire brigade, knowledge of model performance was obtained, resulting in the suggested classification. According to the risk matrices presented in Section 3.2, the defined area where risk reducing measures are recommended to be implemented is associated with the Great Danger level. The presented indexing has already been implemented in a second version mobile application, and is ready for testing within selected Norwegian Fire brigades.

Table 5. Wooden Home Fire Index with associated TTF values and classification.

Index	Color	TTF Class	TTF [min]
Great Danger	Dark red	High	$TTF \leq 4$
Great Danger	Red	Medium high	$4 < TTF \leq 5$
Increased Danger	Orange	Medium	$5 < TTF \leq 6$
Increased Danger	Yellow	Medium low	$6 < TTF \leq 7$
Normal Danger	Turquoise	Low	$7 < TTF$

3.4. The Marshall Fire

In this section, the model is applied outside the cold climate region of Norway to demonstrate the modelling principle for another region. If the model is to be applied to other areas, the most important aspect is the presence of heated homes with wood as an internal cladding material, especially in the ceiling. In such cases the modelling principle is transferable, even to areas less cold than those considered in the present study. However, as seen in the sensitivity analysis for the outdoor weather parameters, the modelled TTF would rely more on the outdoor RH in a warmer climate. Despite the modelling results not necessarily being invalid, warmer climate areas may not be exposed to the same cold climate fire risk. In addition, such areas may typically have quite different traditions for constructing homes, and wood as internal cladding may not be widespread. Nevertheless, without considering the interior of involved building mass, we applied the model to the area of the recent Marshall fire, Boulder County, Colorado, USA on 30 December 2021 [29].

While the Marshall Fire was a WUI fire initially spreading through cured grass, it is interesting to consider the potential in-home drought of involved wooden homes. Potentially, such a large outdoor fire could originate indoors as a rapidly developing wooden home fire. In addition, the FMC of the involved structures is of importance to the post-flashover fire intensity, which affects further fire spread.

Weather data were collected from the nearby Denver airport and used as input for the fire risk model. Model-specific parameters were kept at the recommended values reflecting the generic Norwegian combined kitchen and living room. Figure 14 presents the modelling results. It can be seen that the wooden homes in the area were becoming drier at the beginning of December 2021. By about the 10th of December, the wooden homes had passed the identified critical limit of 4 min, essentially passing into a stage where fire development is known to be extreme based on historical events [12]. The wooden homes then continued to dry out, becoming equally dry as the most severe registered in-home drought conditions, similar to the conditions during the Lærdal fire in Norway on 28 January 2014. The very dry interior of the wooden homes in the Marshall fire likely contributed to fast burnout and increased production of firebrands in the strong winds.

This modelling highlights in-home drought as a potential risk influencing factor in a somewhat different climate than Norway. Further, the area of Boulder County is not unfamiliar in terms of its high wind strengths [29]. Considering the recent findings on conflagration risk depending more on days of strong wind than days of in-home drought, this area likely experiences increased conflagration risk due to the potential combination of very dry wooden homes and strong winds during the winters.

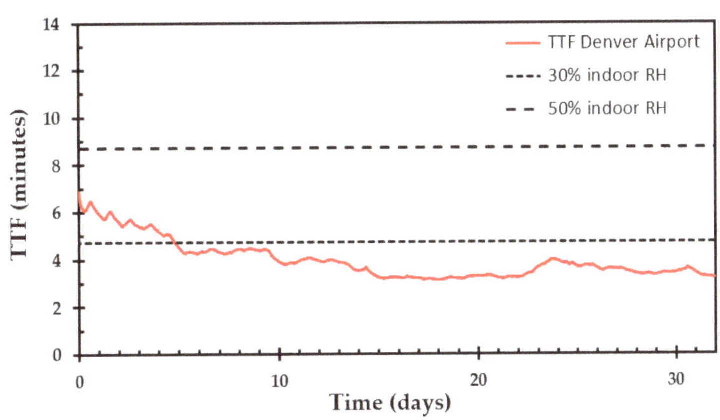

Figure 14. Modelled TTF for Denver during December prior to the Marshall fire. The fire occurred on 30 December. Weather data were taken from Denver airport.

4. Discussion

4.1. The Generic Enclosure

Concerning whether the analysed parameters were considered sensitive or not, a criterion equal to 0.5 times a single TTF class was used, which means that changes within the recommended values could not result in a difference in the 5th percentile TTF equal to more than 0.5 times a TTF risk class. Because the TTF classes are separated into five classes of one minute per class, this means that if a parameter was considered not sensitive, the TTF value changed with less than 0.5 min for the considered change in parameter size. This was the case for nearly all individually considered parameters within ±40% of the recommended value. The exception, however, was the ventilation factor, which, at −40% change, slightly exceeded the 0.5-min criteria. A TTF value of 4 min was used to calculate the percentage deviation, as this has shown to be a critical limit for the low value of TTF [12]. The idea behind this criteria is that minor changes in parameter size should not change the predicted risk class. If they do, the parameter must be considered sensitive, and the issue must be addressed.

While all parameters can be changed beyond the recommendations, this would typically provide unrealistic representations of the generic enclosure. e.g., if the ventilation factor was reduced beyond −40%, significant increases in the 5th percentile TTF were observed as humidity was continuously supplied to the enclosure. Hence, the parameters must be considered in relation to each other.

Indeed, the considered literature and the performed sensitivity analysis have increased the credibility regarding the recommended values and range of key model parameters. The most uncertain parameter, however, is the humidity supply, which appears as a relatively insensitive parameter within the recommended range. This parameter, however, has been recommended at 1 kg/day based on field measurements and model analysis [11]. If the humidity production were changed from 1 to 2 kg/day (100% increase), this would result in a 13.1% increase to the 5th percentile TTF, or 27.3% at 3 kg/day. The reason why this is particularly interesting is due to the literature suggesting higher humidity supplies than are used in the model. Herein, it is important to distinguish between supplied water vapour to the indoor air and the actual water vapour contributing to a rise in indoor RH and FMC. Water vapour is lighter than air, and part of the supplied humidity is likely to rise within the enclosure, escaping (leaking) through small gaps in the ceiling and upper parts of the walls. In addition, most old wooden homes have a kitchen and living room on the first floor with bedrooms on the second floor. A fraction of supplied water vapour is likely to rise within the building through doors and via the staircase to higher-level floors. The ventilation system or ventilation principle is another mechanism for transporting supplied water vapour. Hence, there is probably a difference between supplied water vapour and increased relative humidity and FMC value. For this reason, this parameter was set by changing its value to best fit the field measurements of indoor RH. In contrast, most of the other parameters were explicitly set to describe the particular enclosure. It would be necessary to perform additional field measurements and model analysis to increase the accuracy of the humidity production levels. However, many of the considered quantities are variables, and take different values within different homes, e.g., a particular wooden house with a specific humidity production may be drier than another nearby home exposed to another humidity production. The important aspect is not to explicitly model each wooden home; rather, it is to model a representative average and how this would change with changing ambient temperature and RH. In comparison, the ground and the trees in a forest could have significant local variations in FMC; nonetheless, a generic approach is used to describe the conditions over large areas. In general, the sensitivity analysis performed within this study supports the possibility of computing such an average despite the local variations.

The chosen parameters were recommended based on literature and field studies. Parameters were considered from a conservative point of view, e.g., the humidity production, which is pointed out as the most uncertain parameter, is unlikely to take on much lower values than 1 kg/day; however, it might take higher values. By recommending a lower value compared to a higher value, the model provides lower TTF predictions, indicating increased risk. This could be a problem if the number of high-danger days were too many. The presented modelling does, however, show that the number of days with combined strong wind and low TTF is manageable for all the considered locations, even Røros, the mountain village, which despite increased humidity production would appear with far more days of greater home fire danger than the other considered locations.

4.2. Practicability of the Modelled FMC

In the simplest sense, the modelling of the wooden home fire danger results in the FMC (TTF) indicator of in-home drought, indicating when and to what degree the wooden interior of homes is in a state where it is likely to increase the susceptibility to and negative consequences of an initiating or established fire. Homeowners may use such an indication to implement simple measures, e.g., avoiding candles, deep frying, and unattended electric apparatuses. Additionally, the indicator may be used to notify family members and others about increased risk, such as elderly people living alone. For such use, days of high fire danger may last for several days and occur on multiple occasions. In this case, use is mostly about raising awareness, and potential measures are simple but important.

The other modelling aspect is the coexistence of very dry homes and strong wind, where the combination may be understood as a conflagration danger indicator. This is a major concern of local fire brigades. If the combined occurrence of modelled low TTF and predicted winds results in too many high-risk events, it might not be possible for the fire brigades to address these events, even if the model provides a correct picture of the danger. That is to say, if risk modelling and wind forecasts result in a hundred days of high fire danger per year, the outcome might not result in proactive risk management but rather an acceptance of many days of increased danger. However, if abnormal danger levels are seen, e.g., an expected average of about ten days per year, measures could be taken during predicted days of high conflagration danger, especially to protect dense wooden heritage sites. These might typically be a focus area for the different fire brigades, as they are of historical value and particularly vulnerable; see Section 2.1.

Based on the modelling and the identified manageable number of events needing risk-reducing measures, it appears that the modelling may serve as a tool for identifying increased danger levels. Hence, modelling the indoor FMC (TTF) may be an answer to the Norwegian fire safety regulations supporting a risk-based approach to emergency preparedness. Such identification of upcoming increased danger levels in dense wooden house areas may enable improved risk management, in turn allowing proactive measures to be implemented by the fire departments, e.g., raising awareness among citizens, increasing emergency preparedness, temporarily increasing staff, temporarily staffing unstaffed fire stations, and strategically placing equipment and units to reduce the response time.

The presented results identified many days corresponding to High and Medium high TTF for Lærdal and Røros, i.e., low TTF values. In turn, when considering the coexistence with strong winds, a substantial number of days with high conflagration danger were expected. However, it turned out that Haugesund, with far less days of modelled low TTF, experienced more days of high conflagration danger. This indicates that the number of days with strong wind and low TTF depends more on the number of days with strong wind than the number of days with low TTF. Haugesund is a coastal city where periods with dry in-home conditions occur; when this happens, it is more likely to coexist with strong wind. Historical records support this observation, as the majority of the largest peacetime fires in Norway have been coastal town fires [38].

In the present work, strong winds are associated with the wind classes of Medium to High. In this sense, strong winds reflect critical classes of wind speed. Any critical wind speed is scenario-dependent, e.g., it depends on the fuel, HRR, surroundings, etc. In terms of a city or town fire, the critical wind speed may be winds strong enough to cause flame impingement on adjacent structures. This wind speed may be lower than the strong winds associated with the generation of firebrands and spotting fires. Nonetheless, previous conflagrations [25] and expert opinions [39] suggest that wind speeds close to 10 m/s may be challenging. With respect to in-home drought, the present study associates it with the Medium high and High TTF classes. This is based on observations from Norway, where fires developing very quickly have been associated with TTF values corresponding to these classes [12]. When considering the indoor relative humidity and EMC of wooden surfaces, this corresponds to a wooden surface in equilibrium with an indoor RH equal to or less than 27%, i.e., an equilibrium water content of 5.75 %.

An interesting observation from the sensitivity study was the dependency of model output on outdoor temperatures. At −10 °C, the modelled TTF varied by only 15.7% when the outdoor air humidity changed from 40% to 90% RH. This is an integral part of why Røros mountain village experiences such a high number of days with low TTF values. Røros has a mean temperature during December, January, and February of −8 °C, −11 °C, and −8 °C, respectively. Both before and after this period, the monthly mean temperature can be expected to be −5 °C. From the sensitivity analysis of outdoor weather parameters, Figure 8 shows that these mean temperatures correspond to about three months with a modelled TTF below 4 min, which corresponds to the High TTF risk class. As a mountain village with several months of subzero temperatures, there are long periods of arid in-home conditions; however, the number of days with strong winds during the winter is low.

4.3. Measures of Risk and Risk Communication

In previous work, the FMC was correlated with the TTF to construct an intuitive risk indicator. The idea was to express the FMC risk indicator through a well known quantity within the fire safety community. The correlation was based on a series of one quarter-scale wooden ISO-room experiments [10], similar to the aforementioned one half-scale experiments. Such a correlation becomes relative to the experiments performed, and changing the experiments, e.g., having a basis in a correlation made from the one half-scale experiments would make for a somewhat different correlation. Hence, the use of TTF to express the risk indicator was primarily intended as a practical approach to the communication of the WHFR. However, recent studies [12,34], feedback, and user tests have raised awareness concerning possible misconceptions when using such a specific term with units in minutes. This is especially challenging when model results are compared to historical fires and associated with an observed time to flashover, as in [12]. It is important to note that while the risk indicator expresses a TTF, this is generally not an accurate quantity for attempting to describe the exact time to flashover for wooden homes. Nonetheless, in cases where an enclosure fire develops from a limited initial fuel source and the wooden walls and ceilings become the only fuel for further fire spread until flashover, experimental results suggest that the modelled TTF can indicate the actual TTF, or at least indicate the top point of a skewed normal distribution of expected TTFs if a number of similar tests are repeated. However, this is not a realistic scenario, and other fuels will likely be involved in the fire during the growth phase. In this case, it could be argued that the modelled TTF indicates fire development within an empty wooden enclosure, and including more fuel and more extensive initial fuel sources could result in even less time until flashover than indicated by the model.

Nevertheless, the risk communication approach described in this paper has been implemented in a mobile application and awaits further testing within fire brigades. The modelled TTF was thought to benefit from a well-established fire danger concept, i.e., the wildfire danger system. In turn, this reduces the need for describing and interpreting new concepts, which is likely to limit the number of misinterpretations and simplify the design of the graphical user interface.

4.4. Validity of the Risk Modelling

Computing the risk matrices involves handling and arranging significant amounts of data, e.g., the datasets gathered for a particular location containing outdoor RH, outdoor temperature, and wind speeds, may have missing elements and could need to be treated by handling the associated timestamps. Depending on the duration of the period with missing elements, interpolation can be used to fill the small gaps for modelling purposes. However, if too much time passes without proper data recordings it must be computed as separate series. All handling of data involves the possibility of introducing errors. For this reason, the Matlab scripts responsible for handling the data were validated through a single-step procedure. Small parts of the code were considered separately as the scripts for data analysis were gradually extended by the validated pieces. In addition, two separate scripts using different approaches and hence code were developed to validate the part where low TTF and strong winds were found to coexist. This was an essential part of the data analysis, as the results can support the practicability of the modelling approach. However, handling large amounts of data over a large temporal scale and performing multiple operations on the data sets always carries the chance of introducing possible errors.

Further, it is important to mention that when analysing several years of historical weather data it may be impossible to find the desired combination of weather parameters registered at the exact location. This was a challenge for Lærdal, where outdoor RH and temperature were taken from a unit in Lærdal while recordings of wind velocities had to be taken from Sogndal airport, 20 km west of Lærdal. Considering the local topography, the wind data from Sogndal probably underestimate the actual winds in Lærdal. It is not unlikely that some of these events could be associated with higher wind speeds, and consequently a higher frequency of addressable events. It is important to note that the fire risk modelling primarily relies on predicted weather data interpolated onto a higher-resolution grid by sophisticated models. Hence, the predicted weather data, especially for the near future (hours), are probably more representative of the actual weather at a location than historical weather records at a distant meteorological station.

4.5. Suggestions for Future Research

At this stage, substantial efforts have been put into developing a wooden home fire danger and possibly a conflagration danger indicator. However, there are several aspects in need of further attention and research. More effort should be put into quantifying humidity production and ventilation rates to ensure that the recommended values are representative. Then, the model should preferably be tested for a set of wooden homes with interior wooden panels in the ceiling, and to different degrees on the interior walls. This would serve as a final validation of the FMC modelling and the chosen generic approach. The model has already undergone substantial testing and validation in an earlier phase [11,12]; however, testing arranged for validating the model could be better targeted when the model parameters are known. This could include testing the recorded in-home RH values versus values predicted solely by weather forecasts, which have only been briefly validated previously in [11].

The conflagration danger indicator is at an early stage, indicating a possible conflagration event by considering an initiating event of a wooden home fire starting indoors. However, more substantial modelling efforts are needed to describe the FMC of external wooden panels. When a fire breaches the building envelope and becomes an exterior fire, the wind strength is the most critical factor for initiating a conflagration event. However, if all the neighbouring structures have high FMC values and free water on surfaces and gutters, a conflagration might be less likely to develop. Hence, more research is needed to consider such conditions.

While only a few related studies have been mentioned, notably studies that emphasise the structure and immediate surroundings, other studies have considered urban fire risk from a broader perspective. Noori et al. [40] considered multiple criteria, including static building information, when mapping urban fire risk. When combined with dynamic fire danger indicators, such multi-criteria risk maps may serve as promising tools to improve risk management in the built environment.

5. Conclusions

To support and justify existing approaches for wooden home fire risk modelling, in this paper we considered and analysed homes with wooden internal surfaces, the general concept, input parameters, and model-specific parameters. From a theoretical and experimental point of view, the modelled FMC, often expressed through the TTF indicator, was found to be a reasonable fire risk indicator for homes with sufficient in-home wooden surfaces. An important criterion is that the share of wooden surfaces is managed through the recommended range of the AV ratio parameter. Further, the recommended values and ranges associated with the different parameters likely describe a generic combined kitchen–living room enclosure in a Norwegian wooden home. This assumption was further strengthened through a sensitivity analysis, which identified all of the key parameters in modelling in-home FMC to be relatively insensitive to realistic variations. Considering usability, an analysis of ten years of weather data records revealed that the number of days with high-risk TTF could exceed the practical limit of addressable events for the fire brigades while raising awareness among homeowners. However, when combining days of high-risk TTF with days of strong wind, i.e., conflagration danger, the historical weather data analysis for the considered locations confirmed the assumption of these being infrequent events, i.e., typically 2-10 days each winter for the analysed locations. These days of increased conflagration risk can be addressed by the local fire departments through proactive measures. The fire risk modelling can then serve as a response to the new requirements in the Norwegian Fire and Emergency Regulations requiring systems for identifying days of high risk, in this particular case, risk associated with wooden homes and heritage sites of historical value. Finally, it appears that coastal towns are more prone to conflagrations, as conflagration danger depends more on days of strong wind rather than days of in-home drought. This is in line with the most severe historical fires in Norway since the 1900s.

Author Contributions: Conceptualization, T.L.; Formal analysis, R.D.S.; Funding acquisition, R.D.S. and T.L.; Investigation, T.L.; Methodology, R.D.S. and T.L.; Project administration, R.D.S. and T.L.; Resources, R.D.S. and T.L.; Software, R.D.S.; Supervision, T.L.; Validation, R.D.S. and T.L.; Visualization, R.D.S.; Writing—original draft, R.D.S.; Writing—review and editing, R.D.S. and T.L. All authors have read and agreed to the published version of the manuscript.

Funding: This research was funded by the Research Council of Norway, grant no. 298993, Reducing fire disaster risk through *dynamic risk assessment and management* (DYNAMIC), and supported by Haugaland Kraft Nett, Norwegian Directorate for Cultural Heritage, and Stavanger municipality.

Institutional Review Board Statement: Not applicable.

Informed Consent Statement: Not applicable.

Data Availability Statement: All the weather data are available through https://seklima.met.no/ (accessed 5 July 2023).

Acknowledgments: The authors very much appreciate the valuable guidance on statistical analysis from Sveinung Erland at Western Norway University of Applied Sciences.

Conflicts of Interest: The authors declare no conflict of interest.

Abbreviations

The following abbreviations are used in this manuscript:

WHFR	Wooden Home Fire Risk
RH	Relative Humidity
TTF	Time To Flashover
FMC	Fuel Moisture Content
HRR	Heat Release ate

References

1. Mock, C.; Michael Peck, C.J.; Meddings, D.; Gielen, A.; McKenzie, L. *Burn Prevention: Success Stories and Lessons Learned*; World Health Organization: Geneva, Switzerland, 2011. Available online: https://apps.who.int/iris/bitstream/handle/10665/97938/9789241501187_eng.pdf?sequence=1&isAllowed=y (accessed on 23 June 2023).
2. Villa, V.; Paltrinieri, N.; Khan, F.; Cozzani, V. Chapter 1—A Short Overview of Risk Analysis Background and Recent Developments. In *Dynamic Risk Analysis in the Chemical and Petroleum Industry*; Paltrinieri, N., Khan, F., Eds.; Butterworth-Heinemann: Oxford, UK, 2016; pp. 3–12. [CrossRef]
3. Brushlinsky, N.; Sokolov, S.; Wagner, P.; Messerschmidt, B. *World Fire Statistics*; No. 27; International Association of Fire and Rescue Services: Ljubljana, Slovenia, 2022. Available online: https://ctif.org/sites/default/files/2022-08/CTIF_Report27_ESG.pdf (accessed on 2 June 2023).
4. Log, T. Indoor relative humidity as a fire risk indicator. *Build. Environ.* **2017**, *111*, 238–248. [CrossRef]
5. Directorate for Civil Protection and Emergency Planning. Search in Statistics. Available online: https://www.brannstatistikk.no/brus-ui/search (accessed on 10 April 2023).
6. The Norwegian Ministry of Justice and Public Security. Forskrift om organisering, Bemanning og Utrustning av brann- og Redningsvesen og Nødmeldesentralene. Available online: https://lovdata.no/dokument/LTI/forskrift/2021-09-15-2755 (accessed on 1 June 2023).
7. Pirsko, A.R.; Fons, W.L. *Frequency of Urban Building Fires as Related to Daily Weather Conditions*; Technical Report AFSWP-866; U.S. Department of Agriculture: Washington, DC, USA, 1956. Available online: https://www.fs.usda.gov/psw/publications/documents/cfres_series/cfres_itr_afswp866.pdf (accessed on 23 August 2023).
8. Geving, S.; Thue, J.V. *Fukt i Bygninger*; Sintef–Norsk Byggforskningsinstitutt: Trondheim, Norway, 2002.
9. Log, T. Cold Climate Fire Risk; A Case Study of the Lærdalsøyri Fire. *Fire Technol.* **2014**, *52*, 1815–1843.
10. Kraaijeveld, A.; Gunnarshaug, A.; Schei, B.; Log, T. Burning rate and time to flashover in wooden 1/4 scale compartments as a function of fuel moisture content. In Proceedings of the 8th International Fire Science & Engineering Conference, Interflam, London, UK, 4–6 July 2016; pp. 553–558.
11. Log, T. Modeling Indoor Relative Humidity and Wood Moisture Content as a Proxy for Wooden Home Fire Risk. *Sensors* **2019**, *19*, 5050. [CrossRef] [PubMed]
12. Strand, R.D.; Stokkenes, S.; Kristensen, L.M.; Log, T. Fire Risk Prediction Using Cloud-based Weather Data Services. *J. Ubiquitous Syst. Pervasive Networks* **2021**, *16*, 37–47. [CrossRef]
13. Rahikainen, J.; Keski-Rahkonen, O. Statistical Determination of Ignition Frequency of Structural Fires in Different Premises in Finland. *Fire Technol.* **2004**, *40*, 335–353. [CrossRef]
14. Reducing fire Disaster Risk through Dynamic Risk Assessment and Management (DYNAMIC). Available online: https://www.hvl.no/en/project/2495578/ (accessed on 27 March 2023).
15. Log, T.; Vandvik, V.; Velle, L.; Metallinou, M.M. Reducing Wooden Structure and Wildland-Urban Interface Fire Disaster Risk through Dynamic Risk Assessment and Management. *Appl. Syst. Innov.* **2020**, *3*, 16. [CrossRef]
16. Maria-Monika, M.; Torgrim, L. Cold Climate Structural Fire Danger Rating System? *Challenges* **2018**, *9*, 12. [CrossRef]
17. Tan, S.; Moinuddin, K.; Joseph, P. The Ignition Frequency of Structural Fires in Australia from 2012 to 2019. *Fire* **2023**, *6*, 35. [CrossRef]
18. Tillander, K.; Keski-Rahkonen, O. The Ignition Frequency Of Structural Fires In Finland 1996–1999. *Fire Saf. Sci.* **2003**, *7*, 1051–1062. [CrossRef]
19. Abo El Ezz, A.; Boucher, J.; Cotton-Gagnon, A.; Godbout, A. Framework For Spatial Incident-level Wildfire Risk Modelling to Residential Structures at the Wildland Urban Interface. *Fire Saf. J.* **2022**, *131*, 103625. [CrossRef]
20. Papathoma-Köhle, M.; Schlögl, M.; Garlichs, C.; Diakakis, M.; Mavroulis, S.; Fuchs, S. A Wildfire Vulnerability Index for Buildings. *Sci. Rep.* **2022**, *12*, 6378. [CrossRef] [PubMed]
21. The Directorate for Cultural Heritage. Vil Hindre Brann i Norges 200 Verneverdige Trehusmiljøer. Available online: https://www.riksantikvaren.no/siste-nytt/pressemeldinger/vil-hindre-brann-i-norges-200-verneverdige-trehusmiljoer/ (accessed on 21 June 2023).
22. The Directorate for Cultural Heritage. Innføring i Emnene tett Trehusbebyggelse, Brannvern og Tilskuddsordning for Brannvern. Available online: https://www.riksantikvaren.no/veileder/tilskudd-til-brannsikring-av-tette-trehusomrader/ (accessed on 21 June 2023).
23. Edvardsen, K.I.; Ramstad, T. *Trehus*; Sintef: Trondheim, Norway, 2014.
24. Alapieti, T.; Mikkola, R.; Pasanen, P.; Salonen, H. The Influence of Wooden Interior Materials on Indoor Environment: A Review. *Eur. J. Wood Wood Prod.* **2020**, *78*, 617–634. [CrossRef]
25. Koo, E.; Pagni, P.J.; Weise, D.R.; Woycheese, J.P. Firebrands and spotting in large-scale fires. *Int. J. Wildland Fire* **2010**, *19*, 818–843. [CrossRef]
26. Steen-Hansen, A.; Bøe, A.; Hox, K.; Mikalsen, R.F.; Stensaas, J.; Storesund, K. Evaluation of fire spread in the large Lærdal fire. In Proceedings of the 14th International Conference and Exhibition on Fire and Materials, San Francisco, CA, USA, 2–4 February 2015; pp. 1014–1024.
27. Log, T.; Gjedrem, A.M. A Fire Revealing Coastal Norway's Wildland-Urban Interface Challenges and Possible Low-Cost Sustainable Solutions. *Int. J. Environ. Res.* **2022**, *19*, 3038. [CrossRef] [PubMed]
28. Mjelstad, H. (Meteorologist, Oslo, Norway). Personal communication, 2022.

29. Dougherty, M.T.; Johnson, C. Marshall Fire Investigative Summary and Review. Boulder County. 2023. Available online: https://assets.bouldercounty.gov/wp-content/uploads/2023/06/marshall-fire-investigative-summary.pdf (accessed on 24 August 2023).
30. Drysdale, D. *Fire Dynamics;* John Wiley & Sons Inc: Hoboken, NJ, USA, 2011.
31. Karlsson, B.; Quintiere, J.G. *Enclosure Fire Dynamics;* CRC Press: Boca Raton, FL, USA, 2022.
32. Babrauskas, V.; Peacock, R.D. Heat release rate: The single most important variable in fire hazard. *Fire Saf. J.* **1992**, *18*, 255–272. [CrossRef]
33. Stokkenes, S.; Kristensen, L.; Log, T. Cloud-based Implementation and Validation of a Predictive Fire Risk Indication Model. In Proceedings of the Norwegian Informatics Conference, Narvik, Norway, 25–27 November 2019; pp. 1–12.
34. Strand, R.D.; Kristensen, L.M.; Svendal, T.; Fisketjøn, E.H.; Hussain, A.T. A Mobile Application for Wooden House Fire Risk Notifications Based on Edge Computing. In *Lecture Notes in Networks and Systems* series, Proceedings of the 11th World Conference on Information Systems and Technologies, Pisa, Italy, 4–5 April; Springer: London, UK, 2022; *in press*.
35. Norwegian Meteorological Institute. Historical Weather Data. Frost. Available online: https://frost.met.no/index.html (accessed on 20 June 2023).
36. Norwegian Meteorological Institute. MET Weather API. Predicted Weather Data. Available online: https://api.met.no (accessed on 20 June 2023).
37. Schlunk, L.; Kirkhus, A. *Stue i boliger;* 361.105; Sintef Byggforsk: Trondheim, Norway, 2019. Available online: https://www.byggforsk.no/dokument/140# (accessed on 23 August 2023).
38. Losnegård, G. *Norske ulykker og katastrofar;* Skald: Leikanger, Norway, 2013.
39. Botnen, D. (Chief Fire Officer, Haugesund, Norway). Personal communication, 2023.
40. Noori, S.; Mohammadi, A.; Miguel Ferreira, T.; Ghaffari Gilandeh, A.; Mirahmadzadeh Ardabili, S.J. Modelling and Mapping Urban Vulnerability Index against Potential Structural Fire-Related Risks: An Integrated GIS-MCDM Approach. *Fire* **2023**, *6*, 107. [CrossRef]

fire

Article

Research and Application of Improved Multiple Imputation Based on R Language in Fire Prediction

Jie Wang [1,2,3,*], Meilin Yang [1,2,3], Tianming Li [1,2,3], Xuepeng Jiang [1,2,3] and Kaihua Lu [4]

1 School of Resource and Environmental Engineering, Wuhan University of Science and Technology, Wuhan 430081, China
2 Hubei Research Center of Industrial Safety Engineering Technology, Wuhan 430081, China
3 Safety and Emergency Response Institute, Wuhan University of Science and Technology, Wuhan 430081, China
4 Faculty of Engineering, China University of Geosciences (Wuhan), Wuhan 430074, China
* Correspondence: wangjie87@wust.edu.cn

Abstract: An improved multiple imputation based on R language is proposed to deal with the miss of data in a fire prediction model, which can affect the accuracy of the prediction results. Hazard and operability (HAZOP) is used to accurately find the data related to the research purpose, and exclude data with a missing rate greater than 80% and small differences in characteristics. Then, by changing the *m* value in the mice package under the R language (R-mice), the relevant parameters of the complete filling factor set under different *m* values are obtained. The value of *m* is determined after observing and comparing the parameters. The proposed method fully considers the randomness of filling and the difference between the generated dataset. Taking Hubei Province as an example, the data processed by this method are used as the input of the Bayesian network, and the fire trend is used as the output. The results show that the improved multiple imputation based on R-mice can solve the problem of missing data very well, and have a high prediction effect (AUC = 94.0800). In addition, the results of the predictive reasoning and sensitivity analysis show that the government's supervision has a vital influence on the trend of fires in Hubei Province.

Keywords: R language; mice package; HAZOP; Bayesian network; fire trend

Citation: Wang, J.; Yang, M.; Li, T.; Jiang, X.; Lu, K. Research and Application of Improved Multiple Imputation Based on R Language in Fire Prediction. *Fire* **2023**, *6*, 235. https://doi.org/10.3390/fire6060235

Academic Editor: Tiago Miguel Ferreira

Received: 12 April 2023
Revised: 9 June 2023
Accepted: 11 June 2023
Published: 13 June 2023

1. Introduction

Missing data are a common problem in research, and the main reasons for the deletion are measurement errors, data corruption, and equipment failure [1]. The treatment of missing data can be roughly divided into three categories: not processing, direct deletion, and filling. Among them, non-processing refers to the direct application of incomplete data by machine learning such as Bayesian networks [2] and artificial neural networks [3], they are widely employed to estimate the fire risk of human casualties [4]. However, the requirements for operators are high and the error is large. Direct deletion refers to the deletion of data objects or attributes that contain missing values, but the deletion method applies to completely randomly missing datasets with a missing percentage of less than 5%. Otherwise, the loss of critical information affects the results of the study [5].

The filler method is increasingly recognized by scholars as it preserves key data and obtains more accurate research results [6]. The main methods are the maximum expected value algorithm (EM) [7], regression fill [8], cluster fill, and multiple fill [9] etc., and methods of improvement based on their various theories (e.g., KEMI [10], MIDA [11] etc.). Among them, the multiple-fill method is the most commonly used [12], which can fill the missing data with the appropriate statistical methods according to the pattern of data loss, which not only ensures the correlation between the variables but also effectively solves the uncertainty of data loss [13]. However, most of its applications are used in clinical studies [14], and a few scholars have applied it to the field of fire prediction [15,16]. Most of the ways to deal with missing data in the fire field are direct deletion, such as Liu deletes data with missing key features [17]; Jin et al., retain data containing complete features [18]; and Sattari et al., remove cases where the information required for classification was missing [19]. However, with the rapid development of economy and society, although the risk of fire is decreasing year by year and tends to a stable state, the base is still large and cannot be underestimated. Fire situation assessment and prediction play an important role in reducing its risk, which requires that the error of the research results is as small as possible, and the deletion method may cause larger errors.

The coding of the R-based multi-fill method is detailed in Section 2.2, where the m value defaults to 5 [20], which can be taken as 3 to 5 [21], 5 to 10 [22], or 5 to 20, or even higher [23]; the larger the value, the better, but the size of the calculation amount should also be considered. Therefore, the value of m needs to be discussed on a case-by-case basis, and should not be too large or too small, which may make the filling inaccurate and affect the research results.

To solve the above problems, our work proposes an improved multi-imputation based on the R language. In recent years, the relevant national security departments and enterprises have accumulated a large amount of fire data. Therefore, firstly, the Hazard and Operability Study (HAZOP) is used to qualitatively filter out the categories of data that best affected the trend of fire occurrence from the fire statistics, thereby reducing noise, and then the improved multi-fill was applied. After filling it, it is used as the input of the prediction model and the fire occurrence trend is used as the output. The output can provide a basis for decision-makers to formulate countermeasures to reduce the adverse effects of fires. The idea of this article is shown in Figure 1.

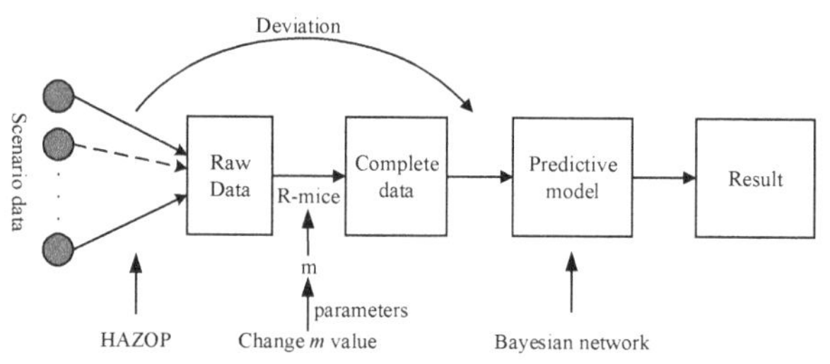

Figure 1. Schematic diagram of improving multiple imputation based on R language.

2. Methodology

2.1. Hazard and Operability (HAZOP)

HAZOP is a qualitative safety evaluation method, guided by keywords (e.g., more, less, part of, no, as well as, reserve, other than) [23], combined with the process parameters (e.g., temperature, pressure, etc.) to find out the deviation of the process state, and then analyze the causes and consequences of the deviation, and finally formulate the prevention and control measures [24], as shown in Figure 2.

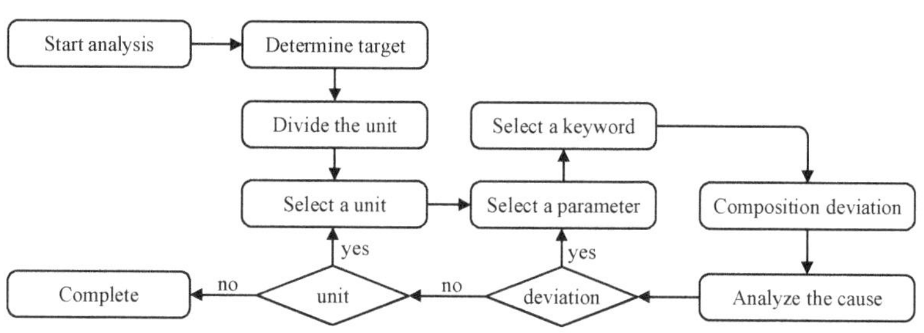

Figure 2. The HAZOP analysis process flow diagram.

The HAZOP analysis is conducted in the form of expert discussion. Before the analysis, the expert decomposes the system into several units according to the research purposes. Each unit has its related functional requirements. There are one or more related parameters, and each parameter corresponds to several guiding words. Select a unit to test the combination of parameters and guiding words that is a deviation. If there is a deviation, analyze its causes and consequences, conduct a risk assessment, and propose the measures to eliminate and control the dangers and reduce losses. After the analysis of all the possible deviations of one unit is completed, transfer to another unit, and repeat the above steps until the analysis of all units is completed.

2.2. Multiple Fills

The multiple-fill method is based on Bayesian estimation theory, and the fills of the missing data are random and all derived from observations. The specific filling steps can be divided into three steps:

1. Choose the appropriate padding method based on the data deletion pattern: monotonic or arbitrary [25], such as predictive mean matching, logistic regression imputation, and professional odds models, *m*-fill missing data to generate *m*-group complete datasets.
2. Statistical analysis of *m*-group datasets.
3. The Rubin method was used to integrate all the analysis results to produce a final inference.

Many pieces of software can implement multiple fillings, such as SOLAS, SAS, and R languages, among others [13]. Compared with other software, R is simple to operate, easy to understand and effective, and is more suitable for scholars to apply it. The steps to implement multiple fillings using the R language and their codes are [26,27]:

library(mice) # Call into the function package
x_1<- *read.spss("data location/file name.sav", to.data.frame = T)* # Import data
x_2<- mice(x_1, *m*, *meth*) # Fill in the data
fit <- *with(x_2, analysis)* # Contains *m* individual statistical analysis results, analysis is used to set the statistical analysis method applied to *m*-filled datasets
pooled <- *pool(fit)* # Contains *m* statistical analysis of the average results
summary(pooled) # Summary

2.3. Bayesian Networks

In solving the practical problems, inferences and judgments need to be made from uncertain knowledge and information, and Bayesian networks (BN) can use a probability theory to deal with the uncertainties caused by conditional correlation [28]. Its network topology is a directed acyclic graph (DAG). The nodes in BN consist of the random variables $X = \{X_1, X_2, \ldots, X_n\}$, non-conditionally independent variables are connected by arrows, and the parent node points to the child node [29]. It has the flexibility to not only learn network structures and parameters through large amounts of data [30], but also to use expert knowledge and data to improve the performance of the model, thus helping us analyze some complex problems (uncertain or missing data) [31], and even support us in developing measures [32].

Let BN = (G, θ), G is the network structure, and θ is the network parameter. The joint probability distribution on X is Equation (1) [33–35]:

$$P(X) = \prod_{i=1}^{n} P(X_i|Pa(X_i)) \tag{1}$$

where $Pa(X_i)$ is all parent nodes of X_i. For arbitrary variables, the joint probability distribution can also be expressed as Equation (2):

$$P(X) = P(X_n|X_1, \ldots, X_{n-1}) \ldots P(X_2|X_1)P(|X_1) \tag{2}$$

A simple Bayesian network: a person with a weak security awareness may be unconsciously littering unextinguished cigarette butts, and if there is combustible material around them, it is highly likely to cause a fire. If the fire is not detected and controlled in time, it will grow larger and quickly spread to the surrounding area, and then improper operation or untimely escape will cause injury or even death; in some cases, such as in Figure 3a, the network structure and condition probability table can be obtained in the input software, as shown in Figure 3b. Its joint probability distribution is $P(X) = P(X_A)$ $P(X_C \mid X_A)$ $P(X_D \mid X_A)$ $P(X_B)$ $P(X_E \mid X_B, X_C)$ $P(X_D \mid X_A, X_E)$. Change either condition, and the probability of being late will change accordingly. For example, when a person has an extremely weak security awareness (True = 100%, Figure 3c), the probability of injury or death will increase from 42% to 68%. Conversely, when a person has a good sense of security (False = 100%, Figure 3d), the probability of injury or death will reduce from 42% to 17%.

No.	A	B	C	D	E
1	T	T	T	T	T
2	T	F	F	F	F
3	F	T	T	T	T
4	F	F	F	F	F
…	…	…	…	…	…

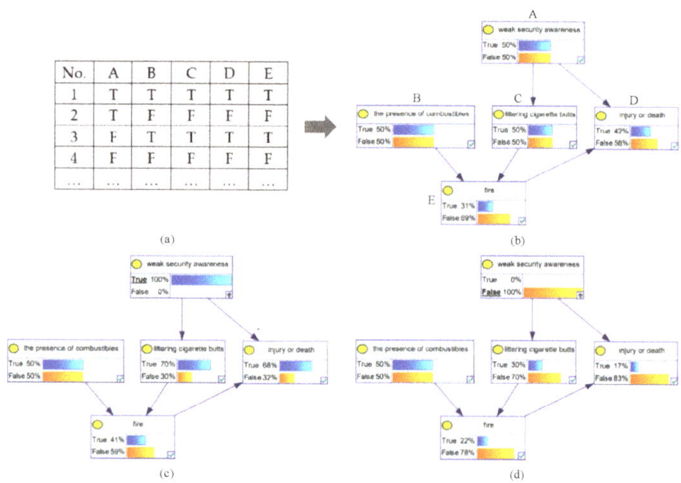

Figure 3. Simple Bayesian network. (**a**) A table of some cases; (**b**) The network structure and condition probability table of a person with moderate security awareness; (**c**) The network structure and condition probability table of a person with extremely weak security awareness; (**d**) The network structure and condition probability table of a person with extremely high security awareness.

3. Instance Calculations

Hubei Province is located in the east of China, with 17 cities and prefectures in total. This article will use an improved multiple imputation based on R-mice to construct a prediction model to predict fire trends x_0 and provide a basis for decision-makers to formulate fire prevention measures. The specific analysis steps are as follows.

3.1. Collection and Processing of Fire Statistics

Through field investigation and consulting the China Fire Rescue Yearbook, we have obtained data on various fire scenarios in Hubei Province from 2016 to 2018, which can be roughly divided into the following 5 categories: fire loss, fire information, fire rescue, fire brigade, and fire equipment [36]. However, not all of them are what we need. Exclude some scenario data that have little influence on the research purposes, which can reduce the complexity and noise of the model, reduce the amount of calculation, and increase the readability and accuracy of the model [37]. Last but not least, all data need to be normalized to facilitate the subsequent analysis [38], as shown in Equation (3):

$$x_i' = \frac{x_i - \min(x_i)}{\max(x_i) - \min(x_i)} \tag{3}$$

The role of HAZOP in this article is to screen out the data we need. The advantage of the HAZOP method is simplicity and ease of operation. Experts from various fields gather together to express their opinions, influence, and inspire each other and discover more problems creatively. The specific process of using HAZOP to screen data is shown in Figure 4, that is, screen out some data irrelevant to the research purpose and data with unobvious characteristics through Condition 1 and Condition 2. The most important thing is in the analysis process, the primary and secondary indicators generated can construct a prediction model structure about the fire trend through correlation analysis, and the deviation can be used as the state of the model node, as shown in Table 1.

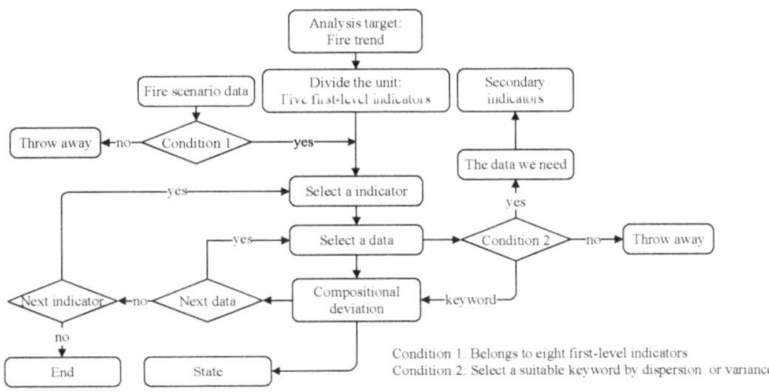

Figure 4. Flow chart of screening scenario data using HAZOP.

Table 1. Prediction model structure about the fire trend.

Test No.	First-Level Indicators	Scenario Data (Illustration)	Abbr.	Deviation
1	Safety awareness y_1	propaganda situation fire propaganda method (internet, slogans, lectures, etc.)	x_1 x_2	frequent, normal, less variety, ordinary, single
2	Adverse effects y_2	number of dead number of injures direct economic loss affected households	x_3 x_4 x_5 x_6	[0, 0.33), [0.33, 0.67), [0.67, 1] [0, 0.33), [0.33, 0.67), [0.67, 1] [0, 0.2), [0.2, 0.4), [0.4, 0.6), [0.6, 0.8), [0.8, 1] [0, 0.2), [0.2, 0.4), [0.4, 0.6), [0.6, 0.8), [0.8, 1]
		Cause of the fire (electrical, production operations, accidental use of fire, smoking, nature, lightning strikes, static electricity, arson, etc.)	x_7	dominant reason
3	Emergency rescue capability y_3	alarm dispatch situation number of combatants number of dispatched vehicles number of rescuers salvage property value joint training situation emergency plan preparation fire research results	x_8 x_9 x_{10} x_{11} x_{12} x_{13} x_{14} x_{15}	[0, 0.2), [0.2, 0.4), [0.4, 0.6), [0.6, 0.8), [0.8, 1] [0, 0.25), [0.25, 0.5), [0.5, 0.75), [0.75, 1] [0, 0.2), [0.2, 0.4), [0.4, 0.6), [0.6, 0.8), [0.8, 1] [0, 0.25), [0.25, 0.5), [0.5, 0.75), [0.75, 1] [0, 0.33), [0.33, 0.67), [0.67, 1] frequent, normal, less sufficient, general, insufficient yes, no
4	Equipment and facilities y_4	number of fire stations number of fire engines number of rescue equipment public fire protection facilities (fire hydrants, smoke alarms, etc.)	x_{16} x_{17} x_{18} x_{19}	[0, 0.25), [0.25, 0.5), [0.5, 0.75), [0.75, 1] [0, 0.25), [0.25, 0.5), [0.5, 0.75), [0.75, 1] [0, 0.2), [0.2, 0.4), [0.4, 0.6), [0.6, 0.8), [0.8, 1] perfect, average, lacking
5	Supervision intensity y_5	organizational structure (informatization, flattening, and standardization) hidden danger rectification rate $((N_{\text{rectifications}})/(N_{\text{discoveries}}))$ fire protection expenditure utilization rate ((use/approve))	x_{20} x_{21} x_{22}	reasonable, unreasonable high, medium, low high, medium, low

3.2. Fill Scenario Data

A total of 51 cases were obtained, with 1683 data and 312 missing data, with a missing rate of 18.54%. When the missing rate is about 20%, compared with the deletion method, mean (mode) method, and regression method, the R-mice filling effect is the best [11]. The specific steps are shown in Section 2.2. In this section, the value of m is determined and a complete dataset is obtained.

First, calculate the total variance t (Equation (4)) of the complete factor set x_1, x_2, x_9~x_{15}, x_{17}~x_{20}, and x_{22} filled under different values of m through code 1 and code 2. The above value of t is smaller the better. The results are shown in Figure 5. The t value of each complete factor set under eleven different m values is presented in the form of a scatter plot. As the m value increases, the t value continues to decrease. Observing the black dashed line in Figure 5, it can be seen that when $m \geq 8$, the change in the t value is small, or even no longer changes. That is, when $m > 8$, the value of m has little effect on the value of t. Therefore, the value of m is 8 in this work.

code 1: **data <-mice(x, m = I, maxit = 100, seed = 2020), i = 1, 2, . . . , n**
code 2: fit <- with(data, lm (x_0~x_1 + x_2 + . . . + x_{21}+ x_{22}))

$$u = \frac{1}{n}\sum_{i=1}^{n}(x_i - \overline{x})^2 \Rightarrow t = u + (1 + \frac{1}{m}) \times u^{\frac{1}{2}} \quad (4)$$

where m refers to the number of interpolated datasets, maxit refers to the number of iterations, seed refers to the number of set seeds, and u refers to the mean squared error (MSE).

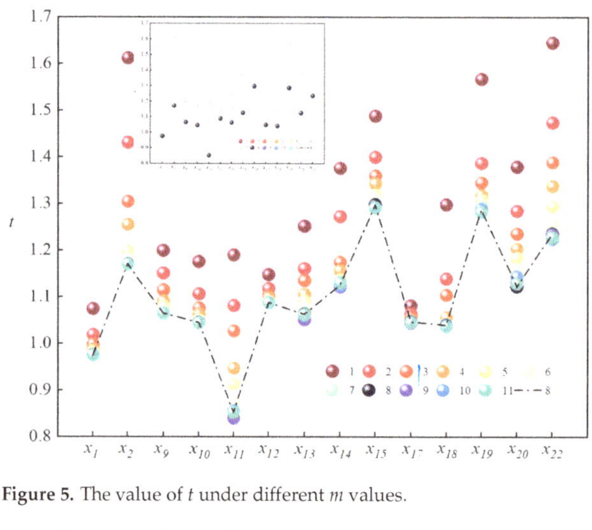

Figure 5. The value of t under different m values.

Then, use code 3 to find eight complete datasets, use code 4 to analyze them, and find the residual standard error (on 57 degrees of freedom), the multiple R-squared, and the F statistic value. It is not difficult to see from the three dashed lines in Figure 6 that the parameters of the fourth dataset are the best in all datasets. Among them, dataset 9 is the dataset with the best parameters when m is 5. Therefore, this article chooses the fourth dataset as the input of the predictive model. Moreover, the results show that determining the value of m by the above method can make the filled data better fit the real data. The impact on the forecast results will be analyzed below.

code 3: data y <- complete(data, y), y = 1, 2, 3, . . . , 8
code 4: fit y <- with(data y, lm (x_0~x_1 + x_2 + . . . + x_{21} + x_{22}))

where the code 3 is to fill in one of the eight filling datasets to the vacant position, and finally form a complete dataset.

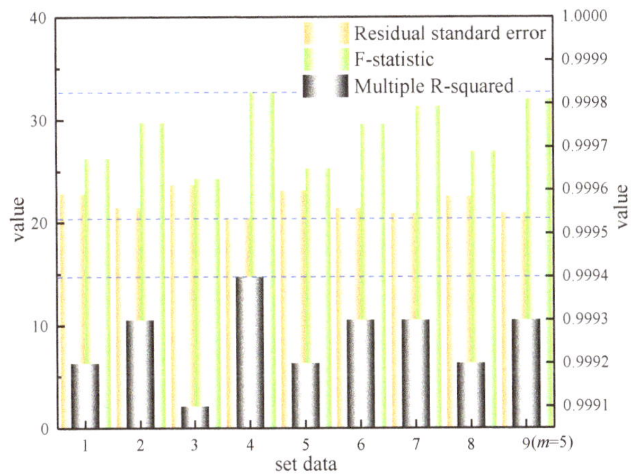

Figure 6. Related parameters of the interpolated dataset.

Based on the filled complete dataset and expert knowledge, the structure of the prediction model is learned, and then the dataset is learned according to the training ratio of 8:2, and the final fire occurrence trend prediction and prediction model is obtained as shown in Figure 7.

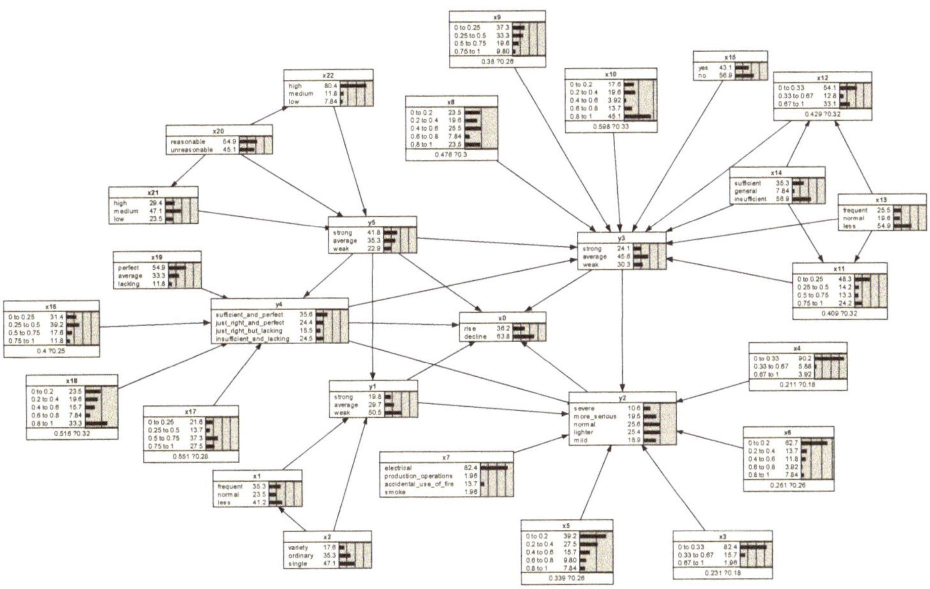

Figure 7. Fire occurrence trend prediction model.

4. Results and Discussion

This section considers the predictive reasoning of the Bayesian network, that is, forward propagation analysis [28,39]. By comprehensively using the fire scenario data and expert knowledge of Hubei Province in 2018, it is input into the prediction model to predict the trend of fire occurrence in Hubei Province in 2019, and compared with actual data to verify the accuracy of the model.

This paper uses the area under the Receiver Operating Characteristic (ROC) curve, that is, the AUC value, to test the prediction effect. Among them, the AUC standard for judging the pros and cons of the prediction model is shown in Table 2 [40,41]. The classification threshold is calculated through the Youden index (Equation (4)). When the predicted probability is greater than the threshold, it is considered to happen; otherwise, it is considered not to happen [42].

$$\text{Youden index} = \text{specificity} + \text{sensitivity} - 1 \quad (5)$$

Table 2. AUC standard for judging the pros and cons of the prediction model.

AUC	[0.5, 0.7]	[0.7, 0.85]	[0.85, 0.95]	[0.95, 1]
Effect	low	fair	good	perfect

4.1. Predictive Reasoning

Through the controlled variable method, three other sets of prediction models are constructed, as shown in Table 3, in which feature screening is to screen out features that have greater weight on the predictor variables from the original data, and the methods include filtering, wrapping, and embedded; this article uses the filtering, because it and HAZOP both screen the data before building the predictive model. According to the analysis and processing of the screening method and m value, the status of each scenario data of each city in Hubei Province in 2018 is obtained then input into the prediction model to find the fire occurrence trend in 2019. The prediction effect is tested using the ROC curve and AUC, as shown in Figure 8.

Table 3. Condition setting of data processing before the model construction.

Text No.	Data Screening Method	m Value
1	/	5
2	Filtering	5
3	HAZOP	5
4	HAZOP	m

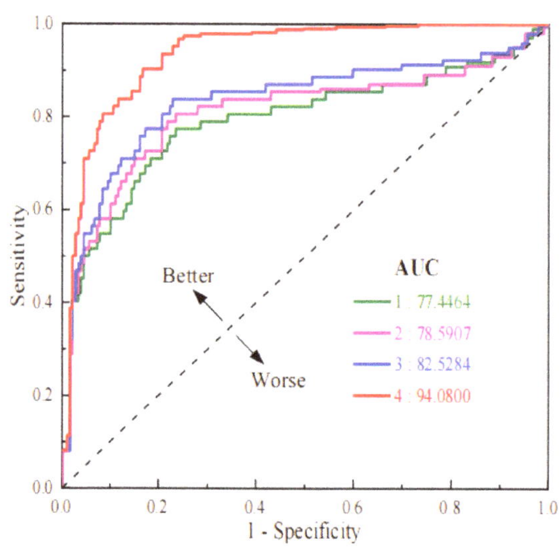

Figure 8. The ROC curve and AUC value of the prediction model.

When m is 5 (the default and the most used), it is better to use HAZOP (3) to screen the data compared with no screening (1) and using the filtering method (2). First, the amount of calculation required is relatively small. Second, it can not only screen quantitative data, but also qualitative data, and the operation difficulty is relatively low. The most important thing is that part of the predictive model construction is completed in the screening process, that is, the state of the node, which greatly reduces the time consumed and improves the efficiency of screening. Finally, the forecasting effect is better (82.52843 > 78.59072 > 77.44641). Therefore, HAZOP is more suitable for data screening.

Compared with m = 5 based on the HAZOP method (3), the prediction effect of m = 8 based on the HAZOP method (4) is better (94.08003 > 82.52844). Although there are differences in other aspects, they are negligible and can be ignored. The results show that improved multiple imputation based on R language can be used for data processing before constructing a prediction model, and it has a good prediction effect (AUC = 94.0800).

Taking full account of the geographical location, population density, and GDP distribution of each city in Hubei Province, combined with the prediction results in Figure 9, a fire risk map of Hubei Province is finally generated, as shown in Figure 9. It can be seen from Figure 9 that the eastern part of Hubei Province is economically developed, although the population is dense, the fire risk is low; the western region is relatively backward in the economy and sparsely populated, but the fire risk is high, especially in the southwest.

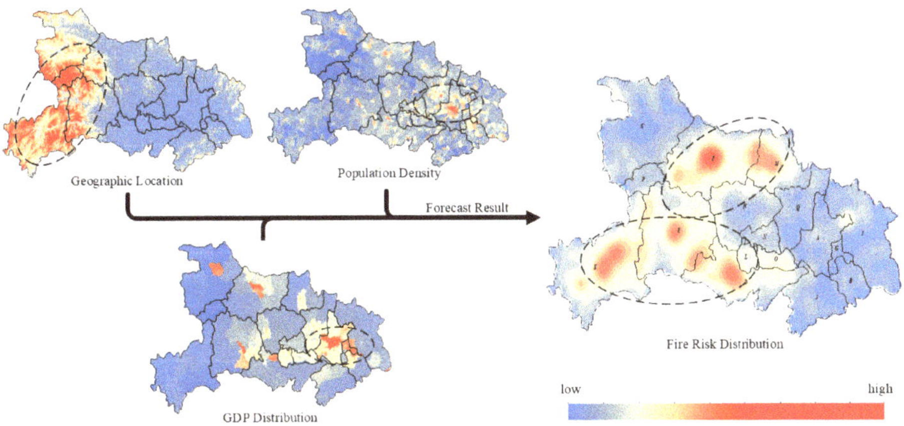

Figure 9. Visualization of fire trend prediction results in Hubei Province.

The reason for this phenomenon may be that compared with the eastern region, the western region has a higher terrain, mostly mountainous areas, and inconvenient transportation. In addition, the processing industry is relatively weak, the development of the high-precision tertiary industry is relatively lagging, the degree of industrialization is low, and the economic relevance is low. This has led to the relatively slow economic development of the western region and the inability to provide sufficient funds and talents for the development of fire rescue services. It is directly manifested in the lack of firefighting funds, weak government supervision, insufficient propaganda, and inadequate facilities and equipment.

4.2. Sensitivity Analysis

Sensitivity analysis refers to the study of the impact degree of changes in one node in the prediction model on other nodes from the perspective of quantitative analysis, which can identify errors in network structure or CPT [28,43], and provide the basis for risk analysis and control measures. In this paper, entropy reduction (Equation (6)) is used to quantify the sensitivity [44], and the nodes were sorted according to the impact of the results on x_0 nodes. The results of the sensitivity analysis are shown in Table 4.

$$ER = H(Y) - H(Y|X) = \sum_{y}\sum_{x} P(y,x)\frac{\log_2[(y,x)]}{P(y)P(x)} \tag{6}$$

where $H(Y)$ is the entropy and $H(Y \mid X)$ is the conditional entropy.

Table 4. The influence of sensitivity analysis results based on entropy reduction on x_0.

Nodes	y_5	y_1	y_4	y_3	y_2
ER	0.7175	0.3238	0.261	0.1376	0.1279
Percentage (%)	45.33	26.77	20.11	9.67	8.93

As shown in Table 4, y_5 is the most significant variable, leading to the maximum entropy reduction of x_0. y_1 and y_4 also have a large impact, with entropy reduction of more than 20%. However, the influence of y_3 and y_2 is relatively small, and the entropy reduction is less than 10%, so they are not suitable to be used as control variables in the prediction. Therefore, y_5, y_1, and y_4 can be used as control variables.

Compared to reality, the result of sensitivity is reasonable. Because if the government's supervision is insufficient, fire hazards cannot be managed in time, firefighting funds cannot be fully utilized, firefighting equipment and facilities will lag behind more and more complex fire situations, and even people's safety awareness will become weak, leading to urban fire risks. The rapid increase cannot be controlled by the emergency rescue capability alone, that is, it cannot improve the state of the target node alone. Because of the importance of supervision and management, the relevant departments should attach great importance to the development and soundness of efficient and resilient supervision mechanisms and systems.

4.3. Measures and Suggestions

The western region should develop processing industries, promote the development of leading industries from resource exporting to processing industries, and increase the added value of products; develop a tertiary industry with a high rate of return, increase the degree of relevance to surrounding industries, cultivate new economic growth points, and drive the development of other industries; and take advantage of talents, develop high-tech industries and emerging industries, and accelerate the pace of transforming scientific and technological achievements into productivity, so that there are surplus funds to develop the firefighting and rescue business, to meet the growing safety needs of the people.

Among them, cities of *F*, *O*, *K*, *D*, *M*, *E*, and *L* should strengthen supervision, promptly discover and supervise the elimination of fire hazards (especially electrical hazards), and archive major hazards and conduct regular inspections; timely check, repair, update, upgrade, and add rescue equipment to ensure that they play their due role in the disaster; and strengthen the publicity, innovative publicity ways, expand the scope of publicity, make fire prevention propaganda ubiquitous, and subtly improve people's safety awareness and skills. While strengthening supervision and propaganda, other cities also need to improve the legal system for fires, increase penalties for deliberate arson, and restrict them.

5. Conclusions

Aiming at the fire prediction model of firefighters, they can accurately obtain the relevant scenario data, predict the fire trend of the next year for the first time, and provide a quantitative basis for effective control measures to reduce the number of fires. This paper constructs a predictive model, analyzes, verifies, and applies the predictive model, and the results show that the model can predict the trend of fire occurrence well. It can not only propose specific measures for a certain city, but also provide a strong basis for a province to formulate accurate and effective fire protection planning, thereby reducing the number of fires, casualties, and property losses, and building a safe living environment. The specific conclusions are as follows:

1. Using the HAZOP method to screen data can complete multiple tasks in a shorter time and make full use of expert knowledge and collected data. This method combines qualitative and quantitative data, which not only makes expert knowledge more convincing but also makes data more meaningful and valuable.
2. Determining the value of m through relevant parameters fully considers the randomness of filling data and the differences between datasets. By continuously changing the value of m and using the relevant parameters as the basis for judgment, the best filling data are selected, so that the complete dataset obtained is closer to the real data, and the prediction effect of the prediction model is greatly improved (AUC: 94.0800 > 82.5284).
3. The forecast results show that the western part of Hubei Province (especially the southwestern part) is a high fire risk area, which is consistent with the actual situation. The reason may be that its geographic location and development strategy has caused its economic development to lag, lack of rich funds to develop the fire protection industry, neglect of safety management, and generally low awareness of public safety. For the above reasons, this article puts forward some suggestions for improvement, see Section 4.3 for details.

Author Contributions: Conceptualization, J.W.; formal analysis, T.L. and M.Y.; investigation, X.J.; writing—original draft preparation, T.L.; writing—review and editing, J.W. and M.Y.; supervision, K.L. All authors have read and agreed to the published version of the manuscript.

Funding: This research was funded by the National Natural Science Foundation of China under Grant Nos. 52076199 and 51806156.

Data Availability Statement: The data presented in this study are available on request from the corresponding author. The data are not publicly available due to privacy.

Acknowledgments: This study was supported by the National Natural Science Foundation of China under Grant Nos. 52076199 and 51806156. The authors gratefully acknowledge all these supports.

Conflicts of Interest: The authors declare no conflict of interest.

References

1. Alipour, M.; La Puma, I.; Picotte, J.; Shamsaei, K.; Rowell, E.; Watts, A.; Kosovic, B.; Ebrahimian, H.; Taciroglu, E. A Multimodal Data Fusion and Deep Learning Framework for Large-Scale Wildfire Surface Fuel Mapping. *Fire* **2023**, *6*, 36. [CrossRef]
2. Tavakol Sadrabadi, M.; Innocente, M.S. Vegetation Cover Type Classification Using Cartographic Data for Prediction of Wildfire Behaviour. *Fire* **2023**, *6*, 76. [CrossRef]
3. Mahamed (Polinova), M.; Wittenberg, L.; Kutiel, H.; Brook, A. Fire Risk Assessment on Wildland–Urban Interface and Adjoined Urban Areas: Estimation Vegetation Ignitability by Artificial Neural Network. *Fire* **2022**, *5*, 184. [CrossRef]
4. Kussul, N.; Fedorov, O.; Yailymov, B.; Pidgorodetska, L.; Kolos, L.; Yailymova, H.; Shelestov, A. Fire Danger Assessment Using Moderate-Spatial Resolution Satellite Data. *Fire* **2023**, *6*, 72. [CrossRef]
5. Łopucki, R.; Kiersztyn, A.; Pitucha, G.; Kitowski, I. Handling missing data in ecological studies: Ignoring gaps in the dataset can distort the inference. *Ecol. Model.* **2022**, *468*, 109964. [CrossRef]
6. Xiong, Z.; Guo, H.; Wu, Y. Review of Missing Data Processing Methods. *Comput. Eng. Appl.* **2021**, *57*, 27–38.
7. Dempster, A.P.; Laird, N.M.; Rubin, D.B. Maximum likelihood from incomplete data via the EM algorithm. *J. R. Stat. Soc. Ser. B Methodol.* **1977**, *39*, 1–22.
8. Little, R.J.A. Hierarchical Logistic Regression Models for Imputation of Unresolved Enumeration Status in Undercount Estimation: Comment. *J. Am. Stat. Assoc.* **1993**, *88*, 1159. [CrossRef]
9. Rubin, D.B.; Service, E.T. Multiple imputations in sample surveys—A phenomenological Bayesian approach to nonresponse. In Proceedings of the Survey Research Methods Section of the American Statistical Association, Alexandria, VA, USA, 2 January 1978.
10. Razavi-Far, R.; Cheng, B.; Saif, M.; Ahmadi, M. Similarity-learning information-fusion schemes for missing data imputation. *Knowl.-Based Syst.* **2020**, *187*, 104805. [CrossRef]
11. Gondara, L.; Wang, K. MIDA: Multiple Imputation Using Denoising Autoencoders. *Adv. Knowl. Discov. Data Min.* **2018**, *91*, 10939.
12. Mohammed, Y.S.; Abdelkader, H.; Pławiak, P.; Hammad, M. A novel model to optimize multiple imputation algorithm for missing data using evolution methods. *Biomed. Signal Process. Control* **2022**, *76*, 103661. [CrossRef]

13. Li, W.; Wang, Y.; Zhang, J.; Li, J.; Huang, B. Application of MICE in R for imputing incomplete multivariate data. *Chin. J. Hosp. Stat.* **2011**, *18*, 309–312.
14. Jakobsen, J.C.; Gluud, C.; Wetterslev, J.; Winkel, P. When and how should multiple imputation be used for handling missing data in randomised clinical trials—A practical guide with flowcharts. *BMC Med. Res. Methodol.* **2017**, *17*, 162. [CrossRef]
15. Cory, W.O.; Bishrant, A.; Simon, P.A.; Paddington, H.; Chen, X.; Thomas, A.M. Predicting Fire Propagation across Heterogeneous Landscapes Using WyoFire: A Monte Carlo-Driven Wildfire Model. *Fire* **2020**, *3*, 71.
16. Malik, A.A.; Nasif, M.S.; Arshad, U.; Mokhtar, A.A.; Tohir, M.Z.; Al-Waked, R. Predictive Modelling of Wind-Influenced Dynamic Fire Spread Probability in Tank Farm Due to Domino Effect by Integrating Numerical Simulation with ANN. *Fire* **2023**, *6*, 85. [CrossRef]
17. Liu, X.; Lu, Y.; Xia, Z.; Li, F.; Zhang, T. A Data Mining Method for Potential Fire Hazard Analysis of Urban Buildings based on Bayesian Network. In Proceedings of the 2nd International Conference on Intelligent Information Processing—IIP'17, Bangkok, Thailand, 17–18 July 2017; ACM Press: New York, NY, USA, 2017; pp. 1–6.
18. Jin, G.; Wang, Q.; Zhu, C.; Feng, Y.; Huang, J.; Hu, X. Urban Fire Situation Forecasting: Deep sequence learning with spatio-temporal dynamics. *Appl. Soft Comput.* **2020**, *97*, 106730. [CrossRef]
19. Sattari, F.; Macciotta, R.; Kurian, D.; Lefsrud, L. Application of Bayesian network and artificial intelligence to reduce accident/incident rates in oil & gas companies. *Saf. Sci.* **2021**, *133*, 104981.
20. Squillante, R.; Santos Fo, D.J.; Maruyama, N.; Junqueira, F.; Moscato, L.A.; Nakamoto, F.Y.; Miyagi, P.E.; Okamoto, J. Modeling accident scenarios from databases with missing data: A probabilistic approach for safety-related systems design. *Saf. Sci.* **2018**, *104*, 119–134. [CrossRef]
21. Carpenter, J.R.; Kenward, M.G. Multiple Imputation And Its Application. *Int. Stat. Rev.* **2014**, *82*, 151–152.
22. Schafer, J.L. Multiple imputation: A primer. *Stat. Methods Med. Res.* **1999**, *8*, 3–15. [CrossRef]
23. Buuren, S.V. *Flexible Imputation of Missing Data*; Chapman and Hall/CRC: New York, NY, USA, 2012.
24. Marhavilas, P.K.; Filippidis, M.; Koulinas, G.K.; Koulouriotis, D.E. An expanded HAZOP-study with fuzzy-AHP (XPA-HAZOP technique): Application in a sour crude-oil processing plant. *Saf. Sci.* **2020**, *124*, 104590. [CrossRef]
25. Dunjo, J.; Fthenakis, V.; Vilchez, J.A.; Arnaldos, J. Hazard and operability (HAZOP) analysis. A literature review. *J. Hazard. Mater.* **2010**, *173*, 19–32. [CrossRef] [PubMed]
26. Dohoo, I.R.; Nielsen, C.R.; Emanuelson, U. Multiple imputation in veterinary epidemiological studies: A case study and simulation. *Prev. Vet. Med.* **2016**, *129*, 35–47. [CrossRef] [PubMed]
27. Flatau Harrison, H.; Griffin, M.A.; Gagne, M.; Andrei, D. Assessing shortened safety climate measures: Simulating a planned missing data design in a field setting. *Saf. Sci.* **2018**, *104*, 189–201. [CrossRef]
28. Enders, C.K. *Applied Missing Data Analysis*, 1st ed.; The Guilford Press: New York, NY, USA, 2010.
29. Lawrence, J.-M.; Ibne Hossain, N.U.; Jaradat, R.; Hamilton, M. Leveraging a Bayesian network approach to model and analyze supplier vulnerability to severe weather risk: A case study of the U.S. pharmaceutical supply chain following Hurricane Maria. *Int. J. Disaster Risk Reduct.* **2020**, *49*, 101607. [CrossRef] [PubMed]
30. Cooper, G.F.; Herskovits, E. A Bayesian method for the induction of probabilistic networks from data. *Mach. Learn.* **1992**, *9*, 309–347. [CrossRef]
31. Dlamini, W.M. Application of Bayesian networks for fire risk mapping using GIS and remote sensing data. *GeoJournal* **2011**, *76*, 283–296. [CrossRef]
32. Sevinc, V.; Kucuk, O.; Goltas, M. A Bayesian network model for prediction and analysis of possible forest fire causes. *For. Ecol. Manag.* **2020**, *457*, 117723. [CrossRef]
33. Leao, T.; Madeira, S.C.; Gromicho, M.; De Carvalho, M.; Carvalho, A.M. Learning dynamic Bayesian networks from time-dependent and time-independent data: Unraveling disease progression in Amyotrophic Lateral Sclerosis. *J. Biomed. Inform.* **2021**, *117*, 103730. [CrossRef]
34. Lee, S.; Kang, J.E.; Park, C.S.; Yoon, D.K.; Yoon, S. Multi-risk assessment of heat waves under intensifying climate change using Bayesian Networks. *Int. J. Disaster Risk Reduct.* **2020**, *50*, 101704. [CrossRef]
35. Baksh, A.-A.; Khan, F.; Gadag, V.; Ferdous, R. Network based approach for predictive accident modelling. *Saf. Sci.* **2015**, *80*, 274–287. [CrossRef]
36. Liu, H.; Guo, Y. Design and implementation of fire prediction model based on scenario data. *Technol. Innov. Appl.* **2018**, *26*, 79–80.
37. Guyon, I.; Elisseeff, A. An Introduction of Variable and Feature Selection. *J. Mach. Learn. Res. Spec. Issue Var. Feature Sel.* **2003**, *3*, 1157–1182.
38. Liang, H.; Zhang, M.; Wang, H. A Neural Network Model for Wildfire Scale Prediction Using Meteorological Factors. *IEEE Access* **2019**, *7*, 176746–176755. [CrossRef]
39. Neil, M. *Risk Assessment and Decision Analysis with Bayesian Networks*, 2nd ed.; Chapman and Hall/CRC: Boca Raton, FL, USA, 2018.
40. Gu, X.; Wu, Z.; Zhang, Y.; Yan, S.; Fu, J.; Du, L. Prediction research of the forest fire in Jiangxi province in the background of climate change. *Acta Ecol. Sin.* **2020**, *40*, 667–677.
41. Bai, H.; Liu, X.; Niu, S.; He, Y. Construction of forest fire prediction model based on Bayesian model averaging method: Taking Dali Prefecture, Yunnan Province of southwestern China as an example. *J. Beijing For. Univ.* **2021**, *43*, 44–52.
42. Chang, Y.; Zhu, Z.; Bu, R.; Chen, H.; Feng, Y.; Li, Y.; Hu, Y.; Wang, Z. Predicting fire occurrence patterns with logistic regression in Heilongjiang Province, China. *Landsc. Ecol.* **2013**, *28*, 1989–2004. [CrossRef]

43. Pollino, C.A.; Woodberry, O.; Nicholson, A.; Korb, K.; Hart, B.T. Parameterisation and evaluation of a Bayesian network for use in an ecological risk assessment. *Environ. Model. Softw.* **2007**, *22*, 1140–1152. [CrossRef]
44. De Iuliis, M.; Kammouh, O.; Cimellaro, G.P.; Tesfamariam, S. Quantifying restoration time of pipelines after earthquakes: Comparison of Bayesian belief networks and fuzzy models. *Int. J. Disaster Risk Reduct.* **2021**, *64*, 102491. [CrossRef]

fire

Article

Modelling and Mapping Urban Vulnerability Index against Potential Structural Fire-Related Risks: An Integrated GIS-MCDM Approach

Sepideh Noori [1], Alireza Mohammadi [1,*], Tiago Miguel Ferreira [2], Ata Ghaffari Gilandeh [1] and Seyed Jamal Mirahmadzadeh Ardabili [3]

[1] Department of Geography and Urban Planning, Faculty of Social Sciences, University of Mohaghegh Ardabili, Ardabil 56199-11367, Iran
[2] College of Arts, Technology and Environment, School of Engineering, University of the West of England (UWE Bristol), Frenchay Campus, Bristol BS16 1QY, UK; tiago.ferreira@uwe.ac.uk
[3] Department of Urban Planning, Islamic Azad University, Tabriz 51579-44533, Iran
* Correspondence: a.mohammadi@uma.ac.ir

Abstract: Identifying the regions with urban vulnerability to potential fire hazards is crucial for designing effective risk mitigation and fire prevention strategies. The present study aims to identify urban areas at risk of fire using 19 evaluation factors across economic, social, and built environment-infrastructure, and prior fire rates dimensions. The methods for "multi-criteria decision making" (MCDM) include the Analytic Hierarchy Process for determining the criteria's importance and weight of the criteria. To demonstrate the applicability of this approach, an urban vulnerability index map of Ardabil city in Iran was created using the Fuzzy-VIKOR approach in a Geographic Information System (GIS). According to the findings, about 9.37 km^2 (31%) of the city, involving roughly 179,000 people, presents a high or very high level of risk. Together with some neighbourhoods with low socioeconomic and environmental conditions, the city centre is the area where the level of risk is more significant. These findings are potentially very meaningful for decision-makers and authorities, providing information that can be used to support decision-making and the implementation of fire risk mitigation strategies in Ardabil city. The results of this research can be used to improve policy, allocate resources, and renew urban areas, including the reconstruction of old, worn-out, and low-income urban areas.

Keywords: risk analysis; structural fire risk; urban vulnerability index; mapping; GIS-MCDM; city

Citation: Noori, S.; Mohammadi, A.; Miguel Ferreira, T.; Ghaffari Gilandeh, A.; Mirahmadzadeh Ardabili, S.J. Modelling and Mapping Urban Vulnerability Index against Potential Structural Fire-Related Risks: An Integrated GIS-MCDM Approach. *Fire* **2023**, *6*, 107. https://doi.org/10.3390/fire6030107

Academic Editor: Grant Williamson

Received: 2 February 2023
Revised: 3 March 2023
Accepted: 6 March 2023
Published: 8 March 2023

1. Introduction

Urban fire risk is the possibility of damage to people's life safety, property loss, and the threat to public security caused by the interaction between fire accidents and urban vulnerability, and the possibility of negative consequences or likely loss such as the breaking up of economic activities and environmental destruction [1]. Urban fire risk is a tremendous challenge to sustainable urban development, especially in low-income countries [2,3]. It causes damage to urban buildings and infrastructure and poses a major hazard to inhabitants' lives and property. The cost of fire damage is disproportionately high in major cities and highly inhabited areas [4]. According to the World Health Organization (WHO), around 3 million fires occur worldwide, with approximately 180,000 people dying each year [5]. Furthermore, most of these catastrophes occur in large cities in low-income countries [6,7] and many economic, social, and environmental conditions in these places raise the risk of possible fires [8]. While total fire prevention is virtually impossible, damages caused by potential building fires can be contained [9]. Modelling and predicting fire dangers is a crucial step in preventing fire damage in urban areas because it employs scientific frameworks to identify hazards [10] and assists local organisations in implementing appropriate geographic data to cope with the damages associated with urban fires [11]. It is critical to identify the risk of fire in urban environments, particularly for rescue agencies in major cities [12]. Discovering and presenting high-risk regions for future fires would assist local organisations in taking effective measures to lower the risk, allocate resources in a more efficient manner, and distribute fire protection infrastructure throughout the city.

While resilience is the capacity of the system to withstand a major disruption within acceptable degradation parameters and to recover within an acceptable time, as well as composite costs and risks, vulnerability is the manifestation of the inherent states of the system that can be subjected to a natural or human-related hazard or be exploited to adversely affect that system. Contrarily, risk is based on probability and is determined by the likelihood and seriousness of unfavourable consequences [13]. In this essay, we take a closer look at fire risk and urban vulnerability to fire. Urban vulnerability is the result of the interaction of a number of disadvantages. Usually, the more vulnerable and distressed areas lack basic services and have a higher number of obsolete buildings, unfavourable social characteristics, vulnerable people, and more prominent social and environmental differences [14,15]. Urban vulnerability in this study refers to the possibility of fires in urban areas where there are poor environmental conditions, disrespect for building engineering requirements, and inadequate urban planning standards. Areas with poor socioeconomic conditions are included as well.

Fires are mostly spatial, meaning that they can be modelled and mapped [16], and that Geographic Information Systems (GIS) can be efficiently used to identify, manage, and anticipate fire events [17]. In fact, modelling and identifying high-risk and sensitive urban areas using spatial metrics [18] is an essential step towards reducing the probability of human and material losses resulting from fire events. Kernel Density Estimation (KDE), Monte Carlo Simulation models (MCS), geographically weighted regression (GWR), and other models of spatial analysis (Cluster analysis) have all been employed in recent years to identify vulnerable urban areas in terms of likely fire outbreaks in urban residential areas [19]. The framework for analysing the geographical patterns of probable fires is provided by identifying vulnerable urban areas and high-risk locations.

Over the last few years, a significant amount of research has been conducted on the spatial analysis and identification of factors impacting the increase of fire danger in urban areas. In the United States, the onset of research in this field dates back to the early 1980s [20,21] with studies primarily focusing on the influence of demographic and socioeconomic characteristics to assess fire risk. According to a Swedish study, the risk of probable fires rises dramatically in urban areas with larger building complexes, particularly when associated with other physical and social vulnerability factors, such as degraded buildings, overcrowded houses, and elderly inhabitants [22].

Some other investigations in Khulna, Bangladesh, found that, along with socioeconomic factors, built environment variables and urban infrastructure such as building quality, distance from high-voltage power plants, distance from fire stations and infrastructure, type of land use, and distance from warehouses and fuel storage or distribution centres all play a role in reducing or increasing the potential risk of fire in urban buildings [23]. According to a study conducted in Nanjing, China, fire risk is also more significant in downtown regions with a high concentration of commercial and economic activity [24]. Another study conducted in Helsinki, Finland, found that although the structure of fire risk distributions is highly variable, socioeconomic, and physical aspects of the urban neighbourhoods (such as building age and quality) have a direct influence on the increase or decrease of fire risk [25]. In addition, a study conducted in Nanjing, China, reported that the risk of fire increased considerably in buildings with mixed land use [26]. Furthermore, a study conducted in the Romanian city of Iași found that the city's outskirts are substantially more sensitive to fire than other metropolitan regions, with factors such as poor income, high population density, and inadequate physical structure all contributing to this urban vulnerability [27]. According to a study conducted in Zanjan, Iran, urban areas with a higher number of tall and old structures have a higher risk of fire [28]. Furthermore, economic-related aspects, such as low household income and high population density, have a critical and direct effect on raising the danger of fire in urban buildings, according to the research of 283 Chinese cities [29]. In a similar study conducted in southern Queensland, Australia, researchers found that in addition to economic and social factors, the distance or proximity to fire stations impacts the degree of sensitivity to potential fire threats in different urban regions [30]. Additionally, a study conducted in Seixal, Portugal, showed that the risk of fire is larger in the old downtowns, which are full of structures with mixed land activities and uses [31]. Deprivation, ethnicity, and sociocultural characteristics may all play a part in lowering or increasing the danger of future urban fires, according to a study conducted in the Midlands of the United Kingdom [32]. Similarly, poverty, population density, and poor building condition were all determined to be key variables in raising the danger of fire in a study conducted in Surabaya, Indonesia [33]. According to research conducted in Melbourne, Australia, while the city's fire risk has followed a complex pattern, the central section of the city has a higher risk of fire due to a mix of economic activities, land use, and property ownership [34]. Moreover, a study conducted in Melbourne, Australia, indicated that the risk of fire is higher in the suburbs with a higher population density and ethnic composition [35]. A similar study conducted in Melbourne validated the significance of high population density in increasing fire rates in urban areas [36]. In recent years, some studies have used GIS and MCDM methods to analyse the risk of residential and structural fires [37–39]. Table S1 in Supplementary File S2, in the supplement, lists and summarises the main findings of some relevant studies addressing fire risk in urban areas.

A review of early research suggests that most of the past research in this field has focused on space–time patterns or the link between variables that influence fire risk. Furthermore, the majority of research was conducted in developed countries or they employed fewer criteria to examine and estimate fire risk [17]. The texture and geometry of historic Middle Eastern cities, particularly in Iran, differ significantly from those of developed countries. However, no study of urban fire risk modelling and zoning utilising GIS approaches and a set of factors has been conducted in Iran. Identifying fire risk in urban areas and GIS-multi-criteria decision making (MCDM) analysis for fire risk mapping are instrumental in supporting informed decision-making and outlining efficient urban vulnerability mitigation strategies [40,41]. Efficient spatial deployment of urban fire stations and emergency services is highly desired to address the risk of modern urban fires [38]. Simple techniques are unable to predict fire risk in various geographic units due to the complexity of fire risk in urban environments. Then, the precise techniques for identifying high-risk locations must be applied. The integrated GIS-MCDA approach provides rapid, effective, and exclusive explanations to complex spatial complications [42]. In this regard, some researchers have employed methods based on the GIS-MCDM approach [43,44]. In this context, the main objective of this study is to apply an integrated GIS-MCDM approach to model and introduce high-risk urban regions in terms of fire occurrence in Ardabil city, located in the northeast of Iran. To model the vulnerability level in a GIS setting, 19 socioeconomic sub-criteria, built environment, facilities, and fire records were defined and applied.

2. Materials and Methods

2.1. Study Area

Over the period between 1990 and 2017, about 20,000 fires were reported in Iran's large and medium-sized cities. For example, in 2017, a fire destroyed a 17-storey Plasco commercial building in central Tehran, killing 25 people, wounding 235 others, and inflicting millions of dollars in damages [45]. Another very relevant Iranian town in terms of fire occurrences in Ardabil city. Between 2015 and 2020, an average of 300 structural fires were reported annually in the historical centre of this city [45]. Located in the northwest area of Iran, the city of Ardabil is the capital of the province with the same name. The town, used as a case study in this work, covers around 76 km^2 and is the house of about 530,000 inhabitants (about 7000 per km^2), according to the most recent census data [45].

Ardabil is divided into 5 administrative districts and 44 neighbourhoods. Regarding fire safety-related infrastructures, there are seven fire stations in the city, whose location is illustrated in Figure 1. Due to its physical and sociodemographic characteristics, fire combat is challenging in this city, particularly in the older parts of the city due to their spatial arrangement [46]. Figure 1 depicts the spatial density map (heatmap) of fire incidents (per hectare) as calculated using the KDE technique in QGIS a free and open-source GIS package [47].

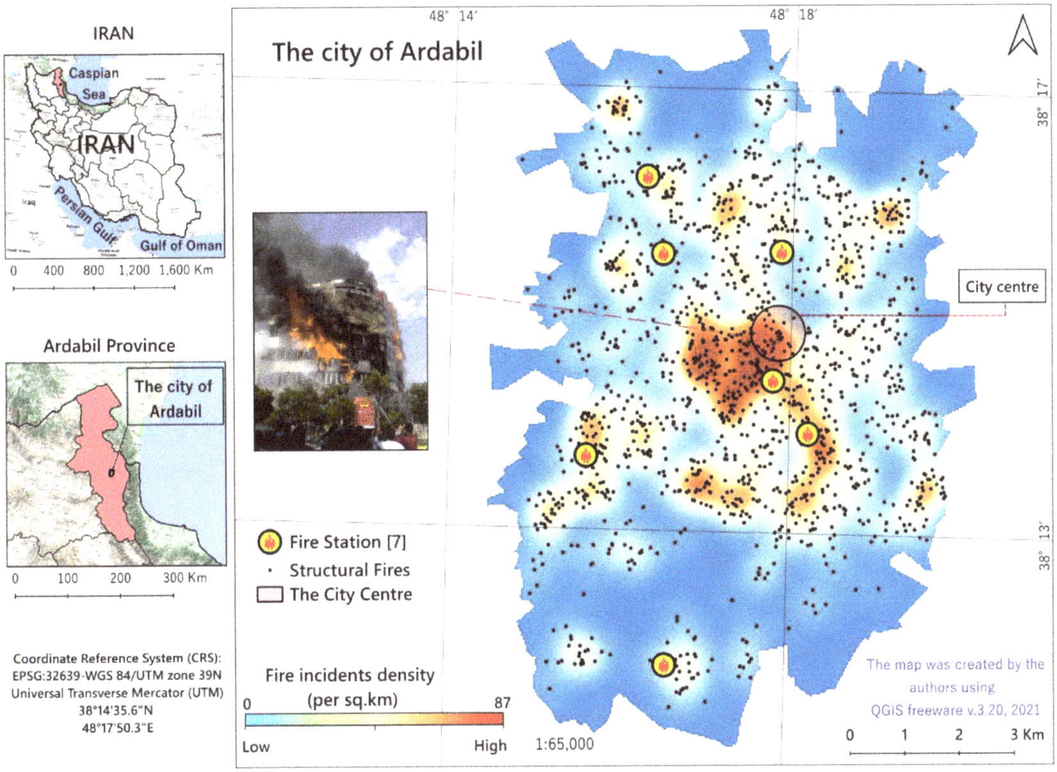

Figure 1. Map of Ardabil, Iran, showing the administrative divisions, fire stations, and the geographical distribution of fire incidents in the study area.

2.2. Data

Geodata sets and study criteria: From the combination of a thorough literature review (summarised in Table S1 of Supplementary File S2) and the objectives defined for the present study, 19 variables were isolated (outlined in Table S2 of Supplementary File S2) and used to evaluate the vulnerability of the buildings (about 250,000 building units) included in the study area. These variables are divided into four categories: (1) socioeconomic, (2) built environment, (3) infrastructure and urban facilities, and (4) previous fire incidence rates. The variables for each category are introduced in the following sections.

Socioeconomic: The Statistics Centre of Iran [45] provided raw data on socioeconomic factors such as population and household, number of elders, children, disabled people, and number of unemployed and illiterate people; see Table S2 in Supplementary File S2 and Figure 2 below.

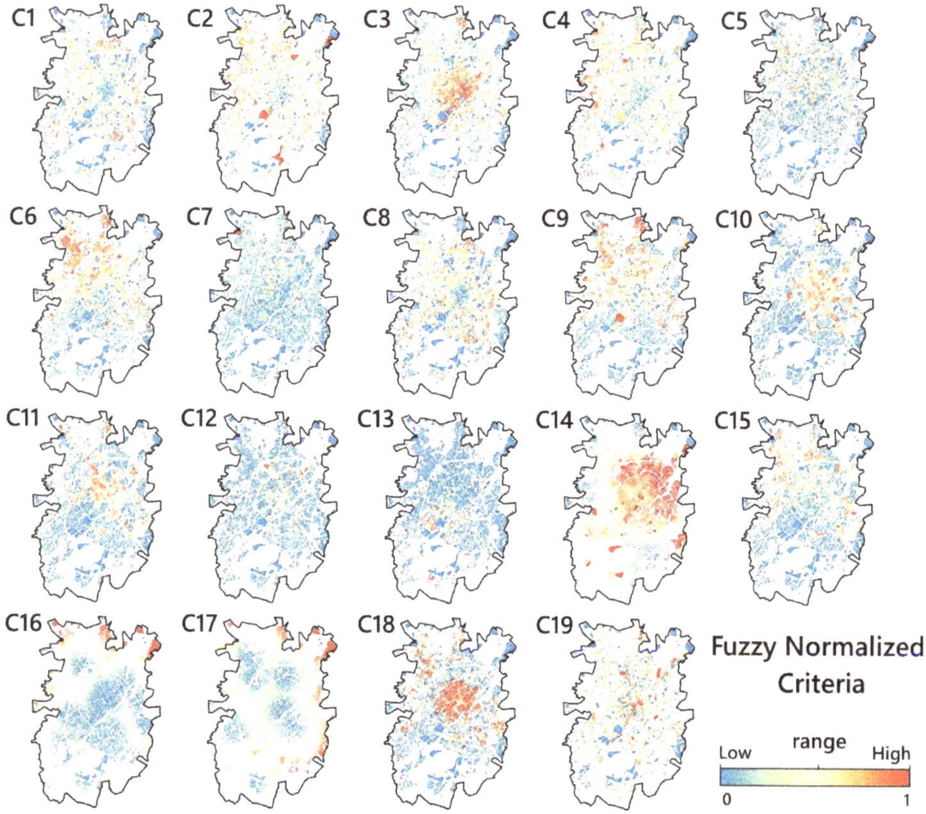

Figure 2. Spatial distribution map of fuzzy normalised criteria. C1: Population density, C2: Household dimension, C3: Old age ratio, C4: The ratio of the 14-year-old group and lower, C5: Disability ratio, C6: Illiteracy rate, C7: Unemployment rate, C8: Residential units' density, C9: The ratio of buildings made of non-durable materials, C10: The ratio of older buildings older than 30 years, C11: The ratio of worn-out and demolishing buildings, C12: Mixed land-use, C13: High-rise buildings ratio, C14: Buildings density with high fire incidence potential, C15: The ratio of small-sized property parts, C16: Euclidean distance from the hydrant valves, C17: Euclidean distance from fire stations, C18: The degree of permeability of the urban texture, C19: Previous fires rate.

Built environment: The type of materials used in the buildings (clay and mud, brick and iron), building quality (reparative, destructive, desolated), and number of storeys were taken from the statistical blocks of the most recent general housing census 2016 [45].

Infrastructures and facilities: Ardabil Municipality provided spatial data on the city road network (including blind roads and roadways less than 6 m wide) and landed segments, while Ardabil City Fire Department provided data on 46 fire hydrants and 7 fire stations in the city.

Location of fire incidents: Data on fire incidents from 2015–2020 were obtained from the files preserved at the Ardabil City Fire Department. The reason for choosing this period is twofold: the completeness, accuracy, and up-to-dateness of the data in this study area. Data that were missing, incomplete, or invalid were omitted from the analysis. Finally, 1488 fires in urban buildings (covering all urban land uses) were chosen as the final data and analysis' foundation. The Empirical Bayes Smoothing method was then used to obtain the rate of fire incidence per 100,000 people. Moreover, GeoDa v. 1.20.0.10 software [48] was used to determine the fire incidence Empirical Bayesian Smoothed (EBS) rates [49].

Within this study, several data from a variety of resources were taken into account. Based on the Iranian deviation systems, the urban blocks are considered as the smallest sector of the urban systems, which covers all information regarding residences, demography, and their specific characteristics. Because small geographical divisions such as urban blocks provide an accurate level for spatial analysis, they were chosen as the spatial basis of the analysis in the study area titled the highest resolution of urban geographical division. Point data were collected in a polygonal feature format layer in 6738 statistical blocks while establishing the UTM-Zone 39N coordinate system. The sub-criteria associated with each criterion utilised in this study (Supplementary File S2) were aggregated in the descriptive data table related to the polygon layer of statistical blocks using raw data and ArcGIS Desktop 10.8, and ArcGIS Pro 3.0.2 packages (ESRI, Redlands, CA, USA, 2022) were used to visualise our final model results [50].

2.3. Methods, Tools, and Procedure

2.3.1. Criteria Ranking and Weighting

In a GIS context, the maps for each criterion were created as raster maps. Because each index had a distinct size, the maps of each criterion were standardised using a fuzzy approach in a GIS environment to overcome this limitation, prepare the data, and execute MCDM methods. Different functions, such as S-shaped or J-shaped, as well as linear functions, are utilised in fuzzy standardisation. According to the nature and the linear relation between our criteria and the probability of fire assurances, the fuzzy S-shaped function was used to standardise benchmark maps in the present study (see Table 1). The fuzzy approach changes all raster layer's values and value rates to the same range of 0 (lowest index value) to 1 (highest index value), in Figure 2. The Fuzzy Overlay function in a GIS system was used to fuzzify and standardise the criteria for analysis [51]. After creating standardised fuzzy maps, the importance of each criterion was established using the numeric pairwise comparison approach by using Thomas L. Saaty's 1–9 Judgement Scale [52] and the opinions of ten experts. The final weight of the criteria was then calculated in the Expert Choice-11 software environment using the Analytic Hierarchy Process (AHP) method [53,54] (see Table 1). The compatibility ratio of the comparisons was calculated using Equations (1) and (2):

$$\mathrm{CR} = \frac{\mathrm{CI}}{\mathrm{RI}} \tag{1}$$

where CI represents the matrix compatibility vector, obtained from the following equation:

$$\mathrm{CI} = \frac{\lambda\mathrm{max} - \mathrm{n}}{\mathrm{n} - 1} \tag{2}$$

where λ_{max} is the largest matrix eigenvalue, RI is a randomness index for the matrix, and its value is proportional to the number of criteria in the matrix, with the number of criteria increasing the value. The pairwise comparison matrix's compatibility ratio should be smaller than 0.1. Otherwise, the preference judgments made are incoherent, and this incoherence should be addressed. Given that the RI ratio was equal to 0.09, the comparisons conducted to establish the importance of the criterion were confirmed. The weights obtained by the AHP method are given in the Table 1. According to experts, not all criteria are equally significant in predicting the likelihood of structural fire incidence, and some criteria, such as C9 (the ratio of buildings made of non-durable materials), C11 (the ratio of worn-out and demolished buildings), and C14 (building density with poor structural quality and a high risk of fire), have a higher importance.

Table 1. Summary statistics of criteria analysis.

Criteria		Statistics					
Symbol	Criterion	Min	Max	Mean	SD	AHP Weights	Fuzzy Membership Function
C1	Population density	0	86.1	2.03	2.22	0.032	linear s-shaped
C2	Household dimension	0	10.7	0.24	0.22	0.022	linear s-shaped
C3	Old age ratio	0	50	3.55	4.40	0.031	linear s-shaped
C4	The ratio of the 14-year-old group and lower	0	50	11.81	8.90	0.035	linear s-shaped
C5	Disability ratio	0	91.73	0.98	2.55	0.036	linear s-shaped
C6	Illiteracy rate	0	91.46	8.23	8.49	0.02	linear s-shaped
C7	Unemployment rate	0	0.34	1.48	2.25	0.019	linear s-shaped
C8	Residential units' density	0	287	58.69	66.59	0.031	linear s-shaped
C9	The ratio of buildings made of non-durable materials	0	100	10.71	19.79	0.069	linear s-shaped
C10	The ratio of older buildings older than 30 years	0	100	13.80	24.12	0.051	linear s-shaped
C11	The ratio of worn-out and demolishing buildings	0	100	19.74	26.85	0.099	linear s-shaped
C12	Mixed land-use	0	0.71	0.04	0.10	0.055	linear s-shaped
C13	High-rise buildings ratio	0	100	2.34	8.48	0.049	linear s-shaped
C14	Buildings density with high fire incidence potential	0.84	4.88	2.88	0.46	0.114	linear s-shaped
C15	The ratio of small-sized property parts	0	100	23.16	30.63	0.033	linear s-shaped
C16	Euclidean distance from the hydrant valves	0	3907.24	949.05	660.7	0.066	linear s-shaped
C17	Euclidean distance from fire stations	0	4091.57	1352.56	686	0.08	linear s-shaped
C18	The degree of permeability of the urban texture	0	100	29.65	38.83	0.079	linear s-shaped
C19	Previous fires rate	0	555	17.87	29.31	0.078	linear s-shaped

2.3.2. Fuzzy-VIKOR Method

VIKOR, as a prevailing MCDM method in the literature, ranks the alternatives based on the distance to the ideal condition [55]. Let $i \in \omega$ represent an alternative or raster cell in the set of alternatives (ω = {1, 2, 3 ... m}) in which m is the last alternative. All cells in the study area are considered an alternative and based on cell value; they have the chance to be evaluated as risk cells for the projected locations. Given that j is a criterion in the set of criteria in which j is the last criterion, x_{ij}, then, is the preference value of alternative i in relation to criterion j. Let f_{ij} be the normalised preference value of alternative i in relation to criterion j, computable according to Equation (3):

$$f_{ij} = \frac{x_{ij}}{\sqrt{\sum_{i=1}^{m} x_{ij}^2}} \quad (3)$$

Using the f_{ij} values, the maps with dissimilar scales and ideal solutions can be converted to the standard maps. The best f_j^* value for the positive and negative criteria is calculated from the following Equation (4):

$$\begin{aligned} f_j^* &= {}^{Max} f_{ij} \text{ if it is a benefit-based function;} \\ f_j^* &= {}^{Min}_{i} f_{ij} \text{ if it is a cost-based function.} \end{aligned} \quad (4)$$

The worst f_j^- value for the positive and negative criteria is calculated from Equation (5):

$$\begin{aligned} f_j^- &= {}^{Min}_{i} f_{ij} \text{ if it is a benefit-based function;} \\ f_j^- &= {}^{Max}_{i} f_{ij} \text{ if it is a cost-based function.} \end{aligned} \quad (5)$$

Let S_i and R_i indicate suitability and regret associated with alternative i, respectively. Then, related values are computable as Equation (6):

$$S_i = \sum_{i=1}^{n} wi \frac{f_j^* - f_{ij}}{f_j^* - f_j^-}$$
$$R_i = Max\left\{ wi \frac{f_j^* - f_{ij}}{f_j^* - f_j^-} \right\} \tag{6}$$

where, w_i represents the weight of the *i*th criterion. The weight of each criterion was calculated through the Delphi method and was applied to each criterion in the GIS environment. The VIKOR value Q_i that represents the maximum group benefits for alternative *i* can then be measured by Equation (7) for each alternative *i*:

$$Q_i = v\left[\frac{S_i - S^-}{S^* - S^-}\right] + (1 - v)\left[\frac{R_i - R^-}{R^* - R^-}\right] \tag{7}$$

where

$$R^* = MaxR_i\ ,\ R^- = MinR_i,\ S^* = MaxS_i\ ,\ S^- = MinS_i$$

refers to the weight of criterion that ensures maximum group utility, and $(1 - v)$ refers to the weight of the minimum regret in dissent. The value of v varies between 0 and 1; however, it is often taken as 0.5.

For an alternative to be preferable, its preference should be confirmed by the associated value of Q_i addition to either of S_i, R_i, Q_n, with the smallest value expressed as the best option among alternatives. In this study, Q_n is the location value of each alternative or cell in GIS. The least-valued alternative (point) is the most appropriate alternative to be selected. In the VIKOR method, if A_1 and A_2 are ranked first and second alternatives, respectively, to specify the value of "Q" (the chance that a fire may occur in each cell), Equation (8) should be satisfied [55]:

$$Q(A_2) - (A_1) \geq \frac{1}{n - 1} \tag{8}$$

In this study, however, the cells have been categorised by the Q_i score of each alternative or cell.

The VIKOR method's conclusions reveal the degree of risk that urban buildings face from a potential fire incidence. In the VIKOR method, the greatest value (high risk) in the output units (cells) is 0 and the lowest value (less risk) is 1. In the final step, we reversed the values for visualisations in the urban vulnerability index map.

The Natural Breaks (Jenks) classifying method approach was used to prioritise probable fire risk into five categories: lower (values = 0.74–1), low (values = 0.6–0.74), moderate (values = 0.48–0.6), high (values = 0.35–0.48), and higher (values = 0.086–0.35) degrees of urban vulnerability. In the Natural Breaks (Jenks) method, the variance within each class is minimised while the variance between classes is maximised [51]. With natural breaks classification, Natural Breaks (Jenks) classes are based on natural groupings inherent in the data. Class breaks are created in a way that best groups similar values together and maximises the differences between classes. The features are divided into classes whose boundaries are set where there are relatively big differences in the data values. Natural breaks are data-specific classifications and not useful for comparing multiple maps built from different underlying information [51,56].

Figure 3 illustrates the methodology utilised in the preparation of the urban vulnerability index map in terms of fire risk in the study area.

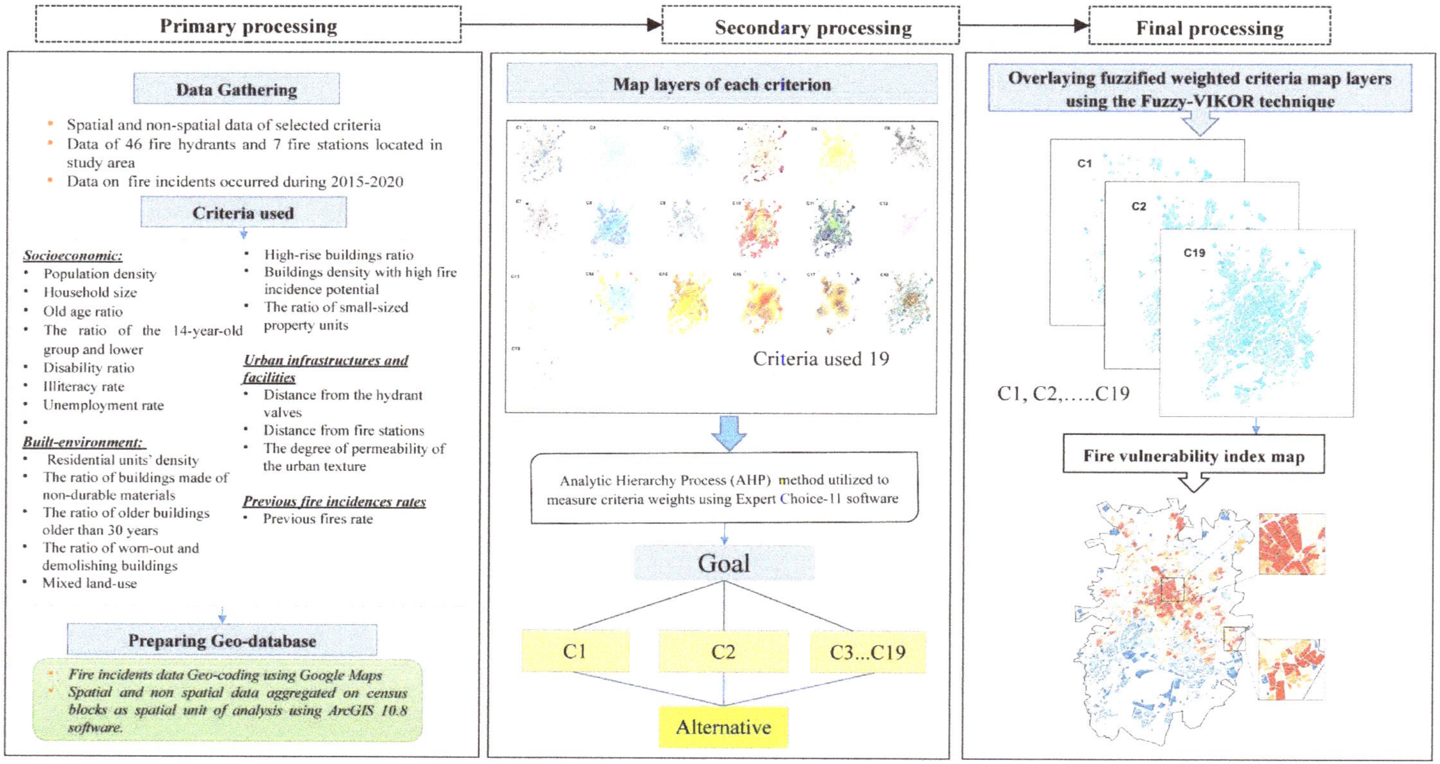

Figure 3. Flowchart illustrating the methodology used in various stages of the research for the preparation of the urban vulnerability index map in terms of fire risk in the study area.

2.3.3. Model Validation

To check the efficiency of the utilized GIS-MCDM model, the Spatial Linear Regression (SLR) approach [57] was used to measure correlation between the urban vulnerability index (result of our proposed model) and the spatial Kernel density of actual fire incidents in the study area using the TerrSet software V.19 (see: [57]). The following stage required mapping two variables using the Bivariate mapping method in ArcGIS Pro in order to visualise the model findings with real incident data. A bivariate map combines various sets of symbols and colours to represent two related but dissimilar variables on a map. It serves as a straightforward approach to depict, graphically and precisely, the link between the two spatially distributed variables. It is also simple to evaluate how two attributes change in respect to one another using this map [51].

3. Results

3.1. Urban Vulnerability Index Map

The aim of this study was to provide an urban vulnerability index map in terms of fire risk using integrated GIS-MCDM methods. The key result of this study is a vulnerability index map which is provided on the basis of our criteria used to map the risk of structural fire in the study area. Figure 4 depicts the vulnerability index map of buildings in different city areas within the urban blocks. Extracting the basic data from the VIKOR model's output map (Table 2) reveals that 639 blocks with a total size of 4.11 km^2 (13.62%) fall into the category of highly vulnerable locations, out of a total area of 30.18 km^2. These blocks are often found throughout the city's northern half, central district, and urban outskirts. A total of 930 blocks with a total size of 5.26 km^2 (17.43%) are classified as very vulnerable regions on this map. The blocks in this category are found throughout much of the city's northern half. As a result, 945 blocks with a total size of 5.31 km^2 (17.59%) are classified as moderately vulnerable. The blocks in this category are still found in most portions of the city's northern half (from east to west and from the centre to the north). The blocks in this category are found throughout much of the city's northern half. According to the findings, 1282 blocks with a total size of 5.97 km^2 (19.78%) fall into the low vulnerability index group. The blocks that fall into these two categories are frequently found at the city's outskirts, far from the downtown area. The blocks in this category are found throughout much of the city's northern half. Accordingly, very vulnerable locations include 2941 blocks with a total size of 9.53 km^2 (31.58%). The majority of these blocks are located on the city's southern side (from east to west and from the centre to south).

Table 2. Summary statistics of the urban vulnerability choropleth map given in Figure 4 based on Natural Breaks classifying method.

Vulnerability Degree	Vulnerability Score	Number of Blocks	Area (sq.km)	Area (%)	Population	Population (%)
Higher	0.086–0.35	639	4.11	13.62	72,471	13.83
High	0.35–0.48	930	5.26	17.43	106,716	20.37
Moderate	0.48–0.6	945	5.31	17.59	104,254	19.90
Low	0.6–0.74	1282	5.97	19.78	103,797	19.81
Lower	0.74–1	2941	9.53	31.58	136,663	26.09

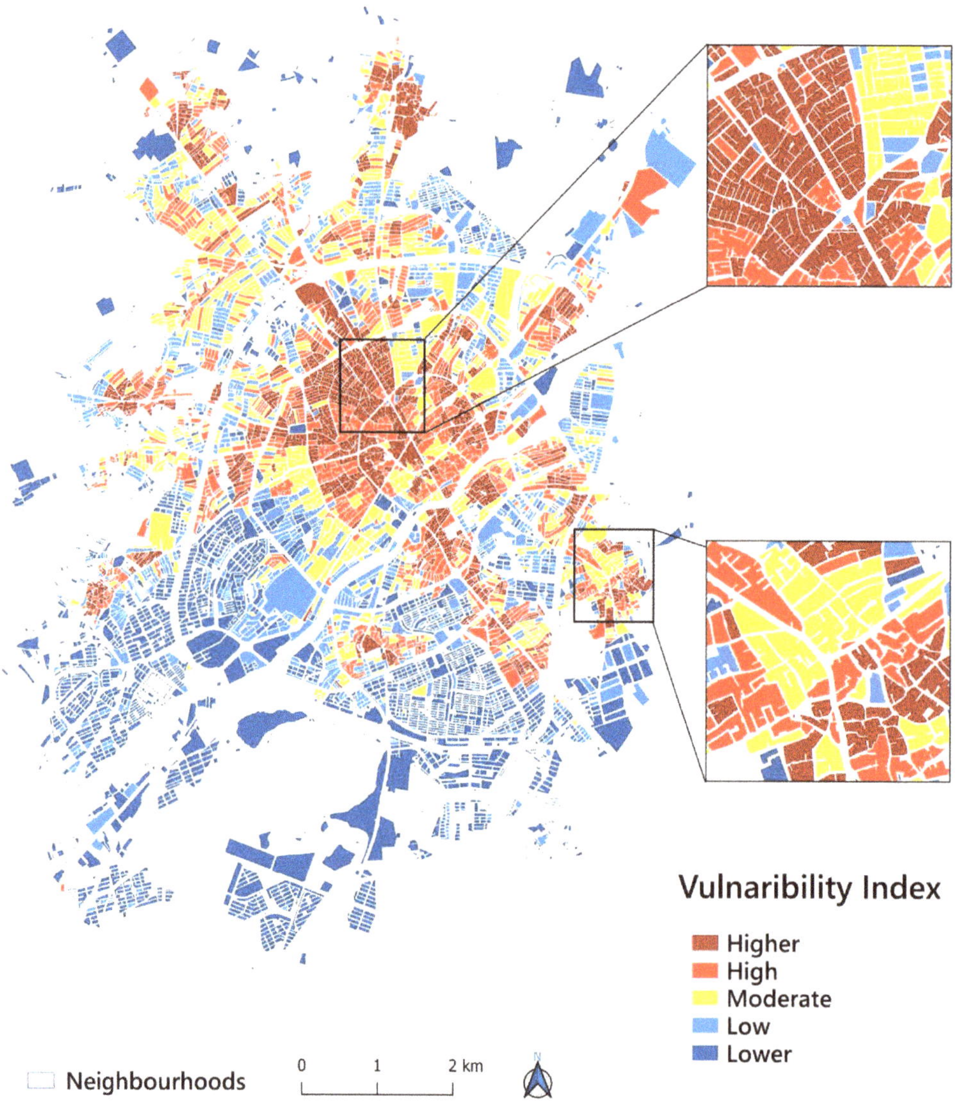

Figure 4. Spatial distribution map of areas with different degrees of urban vulnerability index against the potential structural fires-related risks.

3.2. Model Performance

We used the spatial linear regression correlation test value to compare the urban vulnerability index values with the spatial distribution of the actual fire incidence Kernel density (per hectare) in order to validate the model. The coefficient of determination (r^2) was equal to 100%, and the linear regression correlation's coefficient value was 1 (Figure 5). This significant outcome demonstrates that, using our criteria, the model used in this study was successful in determining the likelihood of fire risk over the sample period (2015–2020) and study area. Furthermore, the small variations between the VIKOR model's output map and the actual fire incident density (per hectare) spatial distribution (Figure 6) indicate a good agreement between our model's measured urban vulnerability index values and the actual fire incident density (per hectare) spatial distribution in terms of the number of fire occurrences in the study area.

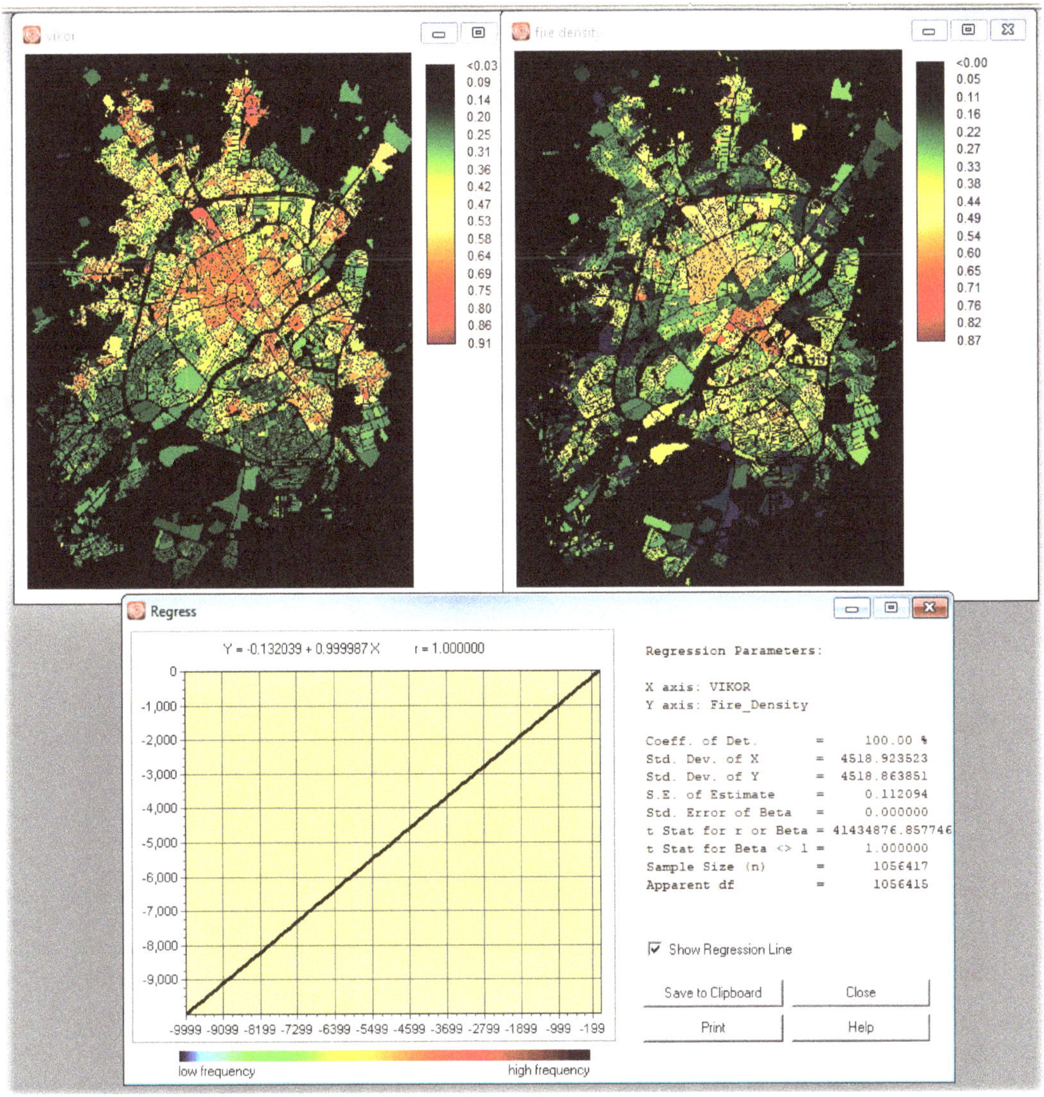

Figure 5. The spatial linear regression scatter plot of the spatial correlation between urban vulnerability against the potential structural fires-related risks index (x-axis) and actual fire incident per hectare values (y-axis). The plot generated in TerrSet (By: Clark Labs, Clark University, Worcester, MA, USA, 2022).

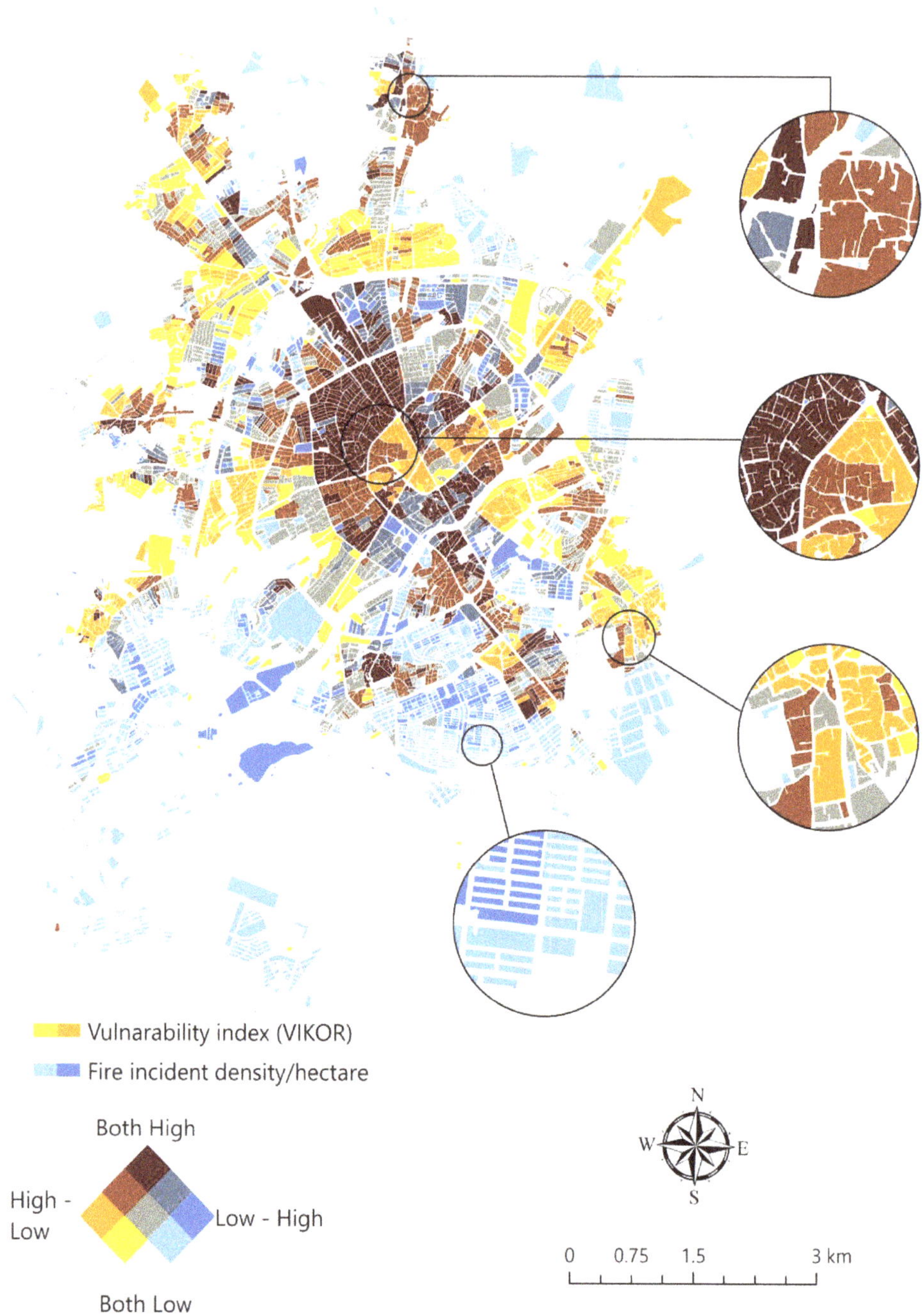

Figure 6. Spatial distribution bivariate map of urban vulnerability against the potential structural fires-related risks index (VIKOR model results) and fire incident per hectare original values. All the maps were generated in ArcGIS Pro 3.0.2 (ESRI, Redlands, CA, USA, 2023).

4. Discussion

The present study aims to model and prioritise urban regions in terms of their vulnerability to potential structural fires-related risks. The Integrated GIS-MCDM approach and 19 socioeconomic and built environment criteria were utilised to model and prioritise vulnerable urban regions. This study contains several important findings.

First, our GIS-MCDM Fuzzy-VIKOR model's output revealed that urban areas with high and extremely high vulnerabilities were increasingly spread from the city centre to the outskirts in the north, east, and west in terms of the spatial pattern (see Figure 4). This result is in line with the findings of Chhetri et al. [26] in southeast Queensland, who have shown that fires are distributed in a central–peripheral pattern. In addition, high-vulnerability areas in terms of potential fire risk frequently correspond to urban districts populated by low-income people (Figure 2 (C1)). The primary features of this portion of the city are the small-sized property (real state) units (Figure 2 (C15)) and high rates of unemployment (Figure 2 (C7)) and illiteracy (Figure 2 (C6)). This finding is in line with those obtained by Chhetri et al. [26,36], who observed that building fires are more common in older neighbourhoods and low-income residential areas on the outskirts of cities. Moreover, this finding is consistent with those of Zhang et al. in Nanjing, China. [3], since they observed a positive, strong, and significant correlation between poor income, unemployment, and illiteracy rates, as well as growing fire incidence rates in various urban areas. According to Rahmawati et al., [29] most poor settlements that are developed unintentionally with high population densities have a higher risk of fire incidence. Additionally, as Ardianto et al. [30] found, in urban areas, the overall socioeconomic and environmental variables have a significant effect in boosting fire incidence rates. In addition to socioeconomic situations, we found that most of the city's old and worn-out buildings are located in high-risk areas (Figure 2 (C11)), with non-durable materials utilised in their construction (Figure 2 (C9)).

Second, this study's results revealed that neighbourhood characteristics can be determinative in lowering or raising the risk of likely structural fire risk, and one component alone cannot play a role. In most vulnerable areas, the physical condition of buildings is relatively low (Figure 2 (C10, C11)), and the population in such places is ageing rapidly (Figure 2 (C3)). Previous studies have found that structural fires are becoming more common and pose a greater danger in urban areas with a high concentration of old and worn-out buildings [20,26,58,59].

Furthermore, we found that the city business district, known locally as the Bazar, is one of the focal points in the category of extremely vulnerable locations, as can be seen in the urban vulnerability index map presented in Figure 4. This place is at the city's most central location and serves a variety of purposes. Previous studies, such as Xia et al. [22], have found that mixed-used developments and commercial structures, particularly in central city areas, directly impacts the fire incidence rates.

Last but not least, the results displayed in Figure 4 make apparent the level of urban vulnerability in the city's outskirts, where spatial access to hydrant valves (Figure 2 (C16)) and fire stations (Figure 2 (C17)) is more reduced. In prior research conducted in Iran, Masoumi et al. [24] found that inadequate spatial access to urban amenities suited for firefighting might increase the probability of fire occurrence in densely populated regions.

Third, the integrated GIS-MCDM approach can be a useful tool for assessing urban vulnerability against probability fire-related risks in cities, as demonstrated in the model performance section (Figures 5 and 6). Because of its adaptability and capacity to interact with human inference and data-based processing, the combination of GIS-MCDM approaches allows for a more accurate prediction of fire risks, as our research revealed. The methods (such as the Point density or Kernel density methods) that ignore the human factor do not provide such a possibility. Moreover, by combining GIS with MCDM methods, it is feasible to combine a number of parameters and obtain more accurate results. In this study, other variables, such as the number of high-rise buildings, the proportion of children aged 14 and under, the population disability rate, the mixed land use coefficient, and the history of previous fires, were included in the integrated GIS-MCDM model as additional variables. However, according to the study's findings based on the opinions of experts, their importance in the final output was lower; in other words, they were less active on a large scale in justifying and explaining the increase or decrease of urban vulnerability. Moreover, although some socioeconomic characteristics, such as population density and built environment variables, such as building age or impermeability and granularity of property components, have a significant influence in raising the risk of structural fire in the study area, as earlier studies [24,26,30] have pointed out, the set of a geographical area's circumstances and features can reduce or increase the risk of future structural fires.

4.1. Policy Implications

Firstly, it is proposed that the city council and municipality emphasise the renovation of ancient buildings, particularly in the downtown area, where the old Bazaar (a local name for the city business district) is located. Secondly, the buildings in the city's integrated villages and worn-out urban textures should be prioritised for rehabilitation. These are densely populated areas where low-income groups dwell, their buildings are of poor quality, and their quality of life is poor. Furthermore, it is critical to facilitate and improve physical access to fire stations for communities that are more vulnerable and have less access. In high-risk and extremely high-risk locations, new stations should be established. It appears that establishing dedicated routes for fire vehicles to approach and depart the city centre can help to lessen the damage caused by potential fires over time.

4.2. Limitations and Futures Research Strategy

The findings of this study have to be seen in the light of some limitations. We were unable to obtain information on household income and expenditures, as well as the status of building insurance. To circumvent this constraint, we examined additional social indices, including illiteracy and unemployment rates, as well as demographic data. The lack of access to urban banks and databases was another disadvantage of this study. Despite these limitations, we think this study has several strengths. The information for this study was gathered from a variety of sources by contacting various organisations. The use of a collection of socioeconomic data, the built environment, and characteristics linked to urban amenities in generating a map of urban vulnerability index to the risk of fire incidence were the study's strengths. We also prepared the final map using integrated MCDM-GIS hybrid approaches, which were less widely employed in earlier research for urban fires. Despite the limitations mentioned, the authors believe the methods utilised in this study will be valuable to academics and policy-makers working in the field of urban fire management.

5. Conclusions

The most vulnerable zones of the urban area analysed herein were identified by resorting to the hybrid MCDM-GIS approach and combining a number of characteristics that determine the fire risk in these areas. The findings revealed that using the hybrid MCDM-GIS approach to identify vulnerable zones in cities might be a useful tool. Urban vulnerability index maps concerning potential structural fires-related risks can assist in identifying elements that enhance fire risk and give useful insights into fire risk estimation and fire service management. Consequently, officials may utilise these maps to take preventative actions and allocate resources and infrastructure more effectively. Future research can look into the spatiotemporal patterns of urban fire phenomena, as well as more appropriate methods, such as spatial regression methods, to explore the relationship between different variables and fire rates and gain a better understanding of fire risk and the factors that influence it in urban areas. To simulate the risk of fire or urban vulnerability, future studies might employ more complex modelling approaches, such as agent-based modelling. We conclude that the methodological approach proposed in this study can be applied successfully to model and map fire risk in urban areas, with the potential to be applied in different urban contexts worldwide. In addition, the obtained results are of great importance to local stockholders, such as the municipality of Ardabil, authorities, and decision-makers, in determining the spatiotemporal pattern of fire risk in the city and developing crisis risk programs.

Supplementary Materials: The following supporting information can be downloaded at: https://www.mdpi.com/article/10.3390/fire6030107/s1, Table: Supplementary File S1; Supplementary File S2. Refs. [60–68] are cited in Supplementary Materials.

Author Contributions: Conceptualisation, A.M. and A.G.G.; methodology, A.M. and S.N.; formal analysis, S.N.; geocoding and cleaning, S.N. and S.J.M.A.; writing, A.M. and S.N.; review, A.G.G., review and editing, T.M.F.; supervision, A.M. and T.M.F. All authors have read and agreed to the published version of the manuscript.

Funding: This research received no external funding.

Institutional Review Board Statement: Not applicable.

Informed Consent Statement: Not applicable.

Data Availability Statement: All the data used in this study to prepare the urban vulnerability index maps against the potential structural fires-related risks are available via Supplementary File S1.

Acknowledgments: We would like to acknowledge the Ardabil City Fire Department for sharing fire incidents data.

Conflicts of Interest: The authors declare no conflict of interest.

References

1. Zhang, Y. Analysis on comprehensive risk assessment for urban fire: The case of Haikou City. *Procedia Eng.* **2013**, *52*, 618–623. [CrossRef]
2. Kiran, K.C.; Corcoran, J. Modelling residential fire incident response times: A spatial analytic approach. *Appl. Geogr.* **2017**, *84*, 64–74.
3. Zhang, X.; Yao, J.; Sila-Nowicka, K.; Jin, Y. Urban fire dynamics and its association with urban growth: Evidence from Nanjing, China. *ISPRS Int. J. Geoinf.* **2020**, *9*, 218. [CrossRef]
4. Tishi, T.R.; Islam, I. Urban fire occurrences in the Dhaka Metropolitan Area. *GeoJournal* **2019**, *84*, 1417–1427. [CrossRef]
5. World Health Organization (WHO). Burns. Published 30 January 2018. Available online: https://www.who.int/news-room/fact-sheets/detail/burns (accessed on 30 January 2023).
6. Bickenbach, J. The world report on disability. *Disabil. Soc.* **2011**, *26*, 655–658. [CrossRef]
7. Murillo, M.; Tutikian, B.F.; Christ, R.; Silva, L.F.; Maschen, M.; Oliveira, M.L. Analysis of the influence of thickness on fire reaction performance in polyisocyanurate core sandwich panels. *J. Mater. Res. Technol.* **2020**, *9*, 9487–9497. [CrossRef]
8. Kaleji, L.K.; Murthy, K. The history of fire fighting in the Iran and the study spatial distribution of the current situation of fire stations and radius its function in Tehran city using GIS. *Asian J. Dev. Matters* **2011**, *5*, 207–218.
9. Ceyhan, E.; Ertuğay, K.; Düzgün, Ş. Exploratory and inferential methods for spatio-temporal analysis of residential fire clustering in urban areas. *Fire Saf. J.* **2013**, *58*, 226–239. [CrossRef]
10. Moshashaei, P.; Alizadeh, S.S. Fire risk assessment: A systematic review of the methodology and functional areas. *Iran. J. Health Saf. Environ.* **2017**, *4*, 654–669.
11. Ziervogel, G.; Pelling, M.; Cartwright, A.; Chu, E.; Deshpande, T.; Harris, L.; Hyams, K.; Kaunda, J.; Klaus, B.; Michael, K.; et al. Inserting rights and justice into urban resilience: A focus on everyday risk. *Environ. Urban* **2017**, *29*, 123–138. [CrossRef]
12. Oliveira, S.; Pereira, J.M.C.; San-Miguel-Ayanz, J.; Lourenço, L. Exploring the spatial patterns of fire density in Southern Europe using Geographically Weighted Regression. *Appl. Geogr.* **2014**, *51*, 143–157. [CrossRef]
13. Aven, T. On Some Recent Definitions and Analysis Frameworks for Risk, Vulnerability, and Resilience. *Risk Anal.* **2011**, *31*, 515–522. [CrossRef] [PubMed]
14. Alguacil, J. Instrumentos para el análisis y políticas para la acción. In Proceedings of the Foro de Debates: Ciudad y Territorio Jornada La Vulnerabilidad Urbana en España, Madrid, Spain, 30 June 2011; Volume 30.
15. Huedo, P.; Ruá, M.J.; Florez-Perez, L.; Agost-Felip, R. Inclusion of Gender Views for the Evaluation and Mitigation of Urban Vulnerability: A Case Study in Castellón. *Sustainability* **2021**, *13*, 10062. [CrossRef]
16. Guldåker, N.; Hallin, P.O. Spatio-temporal patterns of intentional fires, social stress and socio-economic determinants: A case study of Malmö, Sweden. *Fire Saf. J.* **2014**, *70*, 71–80. [CrossRef]
17. Mohammadi, A.; Shahparvari, S.; Kiani, B.; Noori, S.; Chhetri, P. An analysis of Spatio–temporal patterns of fires in an Iranian city. *Indoor Built Environ.* **2022**, *32*, 183–199. [CrossRef]
18. Cicione, A.; Gibson, L.; Wade, C.; Spearpoint, M.; Walls, R.; Rush, D. Towards the Development of a Probabilistic Approach to Informal Settlement Fire Spread Using Ignition Modelling and Spatial Metrics. *Fire* **2020**, *3*, 67. [CrossRef]

19. Runefors, M.; Nilson, F. The Influence of Sociodemographic Factors on the Theoretical Effectiveness of Fire Prevention Interventions on Fatal Residential Fires. *Fire Technol.* **2021**, *57*, 2433–2450. [CrossRef]
20. Shama, S.; Shurid, A.S.; Haque, M.N. Risk Assessment of Accidental Fire Breakdown in a Residential Area of Khulna City, Bangladesh. *J. Eng. Sci.* **2021**, *12*, 109–118. [CrossRef]
21. Todorovic, S. Modelling risk factors in urban residential fires in Helsinki. Published online 2020. Available online: https://eprints.gla.ac.uk/210971/ (accessed on 1 February 2023).
22. Xia, Z.; Li, H.; Chen, Y.; Yu, W. Detecting urban fire high-risk regions using colocation pattern measures. *Sustain. Cities Soc.* **2019**, *49*, 101607. [CrossRef]
23. Bulai, A.T.; Roşu, L.; Bănică, A. Patterns of urban fire occurrence in Iasi City (Romania). *Present Environ. Sustain. Dev.* **2019**, *2*, 87–102. [CrossRef]
24. Masoumi, Z.; van LGenderen, J.; Maleki, J. Fire risk assessment in dense urban areas using information fusion techniques. *ISPRS Int. J. Geoinf.* **2019**, *8*, 579. [CrossRef]
25. Hu, J.; Shu, X.; Xie, S.; Tang, S.; Wu, J.; Deng, B. Socioeconomic determinants of urban fire risk: A city-wide analysis of 283 Chinese cities from 2013 to 2016. *Fire Saf. J.* **2019**, *110*, e102890. [CrossRef]
26. Chhetri, P.; Corcoran, J.; Ahmad, S.; Kiran, K.C. Examining spatio-temporal patterns, drivers and trends of residential fires in South East Queensland, Australia. *Disaster Prev. Manag. Int. J.* **2018**, *27*, 586–603. [CrossRef]
27. Ferreira, T.M.; Vicente, R.; da Silva, J.A.R.M.; Varum, H.; Costa, A.; Maio, R. Urban fire risk: Evaluation and emergency planning. *J. Cult. Herit.* **2016**, *20*, 739–745. [CrossRef]
28. Hastie, C.; Searle, R. Socio-economic and demographic predictors of accidental dwelling fire rates. *Fire Saf. J.* **2016**, *84*, 50–56. [CrossRef]
29. Rahmawati, D.; Pamungkas, A.; Aulia, B.U.; Larasati, K.D.; Rahadyan, G.A.; Dito, A.H. Participatory mapping for urban fire risk reduction in high-density urban settlement. *Procedia-Soc. Behav. Sci.* **2016**, *227*, 395–401. [CrossRef]
30. Ardiantoa, R.; Chhetria, P.; Dunstallb, S. Modelling the Likelihood of Urban Residential Fires Considering Fire History and the Built Environment: A Markov Chain Approach. In Proceedings of the 21st International Congress on Modelling and Simulations, Queensland, Australia, 29 November–4 December 2015. Available online: https://www.mssanz.org.au/ (accessed on 30 January 2023).
31. Shahparvari, S.; Fadaki, M.; Chhetri, P. Spatial accessibility of fire stations for enhancing operational response in Melbourne. *Fire Saf. J.* **2020**, *117*, 103149. [CrossRef]
32. Kiran, K.C. Temporal and spatial patterns of fire incident response time: A case study of residential fires in Brisbane. In Proceedings of the 7th State of Australian Cities Conference, Gold Coast, Australia, 9–11 December 2015. Available online: https://apo.org.au/node/63305 (accessed on 28 January 2023).
33. Špatenková, O.; Virrantaus, K. Discovering spatio-temporal relationships in the distribution of building fires. *Fire Saf. J.* **2013**, *62*, 49–63. [CrossRef]
34. Corcoran, J.; Higgs, G.; Higginson, A. Fire incidence in metropolitan areas: A comparative study of Brisbane (Australia) and Cardiff (United Kingdom). *Appl. Geogr.* **2011**, *31*, 65–75. [CrossRef]
35. Wuschke, K.; Clare, J.; Garis, L. Temporal and geographic clustering of residential structure fires: A theoretical platform for targeted fire prevention. *Fire Saf. J.* **2013**, *62*, 3–12. [CrossRef]
36. Ardianto, R.; Chhetri, P. Modeling Spatial–Temporal dynamics of urban residential fire risk using a Markov Chain technique. *Int. J. Disaster Risk Sci.* **2019**, *10*, 57–73. [CrossRef]
37. Chu, G.Q.; Chen, T.; Sun, Z.H.; Sun, J.H. Probabilistic risk assessment for evacuees in building fires. *Build. Environ.* **2007**, *42*, 1283–1290. [CrossRef]
38. Lee, C.A.; Sung, Y.C.; Lin, Y.S.; Hsiao, G.L.K. Evaluating the severity of building fires with the analytical hierarchy process, big data analysis, and remote sensing. *Nat. Hazards* **2020**, *103*, 1843–1856. [CrossRef]
39. Alkış, S.; Aksoy, E.; Akpınar, K. Risk Assessment of Industrial Fires for Surrounding Vulnerable Facilities Using a Multi-Criteria Decision Support Approach and GIS. *Fire* **2021**, *4*, 53. [CrossRef]
40. Silva, D.; Rodrigues, H.; Ferreira, T.M. Assessment and Mitigation of the Fire Vulnerability and Risk in the Historic City Centre of Aveiro, Portugal. *Fire* **2022**, *5*, 173. [CrossRef]
41. Chen, Y.; Wu, G.; Chen, Y.; Xia, Z. Spatial Location Optimization of Fire Stations with Traffic Status and Urban Functional Areas. *Appl. Spat. Anal. Policy* **2023**, 1–18. [CrossRef]
42. Alizadeh, M.; Hashim, M.; Alizadeh, E.; Shahabi, H.; Karami, M.R.; Pour, A.B.; Pradhan, B.; Zabihi, H. Multi-Criteria Decision Making (MCDM) Model for Seismic Vulnerability Assessment (SVA) of Urban Residential Buildings. *ISPRS Int. J. Geoinf.* **2018**, *7*, 444. [CrossRef]
43. Maniatis, Y.; Doganis, A.; Chatzigeorgiadis, M. Fire Risk Probability Mapping Using Machine Learning Tools and Multi-Criteria Decision Analysis in the GIS Environment: A Case Study in the National Park Forest Dadia-Lefkimi-Soufli, Greece. *Appl. Sci.* **2022**, *12*, 2938. [CrossRef]
44. Jiang, Y.; Lv, A.; Yan, Z.; Yang, Z. A GIS-Based Multi-Criterion Decision-Making Method to Select City Fire Brigade: A Case Study of Wuhan, China. *ISPRS Int. J. Geoinf.* **2021**, *10*, 777. [CrossRef]
45. Statistical Centre of Iran. Ardabil Province Housing and Population Census Reports and Data. Available online: https://www.amar.org.ir/english (accessed on 20 February 2023).

46. Ardabil City Fire Department. Annual Report of Fire Events and Rescue in Ardabil City. Available online: https://fire.ardabilcity.ir/ (accessed on 10 February 2023).
47. QGIS Development Team. QGIS Geographic Information System, Version 3.26. Available online: http://qgis.osgeo.org (accessed on 10 February 2023).
48. Anselin, L.; Syabri, I.; Kho, Y. GeoDa: An introduction to spatial data analysis. In *Geographical Analysis*; Springer: Berlin/Heidelberg, Germany, 2006; Volume 38, pp. 5–22. [CrossRef]
49. Anselin, L.; Lozano, N.; Koschinsky, J. Rate transformations and smoothing. *Urbana* **2006**, *51*, 61801.
50. Environmental Systems Research Institute (ESRI). ArcGIS Professional GIS. Available online: http://http//pro.arcgis.com (accessed on 4 January 2023).
51. ESRI. ArcGIS Pro Help. Available online: https://pro.arcgis.com/en/pro-app/latest/help/main/welcome-to-the-arcgis-pro-app-help.htm (accessed on 1 November 2022).
52. Saaty, T.L. Principles of the analytic hierarchy process. In *Expert Judgment and Expert Systems*; Springer: Berlin/Heidelberg, Germany, 1987; pp. 27–73.
53. Saaty, T.L. An Exposition of the AHP in Reply to the Paper "Remarks on the Analytic Hierarchy Process". *Manag. Sci.* **1990**, *36*, 259–268. [CrossRef]
54. Gumus, S. An evaluation of stakeholder perception differences in forest road assessment factors using the Analytic Hierarchy Process (AHP). *Forests* **2017**, *8*, 165. [CrossRef]
55. Akkas, O.P.; Erten, M.Y.; Cam, E.; Inanc, N. Optimal Site Selection for a Solar Power Plant in the Central Anatolian Region of Turkey. *Int. J. Photoenergy* **2017**, *2017*, 7452715. [CrossRef]
56. De Smith, M.J.; Goodchild, M.F.; Longley, P. *Geospatial Analysis: A Comprehensive Guide to Principles, Techniques and Software Tools*; Troubador publishing Ltd.: Market Harborough, UK, 2007.
57. Clark Labs. The TerrSet Help System. Clark University, 2022. Available online: https://clarklabs.org/download/ (accessed on 2 May 2022).
58. Sternlieb, G.; Burchell, R.; Fires in Abandoned Buildings. The Social and Economic Consequences of Residential Fires Lexington, MA: Lexington Books. 1983. Available online: https://mbln.bibliocommons.com/v2/record/S75C2055156 (accessed on 1 May 2022).
59. Jennings, C.R. Social and economic characteristics as determinants of residential fire risk in urban neighborhoods: A review of the literature. *Fire Saf. J.* **2013**, *62*, 13–19. [CrossRef]
60. Kuran, C.H.A.; Morsut, C.; Kruke, B.I.; Krüger, M.; Segnestam, L.; Orru, K.; Nævestad, T.-O.; Airola, M.; Keränen, J.; Gabel, F.; et al. Vulnerability and vulnerable groups from an intersectionality perspective. *Int. J. Disaster Risk Reduct.* **2020**, *50*, 101826. [CrossRef]
61. Al-Yasiri, H.Q.M. Trends in the illiteracy rate in Iraq and its regional neighborhood for the period 2010–2020. *Misan J. Acad. Stud.* **2021**, *20*. Available online: https://www.iasj.net/iasj/article/215027 (accessed on 20 February 2023).
62. Young, A.O. Cohort Size and Unemployment Rate: New Insights from Nigeria. *Glob. J. Emerg. Mark. Econ.* **2021**, *13*, 122–151. [CrossRef]
63. Ni, S.; Gernay, T. A framework for probabilistic fire loss estimation in concrete building structures. *Struct. Saf.* **2021**, *88*, 102029. [CrossRef]
64. Dadzie, J.; Ding, G.; Runeson, G. Relationship between sustainable technology and building age: Evidence from Australia. *Procedia Eng.* **2017**, *180*, 1131–1138. [CrossRef]
65. Lago, A.; Faridani, H.M.; Trabucco, D. *Damping Technologies for Tall Buildings*, 1st ed.; Butterworth-Heinemann: Woburn, MA, USA, 2018; ISBN 9780128159644.
66. Winandari, M.I.R.; Wijayanto, P.; Faradila. Fire risk based on building density in dense settlement. *IOP Conf.Ser. Earth Environ. Sci.* **2021**, *780*, 12053. [CrossRef]
67. Raškauskaitė, R.; Grigonis, V. An Approach for the Analysis of the Accessibility of Fire Hydrants in Urban Territories. *ISPRS Int. J. Geo-Inf.* **2019**, *8*, 587. [CrossRef]
68. Cruz, M.G.; Hurley, R.J.; Bessell, R.; Sullivan, A.L. Fire behaviour in wheat crops-effect of fuel structure on rate of fire spread. *Int. J. Wildl. Fire* **2020**, *29*, 258–271. [CrossRef]

fire

Article

Effect of Interlayer Materials on Fire Performance of Laminated Glass Used in High-Rise Building: Cone Calorimeter Testing

Md Kamrul Hassan *, Md Rayhan Hasnat, Kai Png Loh, Md Delwar Hossain, Payam Rahnamayiezekavat, Grahame Douglas and Swapan Saha

School of Engineering, Design and Built Environment, Western Sydney University, Penrith, NSW 2751, Australia
* Correspondence: k.hassan@westernsydney.edu.au

Abstract: Laminated glass is prominently used nowadays as building construction material in the façade and architectural glazing of high-rise buildings. On the other hand, the fire safety of the high-rise building with laminated glass is also receiving more attention from the fire safety regulatory authorities and researchers due to recent fire incidents. Different interlayer polymeric materials are used in modern laminated glass to prevent the breakage of the glass façade, which can also increase the fire risk through a lower ignition time, and higher heat release and smoke production. Therefore, further research is required to understand the fire behaviour of laminated glass. In this study, the fire performance of the laminated glass has been investigated using cone calorimeter testing and the effect of different parameters such as glass thickness (6, 10, 12 mm), interlayer materials (PVB, SGP and EVA) and heat flux (25, 50 and 75 kW/m^2) on the fire behaviour of laminated glass has been studied. It is found that the glass thickness, interlayer material and heat flux can significantly influence the reaction-to-fire properties such as peak heat release rate (pHRR), total heat release, time to ignition, and smoke production of laminated glass. In addition, total smoke production (TSP) is also very high for PVB (3.146 m^2) and SGP (3.898 m^2) laminated glass compared to EVA (0.401 m^2) laminated glass and it is affected by these parameters. Finally, a simplified equation is developed to predict the pHRR of laminated glass by correlating the mass loss and external heat flux.

Keywords: laminated glass; fire performance; reaction-to-fire properties fire hazard; smoke hazard

Citation: Hassan, M.K.; Hasnat, M.R.; Loh, K.P.; Hossain, M.D.; Rahnamayiezekavat, P.; Douglas, G.; Saha, S. Effect of Interlayer Materials on Fire Performance of Laminated Glass Used in High-Rise Building: Cone Calorimeter Testing. *Fire* **2023**, *6*, 84. https://doi.org/10.3390/fire6030084

Academic Editor: Tiago Miguel Ferreira

Received: 29 January 2023
Revised: 17 February 2023
Accepted: 21 February 2023
Published: 22 February 2023

1. Introduction

Glass façades have recently been used more frequently in modern high-rise buildings owing to their high artistic, durable, and environmentally friendly qualities [1]. However, when exposed to fire, their breakage and fallout may create a new vent, allowing fresh air entrainment and fire spread, potentially greatly accelerating the development of a compartment fire. When a fire spreads from an interior space to exterior cladding, the glass façade's fire performance is crucial [2]. Modern architecture uses laminated glass instead of monolithic glass due to its limited brittleness. Laminated glass is made up of two or more monolithic layers joined together through polymer interlayers. The most common interlayers are ethylene-vinyl acetate (EVA) and polyvinyl butyral (PVB) [3]. The polymer interlayer allows the element to deform and redistribute the load through the glass panes, which helps to avoid crack propagation and sudden failure of laminated glass [1–3]. This makes laminated glass safer in construction industry applications compared to monolithic glass. In addition, laminated glasses have the ability to reduce the heat transfer, leading to more comfort inside the building. However, the presence of a certain amount of polymer material used as an interlayer in laminated glass makes it combustible, depending on the volume of glass and interlayer material ratio [2], which need to be investigated to identify the fire hazard of laminated glass with different interlayer materials.

To understand the behaviour of laminated glass, many studies have been carried out [4–13]. The mechanism of thermal cracking in glass structures is investigated in [4–6].

It was determined that the thermal gradient is the primary cause of glass breakage in a fire because thermal stress develops between the heated surface and the insulated cooler edge [5]. Harada et al. [7] used a propane radiation panel to test two types of glass (treated float glass and wired glass) under different heat flux and lateral restraint of the glass and measured the time to initial crack and fallout, temperatures and thermal stress. Based on the test data, they developed a simple model to predict the glass cracking and/or breaking. It was concluded that thermal stress varies from 10 to 25 MPa depending on the type of glass and it does not have any effect of the restraint of the glass on crack development. Shields [8,9] conducted full-scale experiments in an ISO 9705 room to investigate the differences in the fire behaviour of single glazing with different fire locations. Wang et al. [10] investigated the effects of thermal breakage on the fire response of single, insulated, and laminated glazing. Their study reported that insulated and laminated glass could last longer than single glass. Debuyser et al. [1] exposed laminated and monolithic glass panes to heat fluxes ranging from 10 to 12 kW/m^2 for around 50 min. The laminated glass panes consisted of three layers of glass bonded with two layers of PVB or SentryGlas interlayer. No ignition or flaming was observed for the seven laminated glass specimens used in their study. However, the heat fluxes used were low compared to what would be expected in a fire or standard tests. More recently, Wang and Hu [6] investigated the performance of laminated glass under fire conditions. They exposed two 600 × 600 mm laminated glass specimens to the thermal exposure. The external heat flux was between 7.7 and 25 kW/m^2. The laminated glass panes consisted of two layers of 6 mm thick glass with a 0.38 mm thick PVB inter-layer. The study recommended the 0.38 mm PVB interlayer and 3 mm glass-pane considering the relatively low construction cost with reasonable fire performance.

Despite the widespread use of non-fire-resistant laminated glass, there is very little information on its fire performance, particularly in reaction-to-fire properties (heat release rate, peak heat release rate, total heat release, mass loss rate) and smoke production rate. Furthermore, most existing research has been conducted in a standard furnace, which provides limited insight into the reaction-to-fire properties and smoke production rate [11]. However, a cone calorimeter, which is useful for analysing smoke production and toxicity, can be used to analyse the reaction-to-fire properties [12,13]. Therefore, further research is required to understand the reaction-to-fire properties and smoke productions of laminated glass using a cone calorimeter.

To address the abovementioned research gaps, the reaction-to-fire properties of the laminated glass have been investigated in the present study using a cone calorimeter. The main parameters considered in this study are glass pane thickness, interlayer materials, and external heat flux. The current study analyses and discusses the impact of those parameters on the reaction-to-fire properties of laminated glass. In addition, the smoke hazard and fire hazard assessment have been discussed in this study. Finally, a simplified formula is provided to predict the peak heat release rate (pHRR) of laminated glass. The outcomes of this study will provide a better understanding of the reaction-to-fire properties and smoke hazards, which will be helpful to train fire safety engineers in designing fire-safe buildings using laminated glass.

2. Materials and Experimental Set-Up

2.1. Materials

In this research study, laminated glasses with different interlayers were purchased from the local market in Australia, which are grade A laminated safety glasses. Laminated glass is made using two glass panes bonded together with a layered material, as shown in Figure 1. The glass panes considered in this study were 3, 5 and 6 mm thick and tested this parameter to observe the effect of the glass pane's thickness on the reaction-fire properties. Three interlayer materials (PVB, SGP and EVA) used in different laminated glasses in building applications were considered in this study. PVB (polyvinyl butyral) is an amorphous polymer that is composed of three monomers such as vinyl butyral

(76–80 wt.%), vinyl alcohol (18 to 22 wt.%) and vinyl acetate (1 to 2 wt.%) [3]. Plasticisers have been incorporated into PVB to improve mechanical properties without affecting the adhesion and optical properties. SGP (SentryGlas® Plus) is an ionomer containing a hydrocarbon backbone with partially or entirely neutralised pendant acid groups. It is stiffer and shows less sensitivity to working temperature than PVB. EVA (ethylene-vinyl acetate) consists of ethylene (10 to 40 wt.%), vinyl acetate (32 to 34 wt.%) and some specific additives [3]. The density and thermal conductivity of the interlayer materials are presented in Table 1.

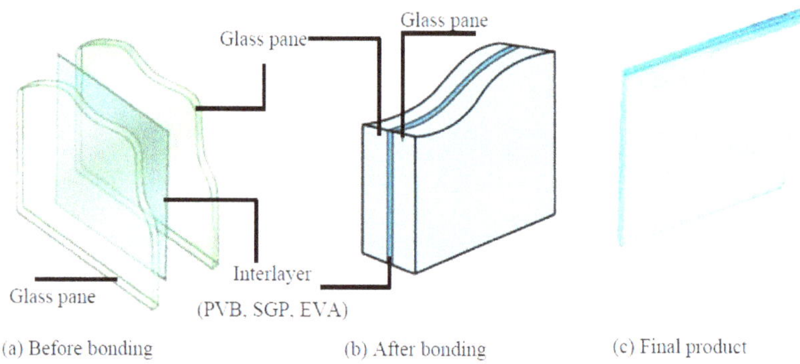

Figure 1. A typical laminated glass with interlayer materials.

Table 1. Test material properties.

Properties	PVB	SGP	EVA
Density of interlayer	915–1070 kg/m^3	950 kg/m^3	945–955 kg/m^3
Density of laminated glass	2274	2242	2402
Thermal conductivity of interlayers	0.20 (W/mK)	0.25 (W/mK)	0.34 (W/mK)

2.2. Test Specimens

A total of 21 samples were considered in this experimental study. The details of each sample are given in Table 2. The parameters that were studied in this study are (1) glass pane thickness (3, 5 and 6 mm); (2) external heat flux (25, 50 and 75 kW/m^2); (3) interlayer material types (PVA, SGP, EVA). Three samples were considered to check the repeatability of each parameter. The last digit of each sample represents the repeating test number for the same configuration. For example, in specimen S6PVB50-1, S represents the sample, 6 represents the nominal thickness value of glass, PVB represents the polymeric interlayer type such as PVB, and the next two digits represent the heat flux considered during testing.

Table 2. Details of test specimens.

Sl No	Sample Label	Glass Pane Thickness (mm)	Interlayer Thickness (mm)	Interlayer Material Type	Heat Flux (kW/m^2)	Sample Size, L × B × t (mm)	pHRR (kW/m^2)
1	S6PVB50-1	3	0.38	PVB	50	100 × 100 × 6.38	237
2	S6PVB50-2	3	0.38	PVB	50	100 × 100 × 6.38	218
3	S6PVB50-3	3	0.38	PVB	50	100 × 100 × 6.38	222
4	S10PVB25-1	5	0.38	PVB	25	100 × 100 × 10.38	44
5	S10PVB25-2	5	0.38	PVB	25	100 × 100 × 10.38	45
6	S10PVB25-3	5	0.38	PVB	25	100 × 100 × 10.38	50
7	S10PVB50-1	5	0.38	PVB	50	100 × 100 × 10.38	156
8	S10PVB50-2	5	0.38	PVB	50	100 × 100 × 10.38	148
9	S10PVB50-3	5	0.38	PVB	50	100 × 100 × 10.38	153
10	S10PVB75-1	5	0.38	PVB	75	100 × 100 × 10.38	203

Table 2. *Cont.*

Sl No	Sample Label	Glass Pane Thickness (mm)	Interlayer Thickness (mm)	Interlayer Material Type	Heat Flux (kW/m^2)	Sample Size, L × B × t (mm)	pHRR (kW/m^2)
11	S10PVB75-2	5	0.38	PVB	75	100 × 100 × 10.38	191
12	S10PVB75-3	5	0.38	PVB	75	100 × 100 × 10.38	197
13	S12PVB50-1	6	1.52	PVB	50	100 × 100 × 13.52	236
14	S12PVB50-2	6	1.52	PVB	50	100 × 100 × 13.52	399
15	S12PVB50-3	6	1.52	PVB	50	100 × 100 × 13.52	318
16	S12SGP50-1	6	1.52	SGP	50	100 × 100 × 13.52	289
17	S12SGP50-2	6	1.52	SGP	50	100 × 100 × 13.52	320
18	S12SGP50-3	6	1.52	SGP	50	100 × 100 × 13.52	293
19	S12EVA50-1	6	1.52	EVA	50	100 × 100 × 13.52	103
20	S12EVA50-2	6	1.52	EVA	50	100 × 100 × 13.52	107
21	S12EVA50-3	6	1.52	EVA	50	100 × 100 × 13.52	106

The first three specimens (S6PVB50-1, S6PVB50-2, S6PVB50-3) were tested at a heat flux of 50 kW/m^2 for the laminated glass with a 3 mm glass pane and 0.38 PVB interlayer. With the same interlayer material (0.38 PVB) and heat flux (50 kW/m^2), another three specimens (S10PVB50-1, S10PVB50-2, S10PVB50-3) were tested to investigate the effect of glass pane thickness (3 mm ad 5 m) on the fire properties. The effect of different heat fluxes (25, 50 and 75 kW/m^2) was also studied, which are labelled as S10PVB25, S10PVB50 and S10PVB75 for 25, 50 and 75 kW/m^2 heat flux, respectively, as shown in Table 2. Another nine specimens (S12PVB50-1, S12PVB50-2, S12PVB50-3, S12SGP50-1, S12SGP50-2, S12SGP50-3, S12EVA50-1, S12EVA50-2 and S12EVA50-3) were tested to analyse the effect of different interlayer materials (PVB, SGP and EVA). The heat flux (50 kW/m^2), interlay thickness (1.52 mm), and glass pane thickness (6 mm) were the same for all nine specimens. The tested specimens of each category are shown in Figure 2.

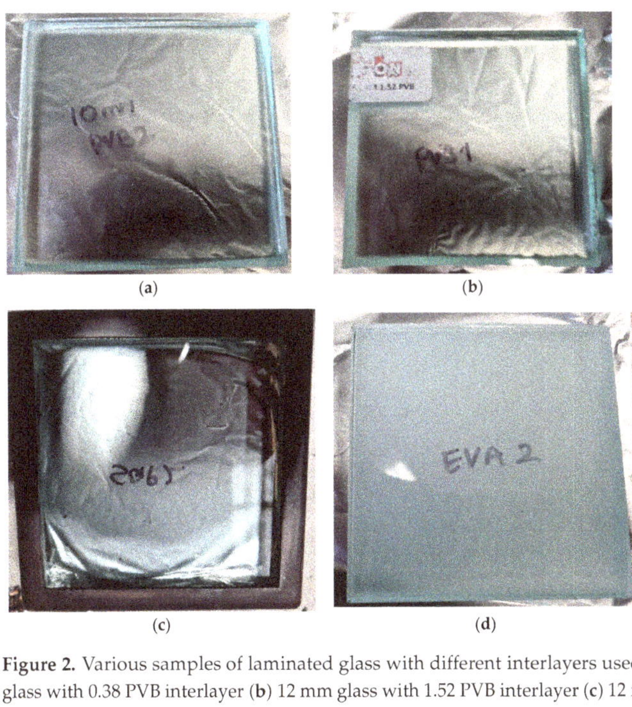

(a) (b) (c) (d)

Figure 2. Various samples of laminated glass with different interlayers used in the test: (**a**) 10 mm glass with 0.38 PVB interlayer (**b**) 12 mm glass with 1.52 PVB interlayer (**c**) 12 mm glass with 1.52 SGP interlayer (**d**) 12 mm glass with 1.52 EVA interlayer.

2.3. Sample Testing Using a Cone Calorimeter

All laminated glass specimens were tested horizontally using a cone calorimeter following AS/NZS 3837. AS/NZS 3837 is the method of testing for heat and smoke release rates of materials and products using an oxygen consumption calorimetry, and it is based on the ISO 5660-1 test standard (Standards Online 2012). In this study, the back and the side of the samples were covered with aluminium foil and then kept over a ceramic fibre backing pad to avoid heat loss [14]. The distance between the cone heater and the top surface of the sample was kept at 25 mm. Similar to other works [15], a spark igniter located 13 mm above the centre of the specimen was used to pilot the ignition. Laminated glass samples were placed horizontally on the specimen holder. The cone heater consists of 5 kW electrical heating element, and the temperature of the heater is controlled by three type k thermocouples inserted inspires. It is worth mentioning that the target heat flux, such as 25 kW/m^2, 50 kW/m^2 and 75 kW/m^2, is kept constant before starting any test. The test data, such as heat release rate (HRR), smoke production rate (SPR) and carbon monoxide production (COP), were recorded at 10-second intervals throughout each test. A typical cone calorimeter horizontal test set-up for a laminated glass specimen is shown in Figure 3.

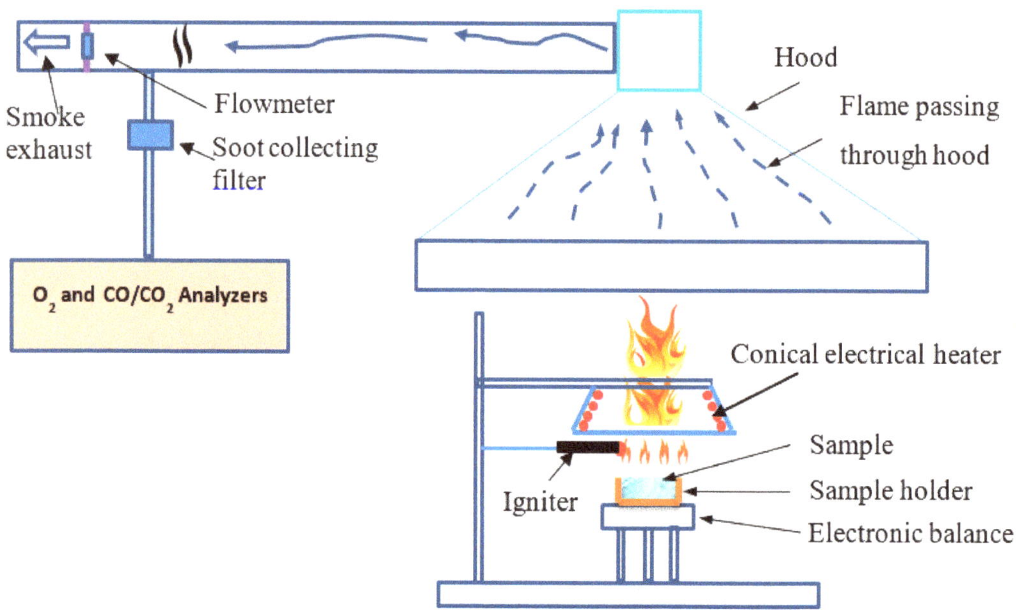

Figure 3. Schematic diagram of cone calorimeter test set-up used for laminated glass.

2.4. Data Measurement Details

In this study, the heat release rate (HRR), total heat release (THR), effective heat of combustion (EHC), mass loss rate (MLR), smoke production (TSP) and specific extinction area (SEA) are measured. The HRR was calculated according to oxygen depletion using Equation (1) [16].

$$HRR = 13.1 \times 10^3 \times 1.10C \times \frac{\sqrt{\Delta P}}{T_e} \times \frac{(0.2095 - Xo_2)}{(1.105 - 1.5Xo_2)} \quad (1)$$

where C is the orifice flow meter calibration constant, ΔP is the pressure drop at the orifice meter, and T_e is the absolute temperature of the gas at the orifice meter [16]. The pHRR was determined by calculating the mean values for each sample and for each heat flux. The mass loss rate (MLR) was also measured during each test, and the average values were calculated. Total heat release (THR) was measured while testing (Equation (2)). Effective heat of combustion (EHC) can be calculated with Equation (3).

$$\mathrm{THR} = \int_{0}^{\infty} \mathrm{HRR}(t)dt \tag{2}$$

$$\mathrm{EHC} = \frac{\mathrm{HRR}}{\mathrm{MLR}} \tag{3}$$

$$\mathrm{pHRR} = \mathrm{pMLR} \times \mathrm{EHC} \tag{4}$$

3. Results and Discussion

For studying the flammability of laminated glass materials, pHRR and time to ignition parameters are the key parameters used to assess fire safety [17,18]. These parameters are also helpful for correlating the full-scale and bench-scale tests.

3.1. *Effect of Glass Pane Thickness*

The effect of laminated glass thickness was studied by analysing the samples with 3 mm and 5 mm glass pane thickness with the same interlayer PVB thickness at 50 kW/m^2 heat flux. The heat release rate of each sample of 6 mm laminated glass and 10 mm laminated glass is shown in Figure 4a–c, and the average heat release rate of 6 mm and 10 mm laminated glass with 0.38 PVB interlayer is shown in Figure 4a. It can be seen from Table 2 and Figure 4a and that the maximum and minimum heat release rate for 6 mm laminated glass samples were observed as 222 kW/m^2 and 175.05 kW/m^2, respectively. The average heat release rate of 6 mm laminated glass samples with 0.38 mm PVB interlayer was 194.45 kW/m^2, as shown in Figure 4c. In the case of three 10 mm laminated glass thickness samples, the maximum heat release rate was observed at 153 kW/m^2. The average heat release rate of 10 mm laminated glass samples with 0.38 PBV interlayer is 137.33 kW/m^2, as shown in Figure 4c. The 6 mm thick laminated glass showed a higher average heat release rate than the 10 mm thick laminated glass sample at 50 kW/m^2 heat flux. For both types, there was a 0.38 mm thick PVB interlayer. Therefore, the glass pane thickness greatly influenced the heat release rate. It was also observed that the 3 mm glass pane had started to crack earlier. As a result, ignition occurred earlier for a 3 mm glass pane when compared to a 5 mm thick glass pane, as shown in Figure 5. Similar works have been carried out on PMMA by Paul et al. [19], where the HRR values decreased with sample thickness. Similarly, in another work on EPS and XPS insulations by Weiguang et al. [20], the pHRR value gradually decreased with a thickness increase for the EPS sample.

The top glass layer of the 6 mm samples cracked in 30 to 40 s, whereas the top glass layer of the 10 mm samples cracked in 57 to 108 s (see Figure 5a,b). These findings imply that as thickness increases, cracking time can be delayed. When the maximum normal stress of materials reaches the materials' ultimate strength, cracks occur [14]. Therefore, the maximum stress required to reach the ultimate strength of 10 mm samples is assumed to take longer than that of 6mm samples. Moreover, thicker samples (10 mm) had fewer cracks than thinner samples (6 mm). This is because the crack was first initiated at the edge for both samples and then propagated very quickly.

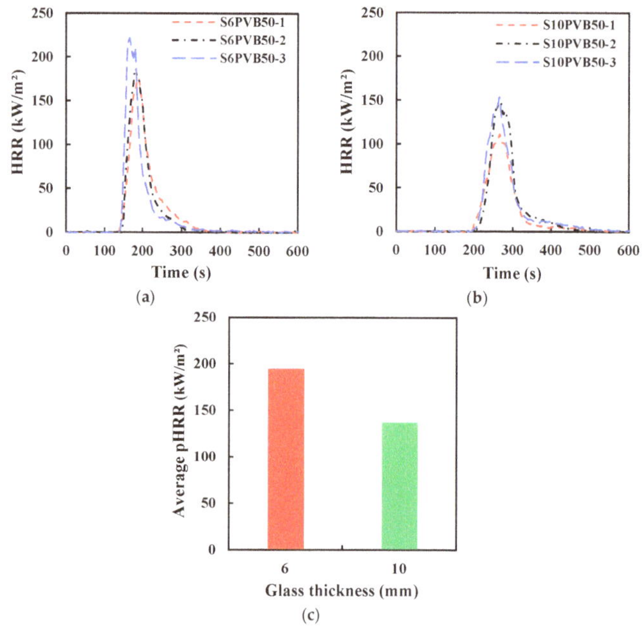

Figure 4. Effect of glass thicknesses on HRR curves for 6 and 10 mm glass with 0.38 PVB interlayer at 50 kW/m^2 heat flux: (**a**) HRR for 6 mm glass pane (**b**) HRR for10 mm glass pane and (**c**) average Peak heat release rate (pHRR).

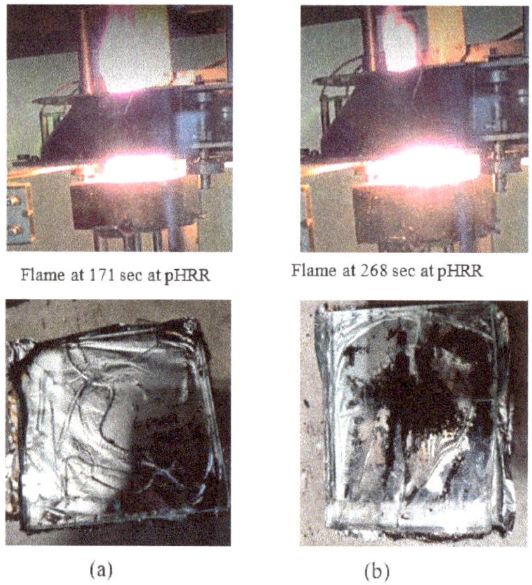

Figure 5. Observation during and after fire of (**a**) 6 mm and (**b**) 10 mm glass with 0.38 PVB interlayer at 50 kW/m^2 heat flux.

3.2. Effect of Heat Flux

Three 10 mm laminated glass samples with 0.38 mm PVB interlayer were tested for each heat flux exposure of 25, 50 and 75 kW/m^2. The average heat release rate of the 10 mm laminated glass with different heat fluxes (25 kW/m^2, 50 kW/m^2, 75 kW/m^2) is shown in Figure 6a. At 25 kW/m^2 heat flux, the peak heat release rate was 40.52 kW/m^2, whereas at 50 kW/m^2 heat flux, the peak heat release rate was 137.33 kW/m^2. The peak heat release rate at 75 kW/m^2 heat flux was 197 kW/m^2. A gradual increase was observed in the heat release rate (HRR), with an increase in the heat flux from 25 to 75 kW/m^2, as shown in Figure 6b. The key observations indicate the glass had started to crack earlier at 75 kW/m^2 of heat irradiance when compared to 50 kW/m^2 and 25 kW/m^2. This resulted in the samples being exposed to 25 kW/m^2 of heat irradiance to record a lower peak heat release rate, as the interlayer had reacted within the glass but did not emit a large amount of gas. This behaviour indicates that the time it takes for the glass to crack on the exposed surface and how the gasses from the reaction of the interlayer are emitted will affect the heat release of the laminated glass. It was stated by Babrauskas [5] that in a broader sense, for many products, the HRR value is linearly proportional to the external heat flux, though the deviation from the linearity at very high and low heat flux is common for many materials. A similar response was observed in a previous study related to XPS polymer [15], where the peak heat release rate (pHRR) increased to 492 kW/m^2 at 50 kW/m^2 from 423 kW/m^2 at 35 kW/m^2. However, the trend was the opposite in the case of EPS. Similar work was conducted by Rickard et al. [11] by imposing radiant heat flux on the PVB interlayer laminated glass samples. It was found that the HRR value increased at 50 kW/m^2 compared to 25 kW/m^2 and again slightly decreased at 75 kW/m^2. The trend of time-to-ignition is the same as studied by Rickard et al. [11], as shown in Figure 6c. Though the interlayer thickness was 1.2 mm in Rickard's study, the required time-to-ignition at higher heat flux was almost similar. At lower heat flux (25 kW/m^2), the effect of the thickness variation was acute; thus, the ignition was delayed significantly (Figure 6c).

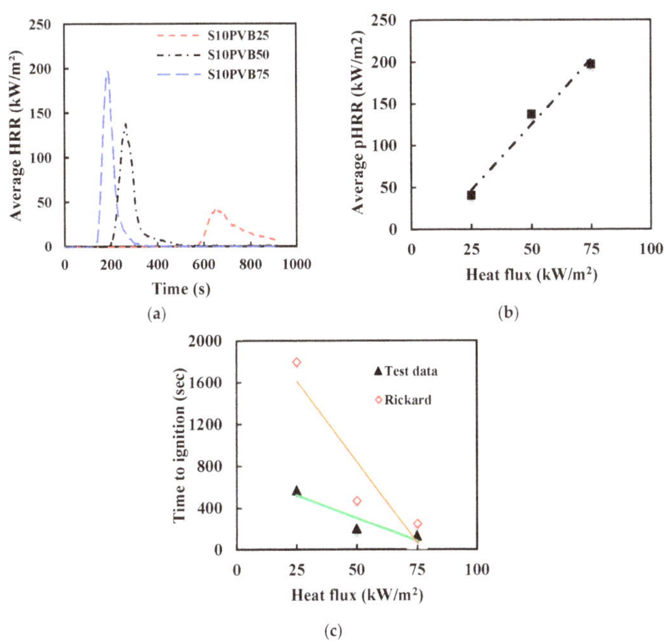

Figure 6. Effect of external heat flux on (**a**) HRR curves, (**b**) pHRR and (**c**) time to ignition of 10 mm PVB laminated glass samples.

The fire hazard increases with the increase in external heat flux, as seen in Figure 7. The burnout was high for the sample tested at 75 kW/m^2 compared to the 25 kW/m^2 heat flux. Within 185 s, the flame reached its peak state when mass loss and pyrolysis accelerated greatly. The pHRR was also significant in that case. In the case of 25 kW/m^2 heat flux, the cracking was initiated at the corner; after the fire test, only the crack existed (Figure 7a). There was no significant burning, which indicates very low mass loss during the burning process at 25 kW/m^2 heat flux.

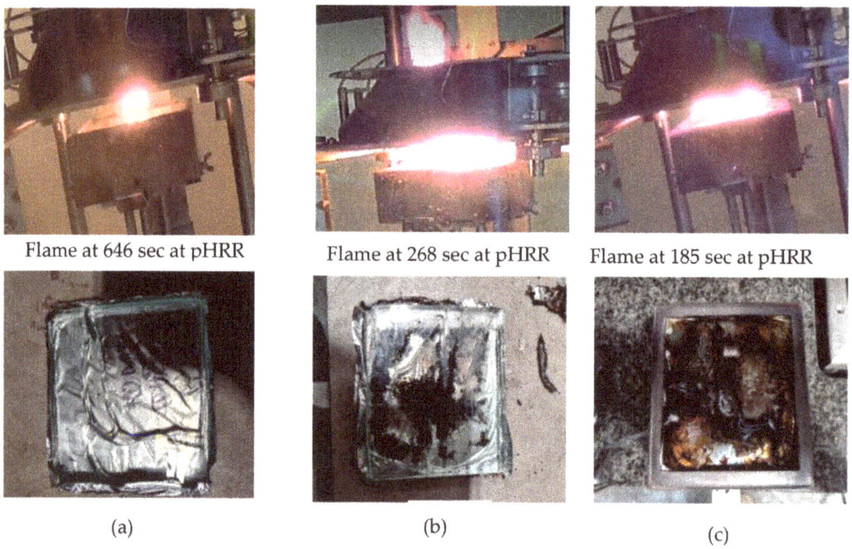

Figure 7. Observation during and after fire of 10 mm glass with 0.38 PVB interlayer with different heat fluxes: (**a**) 25 kW/m^2 (**b**) 50 kW/m^2 (**c**) 75 kW/m^2.

3.3. *Effect of Interlayer Materials of Laminated Glasses*

The effect of interlayer materials (PVB, SGP and EVA) of the laminated glass with a 6 mm glass pane at 50 kW/m^2 heat flux was investigated. For each category of the interlayer, three samples were considered. The heat release rate of each sample of PVB, SGP and EVA laminated glass is shown in Figure 8a–c, and the average peak heat release rates of PVB, SGP and EVA laminated glass are shown in Figure 8d. For PVB interlayer samples, the maximum heat release rate was 399 kW/m^2, as shown in Figure 8a, and the average value of peak heat release rate was 313.5 kW/m^2, as shown in Figure 8d for the second sample. Ignition of the first sample was observed to have started along the edges of the sample rather than along the crack (Figure 9). This phenomenon led to a slower spread of flame along the cracks, slower flame extinguishment and a lower peak heat release rate of 236 kW/m^2, which was 935 s after ignition. The peak heat release rate for the first and second samples was 236 kW/m^2 and 399 kW/m^2, with an average of 313.5 kW/m^2. The average time it took for the two samples to reach their peak heat release rate from the flaming starting time was about 77 s.

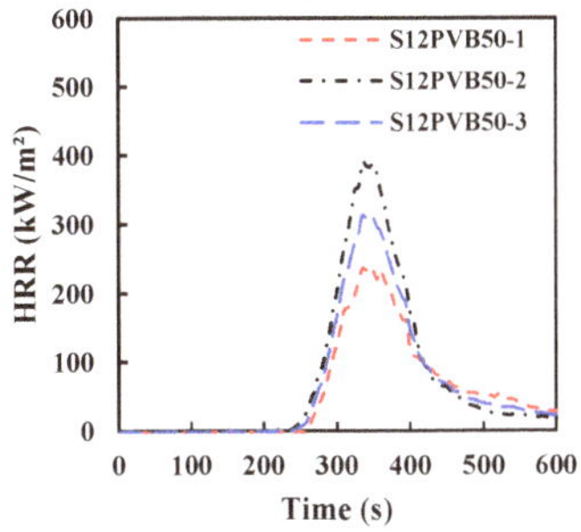

(**a**) 12 mm glass pane with 1.52 PVB interlayer

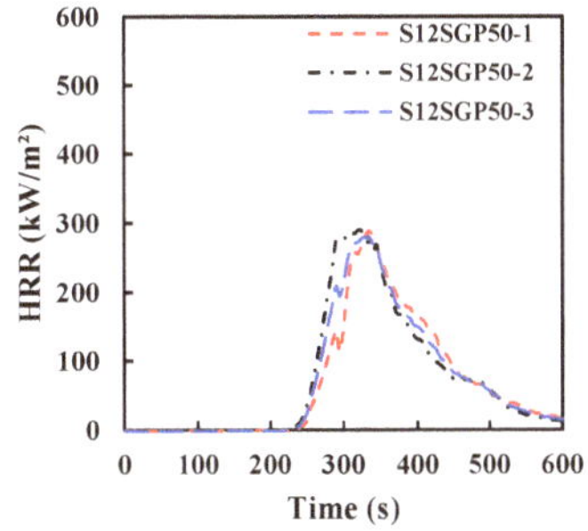

(**b**) 12 mm glass pane with 1.52 SGP interlayer

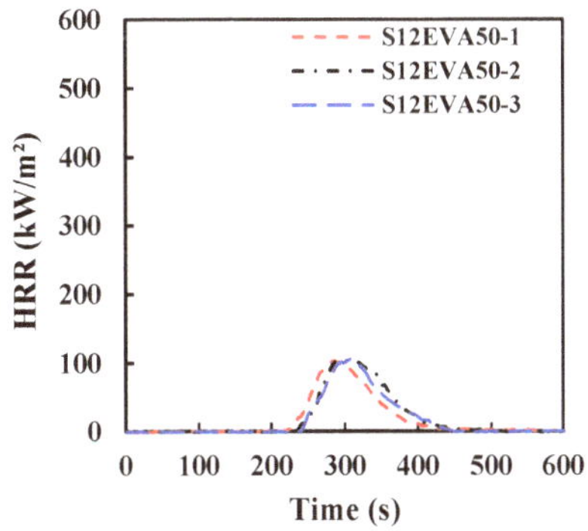

(**c**) 12 mm glass pane with 1.52 EVA interlayer

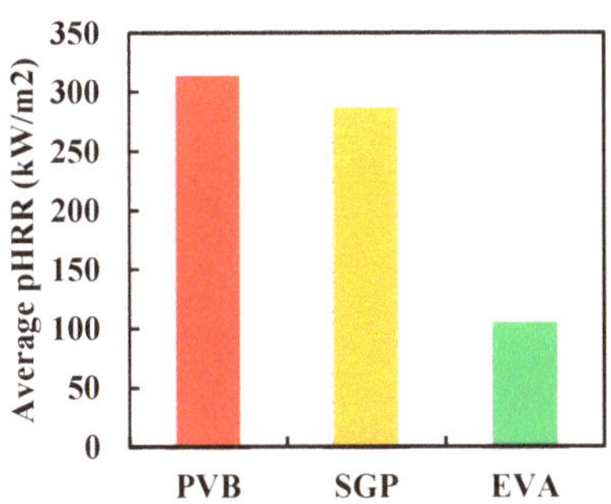

(**d**) Average pHRR

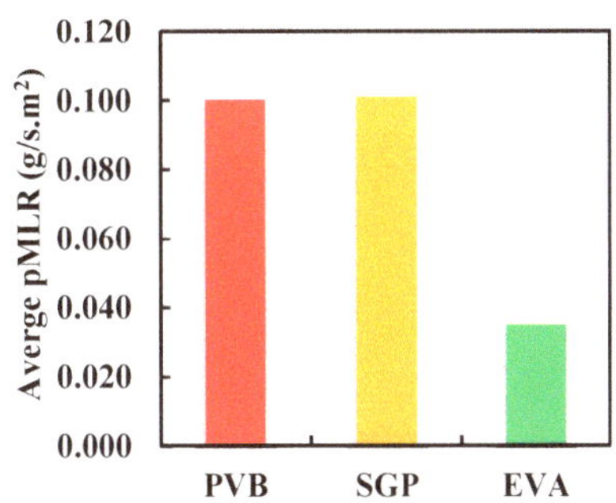

(**e**) Average MLR

Figure 8. Effect of interlayer materials on (**a**–**c**) HRR curves, (**d**) average pHRR values and (**e**) average MLR of 12 mm glass tested at 50 kW/m^2 heat flux.

In the case of 12 mm SGP laminated glass, the maximum heat release rate was 290 kW/m^2, as shown in Figure 8b, and the average heat release rate was 286.23 kW/m^2, as shown in Figure 8d. However, for 12 mm EVA laminated glass samples, the peak heat release rate was 107 kW/m^2, as shown in Figure 8c, and the average heat release rate was 105 kW/m^2, as shown in Figure 8d. Therefore, the peak mass loss rate of EVA laminated glass is remarkably different compared to PVB and SGP laminated glasses, as shown in Figure 8e. The pHRR is directly related to the mass loss rate (Equation (4)) [21]. The higher the mass loss rate, the greater the pHRR value. The reason is due to a high mass loss; as there will be more pyrolysis gas generation, the more combustible gas will be there to accelerate the heat generation.

Interlayers also played an important role in crack development. The PVB samples cracked between 35 and 127 s (see Figure 9a), whereas the SGP samples cracked between 60 and 87 s (see Figure 9b). In the case of EVA samples, the crack appears between 84 and 154 s (see Figure 9c). As a result, EVA samples provide greater stress crack resistance than the other interlayer samples. This longer cracking time may be due to EVA's higher stiffness compared to PVB, which is almost similar to that of SGP [3]. Nearly 100 s before ignition, pyrolysis gases were detected. The flame reached its peak moment at 335 s after the test started in case of PVB interlayer (see Figure 9a), whereas it was at 296 s for the EVA sample (see Figure 9c).

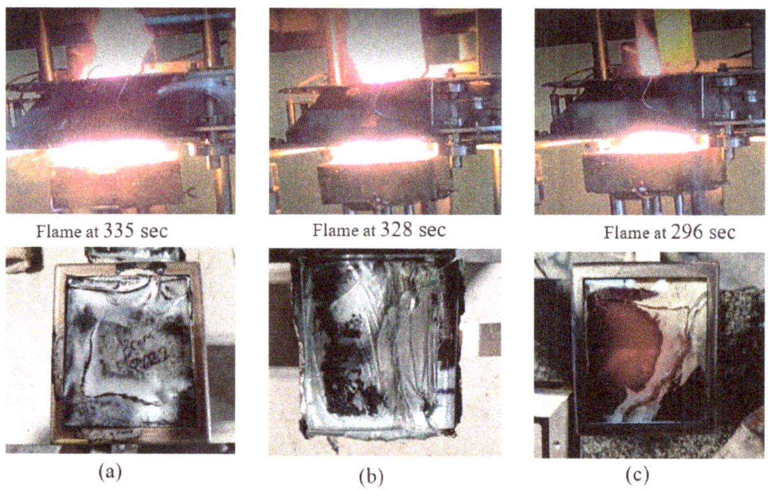

Figure 9. Observation during and after fire of 12 mm glass at 50 kW/m^2 heat flux with 1.52 mm thick (**a**) PVB, (**b**) SGP and (**c**) EVA interlayers.

3.4. Discussions

The fire performance of any combustible and non-combustible materials is evaluated based on the reaction-to-fire properties, fire hazard and smoke hazard [12–15]. In this study, the fire performance of PVB, SGP and EVA interlayer laminated glasses is also discussed based on the reaction-to-fire properties, fire hazard and smoke hazard. The measured reaction-to-fire properties, fire and smoke hazards of PVB, SGP and EVA interlayer laminated glasses are listed with averaged calculated values in Table 3. The details of these factors are discussed in the below sub-sections.

Table 3. Average reaction-to-fire properties of laminated glass samples.

Test Specimens	S6PVB50	S10PVB50	S10PVB25	S10PVB75	S12PVB50	S12SGP50	S12EVA50
Time to ignition (s)	130	200	570	130	225	225	225
Time to flameout (s)	142	219	598	143	261	243	250
Total burning time (s)	435	551	1413	408	1208	783	510
pHRR (kW/m^2)	194.45	153	40.52	197	313.5	289.5	105.2
Time to pHRR (s)	171	268	646	185	335	328	296
Average THR (MJ/m^2)	11.27	11.21	8.48	11.19	42.28	42.79	10.26
Average mass loss rate ($g/s{\cdot}m^2$)	0.063	0.043	0.015	0.064	0.100	0.101	0.035

3.4.1. Ignition Time

It is found that the ignition time is greatly influenced by the thickness of the glass pane as well as by the different external heat fluxes. It is noticeable from Figure 10a that the ignition time is the same (225 s) for all three types of interlayers (PVB, SGP, EVA). However, the flame-out time was 250 s and 243 s in the case of EVA and SGP, respectively, whereas it was slightly longer for PVB samples (261 s) (Figure 10b). The flameout time also increased with the increase in thickness but gradually decreased with the increase in external heat flux. EVA interlayer samples reached pHRR more quickly than the others (Figure 10c). It took 296 s for the EVA sample to reach pHRR, whereas it took 328 s and 335 s for the SGP and PVB samples, respectively. The ignition time for the 6 mm glass sample was 130 s and increased to 200 s for the 10 mm thick laminated glass sample with the same heat flux exposure of 50 kW/m^2 in Figure 10a. The thinner glass pane cracked earlier, allowing volatile gases to emit earlier and ultimately resulting in an earlier ignition than that of the 10 mm thick glass samples. The glass pane crack was due to thermal stress built because of the temperature gradient. The 3 mm glass pane had started to crack earlier as the inner portion expanded more quickly than the 5 mm glass pane. Before ignition, the mass loss rate was significantly higher than any other time, which subsequently contributed to the pyrolysis and ignition. In the case of the 10 mm glass sample, the mass loss rate was lower than the 6 mm glass sample, which delayed the time to evolve the pyrolysis gas to contribute to the ignition process. The time required for the 12 mm glass samples to reach their peak heat release rate from the flaming starting time was varied. The average time was 316 s from the test's beginning. It was observed that flaming started occurring at the bottom of the case before stopping and intensifying on the exposed face. In Rickard et al.'s [11] work, the authors pointed out that the ignition time altered in various samples due to changes in the transmissivity of heat flux. As the glass pane cracked and the interlayer started bubbling, it influenced the transmissivity of the glass and interlayer. Their experiment found that after glass cracking and interlayer bubbling, the measured heat flux on the sample was greatly reduced in the case of a lower external heat flux of 25 kW/m^2. At 75 kW/m^2, the measured heat flux on the sample was less reduced than in the earlier case. It was mentioned that the presence of gas also influenced the measured heat flux on the sample. Weiguang et al. [20] used a cone calorimeter to observe the fire performance of polystyrene (PS) foam with different external kW/m^2, where it was only 23.5 s at 45 kW/m^2. Another reason was mentioned by Klasen et al. [22], who reported that the re-radiation of the sample altered with the increase in sample temperature.

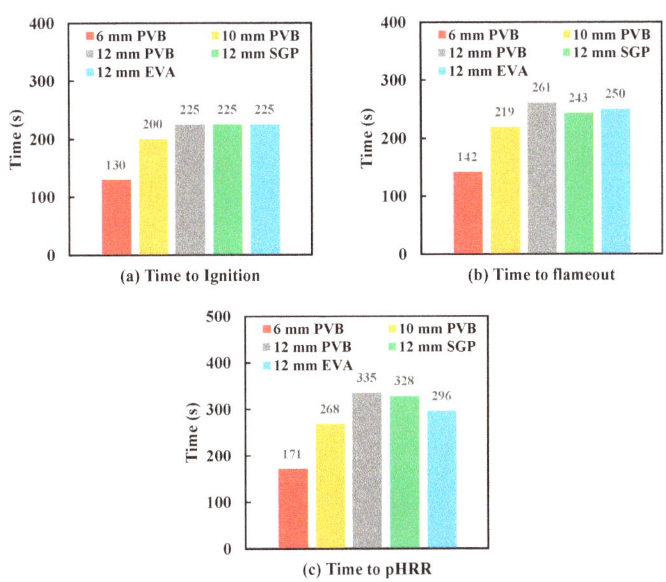

Figure 10. Effect of thickness and interlayer materials on (**a**) time to ignition (**b**) time to flameout and (**c**) time to pHRR of 6, 10 and 12 mm glass tested at 50 kW/m^2 heat flux.

3.4.2. Total Heat Release

The total heat release of all samples is determined to quantify the total fire load of each specimen at the end of the test, which will provide an indication of the fire hazard of laminated glass [23]. The total heat release obtained from each test at the heat flux of 50 kW/m^2 is shown in Figure 11. It can be seen from Table 3 that the values of total heat release were almost the same for both the 6 mm laminated glass with 0.38 mm PVB interlay (11.27 MJ/m^2) and the 10 mm laminated glass with 0.38 mm PVB interlay (11.21 MJ/m^2). It was also the same (11.19 MJ/m^2) for the 10 mm laminated glass with 0.38 mm PVB interlay when tested at the heat flux of 75 kW/m^2, see Table 3, but lower THR (8.48 MJ/m^2) was observed when testing the same sample at the heat flux of 25 kW/m^2. However, the average THR value was significantly high for 13 mm laminated glass with 1.52 mm PVB interlayer (42.28 MJ/m^2) and 13 mm laminated glass with 1.52 mm SGP interlayer (42.79 MJ/m^2) (see Table 3 and Figure 11). On the other hand, the average THR value was significantly low (10.26 MJ/m^2) for 13 mm laminated glass with 1.52 EVA interlayer samples (see Table 3 and Figure 11). It implies that based on THR, EVA possesses less fire load than the other two interlayers (PVB and SGP).

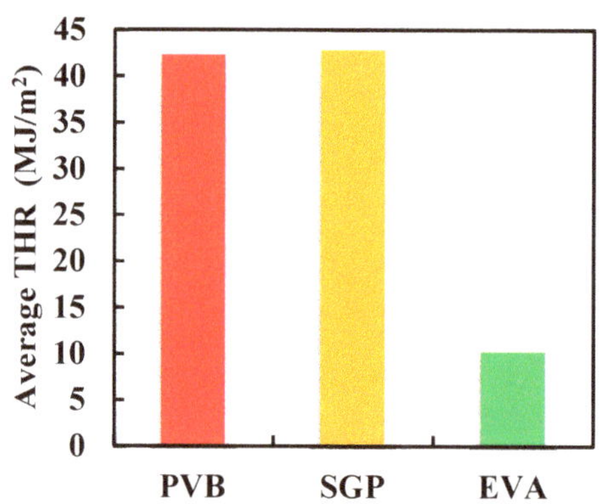

Figure 11. THR of different interlayer 12 mm laminated glass at 50 kW/m^2.

3.4.3. Mass Loss Rate

The mass loss rate (MLR) obtained during the testing of each sample at the heat flux of 50 kW/m^2 is shown in Table 3. It can be seen from Figure 12a,b that the MLR value of 6 mm laminated glass with 0.38 mm PVB interlayer was higher than that of 10 mm laminated glass with 0.38 mm PVB interlayer. The average MLR for 6 mm laminated glass was 0.063 g/s·m^2, whereas it was 0.043 g/s·m^2 for 10 mm laminated glass (see Table 3). It can be seen from Figure 12d that the maximum MLR was observed for 12 mm laminated glass with 1.52 mm SGP interlayer compared to other 12 mm PVB (Figure 12c) and EVA (Figure 12e) laminated glass samples with the same interlayer thickness. The average MLR of 13 mm PVB, SGP and EVA laminated glass samples with 1.52 mm interlayer was 0.100, 0.101, 0.035 g/s·m^2, respectively) (see Table 3). It indicates that the fire spread for EVA laminated glass will be less compared to PVB and SGP laminated glass. It can also be seen from Table 3 that the average MLR value of 10 mm PVB laminated glass increases from 0.015 to 0.064 g/s·m^2 with an increase in the heat flux from 25 kW/m^2 to 75 kW/m^2.

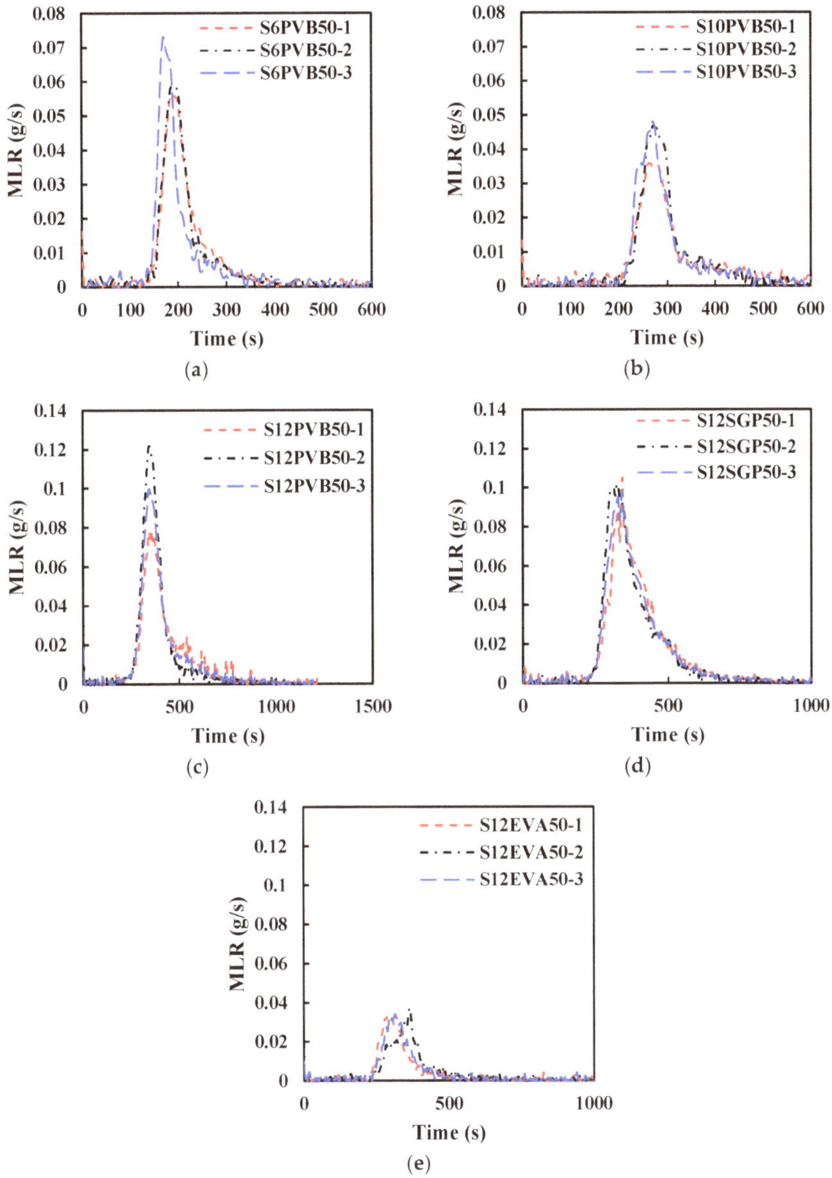

Figure 12. Mass loss rate at 50 kW/m^2: (**a**) 6 mm glass with 0.38 PVB interlayer (**b**) 10 mm glass with 0.38 PVB interlayer (**c**) 12 mm glass with 1.52 PVB interlayer (**d**) 12 mm glass with 1.52 SGP interlayer and (**e**) 12 mm glass with 1.52 EVA interlayer.

pMLR was also greatly influenced by external heat fluxes, as stated in Section 3.2. When the heat flux increased to 50 kW/m^2, the mass loss rate (MLR) also increased as the pyrolysis process accelerated with higher heat flux (see Figure 13). Thus, the ignition time rapidly decreases with higher external heat flux, as shown in Figure 6c. At 75 kW/m^2, the MLR increased rapidly to 0.064 g/s·m^2 for the same reason stated earlier.

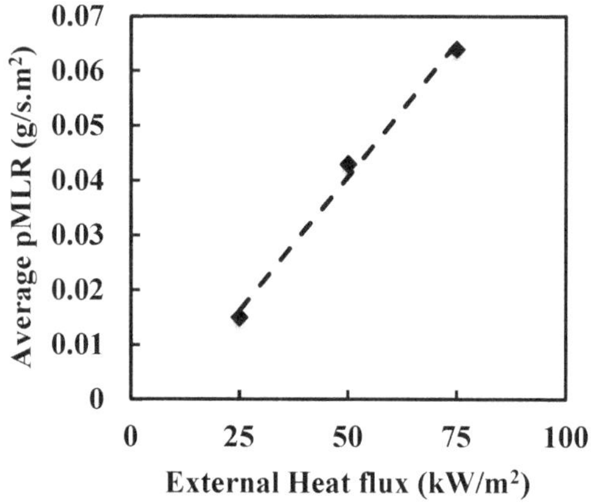

Figure 13. Effect of external heat fluxes on 10 mm PVB laminated glass.

3.4.4. Fire Hazard Assessment

Some common fire performance indicators such as fire growth index (FGI), fire performance index (FPI) and maximum average rate of heat emission (MARHE) were calculated based on cone calorimeter test data to assess the fire hazard of PVB, EVA and SGP interlayer laminated glass samples. FGI and FPI values are determined using Equation (5) and Equation (6), respectively, based on the pHRR and time to reach pHRR (t_{pHRR}) of the laminated glass samples.

$$FGI = \frac{\mathrm{pHRR}}{t_{\mathrm{pHRR}}} \tag{5}$$

$$\mathrm{FPI} = \frac{t_{\mathrm{ign}}}{\mathrm{pHRR}} \tag{6}$$

The fire hazard performance assessment properties of tested samples are shown in Figure 14. In fire performance studies, a low FGI rating means excellent fire retardant and the higher the FPI value, the better the flame resistance property [24,25]. It can be seen from Table 4 that the FGI and FPI values of laminated glass change with changes in the glass pane thickness, heat release rate and interlayer materials. Based on the FIGRA and FPI bars shown in Figure 14a, the rank of the fire and flame resistance of the investigated laminated glass samples with the same 1.52 mm interlayer can be ranked as EVA > SGP > PVB. It indicates that the fire hazard of EVA laminated glass is comparatively lower compared to PVB and SGP laminated glasses. PVB laminated glass showed the most fire hazard among the three types of interlayers. However, 5 mm glass pane thickness on both sides of PVB improved the performance of laminated glass compared to 3 mm glass pane thickness. On the other hand, a linear correlation was observed between the external heat-flux exposure and the fire performance of the PVB laminated glass samples.

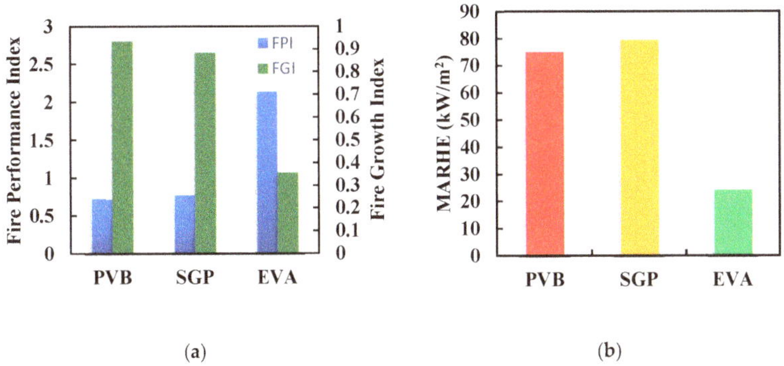

Figure 14. (**a**) FPI, FGI and (**b**) MARHE of different interlayer laminated glass at 50 kW/m^2.

Table 4. Fire hazard performance assessment properties of laminated glass samples.

Test Specimens	S6PVB50	S10PVB25	S10PVB50	S10PVB75	S12PVB50	S12SGP50	S12EVA50
FGI	1.290	0.072	0.575	1.063	0.936	0.883	0.355
FPI	0.576	12.284	1.311	0.661	0.718	0.777	2.138
MARHE (kW/m^2)	44.12	8.47	29.48	42.79	74.71	79.20	24.37

MARHE is used to determine the average heat generated during each combustion period, which is measured according to Equation (7). This parameter is defined as a maximum value of the cumulative heat release over time (highest average heat release rate). Therefore, it can be regarded as an excellent indicator of fire development in controlled circumstances [26].

$$\text{MARHE} = \max\left|\frac{\int_{n}^{t_{n+1}} \overline{\dot{q}''}(t)dt}{t_{n+1} - t_n}\right| \tag{7}$$

Here, $\overline{\dot{q}''}(t)$ is the mean heat release rate at time, t.

The trends of the MARHE ranking seem akin to those of FGI ranking, as shown in Table 4. The MARHE value of PVB and SGP laminated glass is more than three (03) times higher compared to EVA laminated glass with 1.52 mm interlayer (see Figure 14b). The MARHE values of PVB, SGP and EVA laminated glass samples tested at the heat flux of 50 kW/m^2 are 74.71, 79.20 and 24.37 kW/m^2, respectively.

3.4.5. Smoke Hazard

Total smoke production (TSP) and specific extinction area (SEA) are determined to investigate the smoke hazard potential of the different interlayer samples with varying thicknesses and heat flux. The time-averaged values of the SEA of the samples were considered the peak values of SEA, which are sensitive to mass changing of the sample. Hence, the average SEA indicates a better understanding of the overall smoking behaviour of the samples. The values of TSP and average SEA were reduced by almost 50% when the thickness of the glass pane was increased from 3 mm to 5 mm (Table 5).

Table 5. Smoke hazard properties of laminated glass samples.

Test Specimens	S6PVB50	S10PVB25	S10PVB50	S10PVB75	S12PVB50	S12SGP50	S12EVA50
TSP (m^2)	0.821	2.364	0.545	1.009	3.146	3.898	0.401
Avg. SEA (m^2/kg)	175.395	710.07	76.63	175.913	221.93	254.84	56.073
CO yield (kg/kg)	0.097	0.051	0.104	0.126	0.088	0.087	0.127
CO_2 yield (kg/kg)	1.563	1.600	1.729	1.606	1.734	1.836	1.735

For the same glass thickness and interlayer material, the TSP and avg. SEA values were significantly reduced at 50 kW/m^2 though slightly increased at 75 kW/m^2 heat flux. This result implies that there was remarkable smoke generation at 25 kW/m^2 heat flux, and it became lowest at 50 kW/m^2. Considering three interlayer samples with the same thickness from Table 5, it is identified that there was higher total smoke production in the case of the SGP interlayer laminated glass sample 3.898 m^2, whereas only 0.401 m^2 smoke was generated in total for the EVA interlayer sample. The same trend also followed in the case of avg. SEA values. The CO yield values were changed with thickness increase, whereas it was gradually increased with heat flux increase, as shown in Table 5. However, it was the same for SGP and PVB samples but increased for EVA samples (Figure 15). In the case of CO_2 yield, the value was the same for PVB and EVA samples but slightly increased for SGP samples.

It can be seen from Figure 16a,b that the smoke production rate (SPR) values were decreased with an increase in the glass pane thickness. The SPR values were almost the same for both PVB and SGP interlayers, whereas the SPR values for EVA samples were significantly reduced (less than 0.01 m^2/s), as shown in Figure 16c–e.

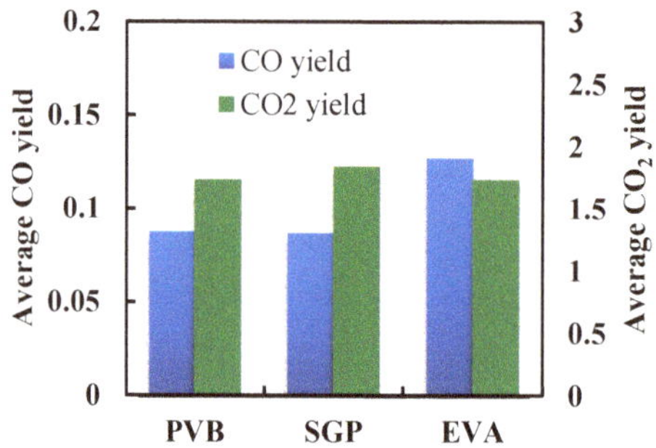

Figure 15. CO yield and CO_2 yield of different interlayer laminated glass at 50 kW/m^2.

The carbon monoxide production (COP) decreased with the glass thickness increase presented in Figure 17a,b from 0.004 g/s to less than 0.003 g/s. With the same thickness and heat flux of 50 kW/m^2, the values of COP were maximum for SGP (0.006 g/s), whereas it was significantly reduced to around 0.002 g/s for EVA samples, as presented in Figure 17c–e. These results indicate the fire hazard potential of three types of interlayers containing laminated glass with varying thickness and heat flux. EVA interlayer samples exhibited less hazard potential than the others.

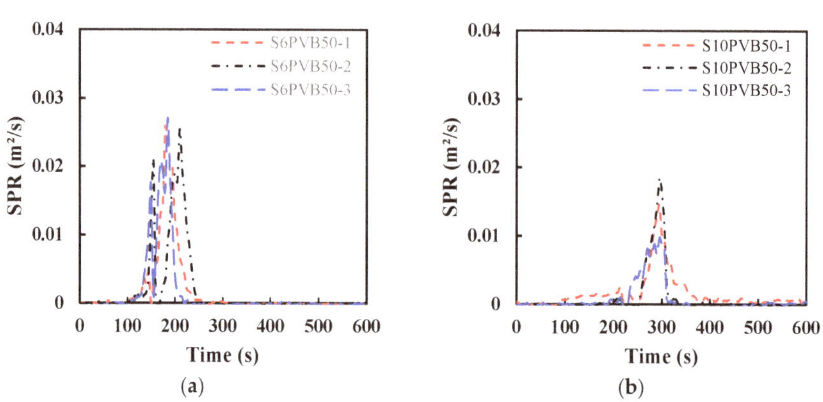

Figure 16. *Cont.*

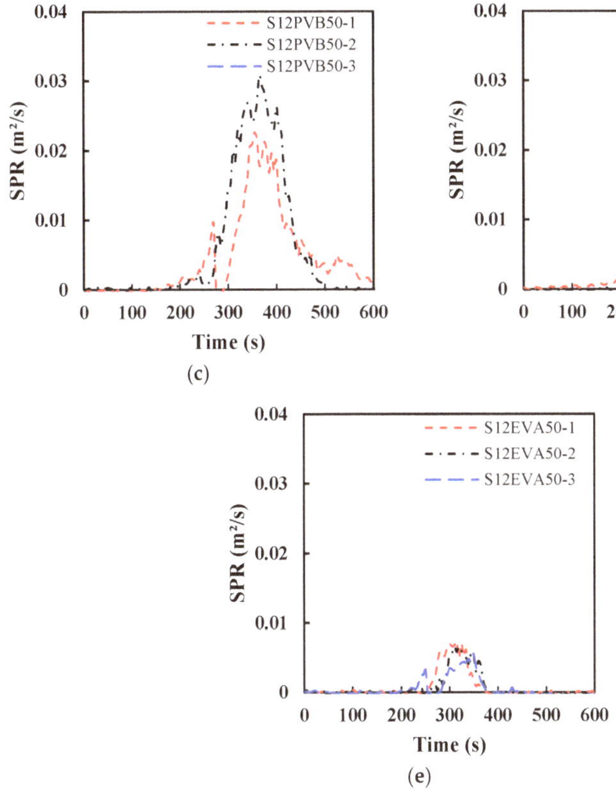

Figure 16. Smoke production rate at 50 kW/m^2: (**a**) 6 mm glass with 0.38 PVB interlayer (**b**) 10 mm glass with 0.38 PVB interlayer (**c**) 12 mm glass with 1.52 PVB interlayer (**d**) 12 mm glass with 1.52 SGP interlayer and (**e**) 12 mm glass with 1.52 EVA interlayer.

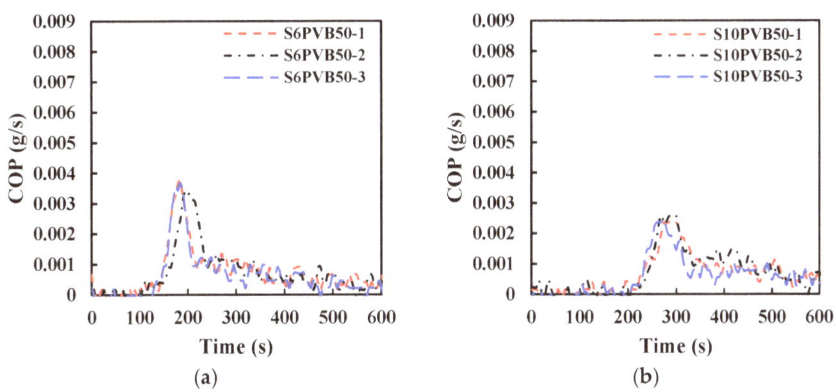

Figure 17. *Cont.*

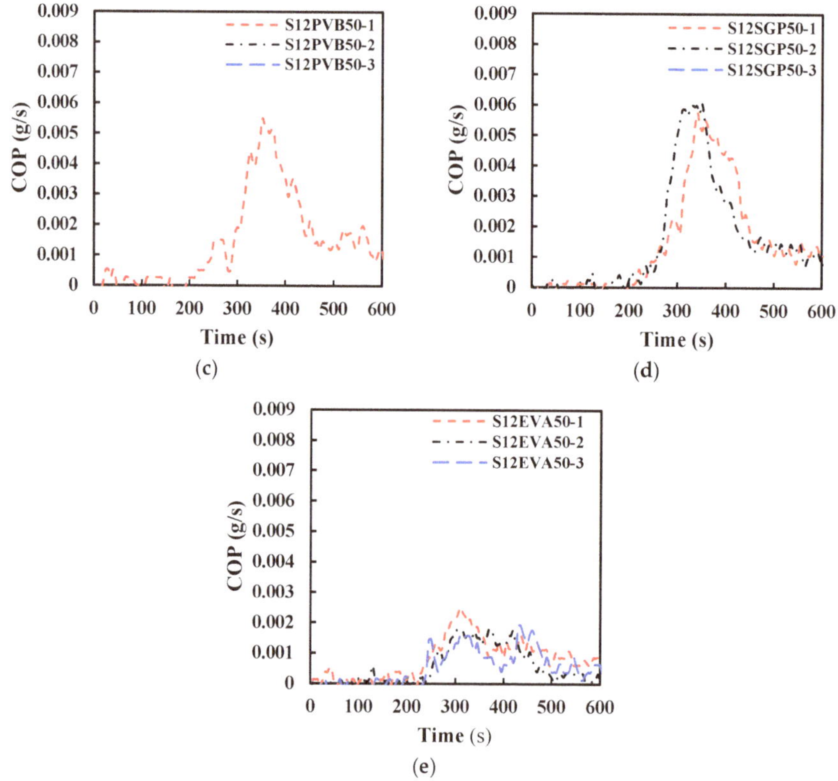

Figure 17. CO production rate of laminated glass at 50 kW/m^2: (**a**) 6 mm glass with 0.38 PVB interlayer (**b**) 10 mm glass with 0.38 PVB interlayer (**c**) 12 mm glass with 1.52 PVB interlayer (**d**) 12 mm glass with 1.52 SGP interlayer and (**e**) 12 mm glass with 1.52 EVA interlayer.

4. Phenomenological Modelling

A phenomenological model is developed in this study based on the test data and using regression analysis to predict the pHRR values of the laminated glass samples. The pHRR value of a material is considered as one of the main flammability parameters for fire hazard [22]. Fire hazard assessment of material is very important for construction applications. By predicting the pHRR values, it will be possible to identify the fire hazard status of the material. The researchers have developed several models to predict pHRR for flame retardancy identification [22,27]. In this research, a simplified equation (Equation (8)) is developed based on the mass loss, effective heat of combustion and external heat flux. The predicted pHRR is determined using Equation (8). It can be seen from Table 3 that the pHRR value depends on the mass loss rate, which is also dependent on the external heat flux and type of materials. It is worth noting that EHC is greatly influenced by the type of materials than the heat flux.

$$\text{pHRR} = 2.2e^{0.076A} \tag{8}$$

where A is the correlating factor determined using Equation (9), developed based on the regression analysis by considering the effect of mass loss, effective heat of combustion, external heat flux and type of interlayer material, as shown in Figure 18.

$$A = 3.54 \times B \times q_{\text{ext}} \times \text{EHC}^{0.1}\text{MLR}^{0.17}\, e^{-0.02q_{\text{ex}}} \tag{9}$$

where B is a factor for interlayer materials (Equation (10)), EHC is effective heat of combustion, MLR is mass loss rate, and q_{ext} is external heat flux.

$$B = \begin{cases} 0.95 & \text{for PVB interlayer} \\ 0.95 & \text{for SGP interlayer} \\ 0.90 & \text{for EVA interlayer} \end{cases} \quad (10)$$

The calculated value of pHRR determined using Equation (8) is plotted in Figure 19 against the experimental values of 21 test data. It can be seen that the predicted value is very close to the test data. The average value is 1.025 and stadandard deviation is 0.079. The average prediction error is 7.7%. It is worth noting that the prediction error of 1 datum out of 21 tested samples is 17% and the prediction error of the remaining test samples is within the limit of 10%, which indicates a very good prediction accuracy.

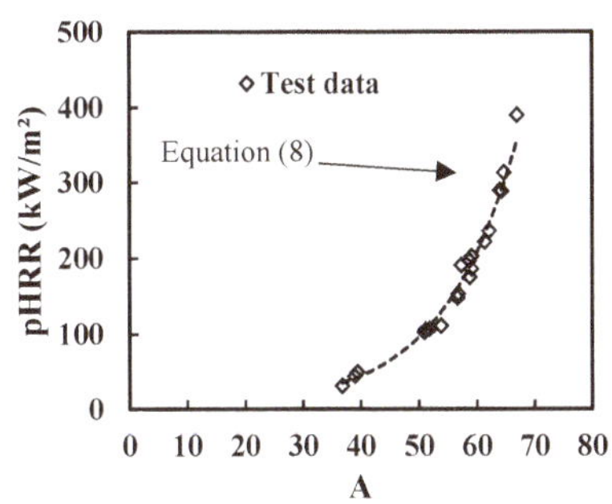

Figure 18. pHRR vs correlating factor A for the laminated glass samples.

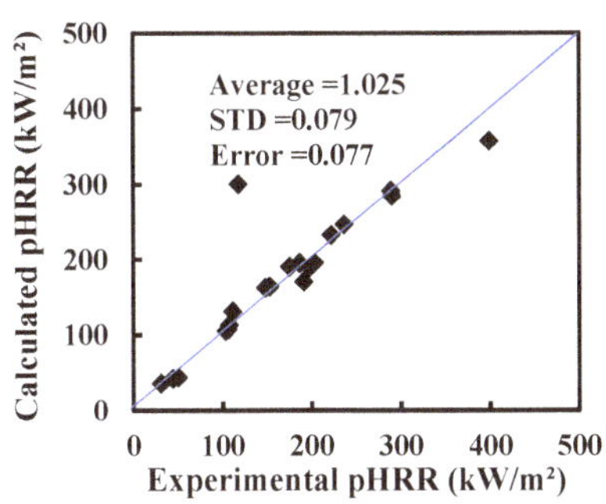

Figure 19. Calculated values vs experimental values of pHRR for laminated glass samples.

5. Conclusions

This study investigated the reaction-to-fire performance of laminated glass by considering different thicknesses of the glass pane, heat fluxes, and types of interlayer materials. The effects of the interlayer thickness on the fire hazard properties of the laminated glass were also investigated. The reaction-to-fire properties, fire hazard and smoke hazard were analysed. The following conclusion can be drawn based on the test results of the current study.

➢ The ignition time of laminated glass is increased to 200 s for 10 mm thick laminated glass, which was 130 s in the case of the 6 mm glass sample. The ignition time is significantly dropped in the case of higher heat exposures (130 s at 75 kW/m^2), while it was 570 s at 25 kW/m^2 heat flux. The interlayer materials (PVB, SGP, EVA) do not influence the ignition time of the laminated glass, as in all cases with 50 kW/m^2 heat flux, the ignition time was 225 s.

- The peak heat release rate (pHRR) of laminated glass decreases with the increase in the thickness of the glass pane, and early cracks are observed for thinner glass pane compared to thicker glass panes. The higher pHRR is observed for PVB laminated glass (313.5 kW/m^2) and the lower pHRR is observed for EVA laminated glass (105 kW/m^2). This variation mainly depends on the glass pane thickness and interlayer materials.
- The average total heat release (THR) of EVA laminated glass is significantly low (10.26 MJ/m^2) compared to PVB and SGP laminated glasses (42.28 MJ/m^2 for PVB interlayer and 42.79 MJ/m^2 for SGP interlayer). This could be due to the lower MLR of EVA laminated glass compared to PVB and SGP laminated glass. It indicates that the fire spread for EVA laminated glass will be less compared to PVB and SGP laminated glass.
- The cracking time is higher for low-thickness glass panes. A thinner glass pane exposed to a higher heat irradiance can crack earlier than a thicker glass pane. The early cracking exposes the interlayer to heat irradiance so that fire hazards can be increased. EVA samples provide greater stress crack resistance (at 154 s) than the other interlayer samples (87 s and 127 s for SGP and PVB interlayer laminated glass samples respectively). This longer cracking time may be due to EVA's higher thermal inertia than PVB and SGP.
- The values of total smoke production (TSP) and specific extinction area (SEA) were reduced by almost half of the total value when the thickness of the glass panes was increased from 3 mm to 5 mm. EVA laminated glass produced less smoke and exhibited less smoke hazard potential than PVB and SGP laminated glass.

Further research can be conducted to investigate the high temperature properties, char formation of burning materials and material characterisation of interlayer materials and glass materials. The proposed model to predict the value pHRR of laminated glass can be validated further with more test data such as thickness.

Author Contributions: Conceptualisation, M.K.H. and P.R.; formal analysis, M.K.H. and M.R.H.; investigation, K.P.L.; methodology, M.K.H.; supervision, P.R.; writing—original draft, M.K.H. and M.R.H.; writing—review and editing, K.P.L., M.D.H., P.R., G.D. and S.S. All authors have read and agreed to the published version of the manuscript.

Funding: This research received no external funding.

Institutional Review Board Statement: Not applicable.

Informed Consent Statement: Not applicable.

Data Availability Statement: Data is available upon the request.

Acknowledgments: The authors would like to acknowledge the support provided by the fire testing lab of Warringtonfire to perform all tests. The authors sincerely appreciate all financial and technical support by Western Sydney University and Warringtonfire.

Conflicts of Interest: The authors declare no conflict of interest.

References

1. Debuyser, M.; Sjöström, J.; Lange, D.; Honfi, D.; Sonck, D.; Belis, J. Behaviour of monolithic and laminated glass exposed to radiant heating. *Constr. Build. Mater.* **2017**, *130*, 212–229. [CrossRef]
2. Bedon, C.; Zhang, X.; Santos, F.; Honfi, D.; Kozłowski, M.; Arrigoni, M.; Figuli, L.; Lange, D. Performance of structural glass facades under extreme loads–Design methods, existing research, current issues and trends. *Constr. Build. Mater.* **2018**, *163*, 921–937. [CrossRef]
3. Martin, M.; Centelles, X.; Sole, A.; Barreneche, C.; Fernández, A.; Cabeza, L. Polymeric interlayer materials for laminated glass: A review. *Constr. Build. Mater.* **2020**, *230*, 116897. [CrossRef]
4. Emmons, H. The needed fire science. In *Proceedings of the Fire Safety Science-Proceedings of the First International Symposium*; IAFSS: Berkeley, CA, USA, 1986; pp. 33–53. Available online: https://publications.iafss.org/publications/fss/1/33/view/fss_1-33.pdf (accessed on 28 January 2023).
5. Babrauskas, V. *Glass Breakage in Fires*; Fire Science and Technology Inc.: Issaquah, WA, USA, 2011.
6. Wang, Y.; Hu, J. Performance of laminated glazing under fire conditions. *Compos. Struct.* **2019**, *223*, 110903. [CrossRef]

7. Harada, K.; Enomoto, A.; Uede, K.; Wakamatsu, T. An experimental study on glass cracking and fallout by radiant heat exposure. *Fire Saf. Sci.* **2000**, *6*, 1063–1074. [CrossRef]
8. Shields, T.J.; Silcock, G.; Flood, M. Performance of a single glazing assembly exposed to a fire in the centre of an enclosure. *Fire Mater.* **2002**, *26*, 51–75. [CrossRef]
9. Shields, T.; Silcock, G.; Flood, M. Performance of a single glazing assembly exposed to enclosure corner fires of increasing severity. *Fire Mater.* **2001**, *25*, 123–152. [CrossRef]
10. Wang, Y.; Wang, Q.; Wen, J.; Sun, J.; Liew, K. Investigation of thermal breakage and heat transfer in single, insulated and laminated glazing under fire conditions. *Appl. Therm. Eng.* **2017**, *125*, 662–672. [CrossRef]
11. Rickard, I.; Spearpoint, M.; Lay, S. The performance of laminated glass subjected to constant heat fluxes related to building fires. *Fire Mater.* **2021**, *45*, 283–295. [CrossRef]
12. MHossain, D.; Hassan, M.; Akl, M.; Pathirana, S.; Rahnamayiezekavat, P.; Douglas, G.; Bhat, T.; Saha, S. Fire Behaviour of Insulation Panels Commonly Used in High-Rise Buildings. *Fire* **2022**, *5*, 81. [CrossRef]
13. Hossain, M.; Hassan, M.; Yuen, A.; He, Y.; Saha, S.; Hittini, W. Flame behaviour, fire hazard and fire testing approach for lightweight composite claddings—A review. *J. Struct. Fire Eng.* **2021**, *12*, 257–292. [CrossRef]
14. Hossain, D.; Hassan, K.; Yuen, A.C.Y.; Wang, C.; Hittini, W. Influencing factors in small-scale fire testing of aluminium composite panels. In Proceedings of the 12th Asia-Oceania Symposium on Fire Science and Technology (AOSFST 2021), Virtual, 7–9 December 2021; The University of Queensland: Brisbane, Australia, 2021.
15. Hossain, M.D.; Saha, S.; Hassan, M.K.; Yuen, A.C.Y.; Wang, C.; Hittini, W.; George, L.; Wuhrer, R. Testing of aluminium composite panels in a cone calorimeter: A new specimen preparation method. *Polym. Test.* **2022**, *106*, 107454. [CrossRef]
16. *ISO 5660-2:3025*; Reaction-to-Fire Tests—Heat Release, Smoke Production and Mass Loss Rate. BSI Standards Publication: London, UK, 2019.
17. Babrauskas, V. Comparative Rates of Heat Release from Five Different Types of Test Apparatuses. *J. Fire Sci.* **1986**, *4*, 148–159. [CrossRef]
18. Ostman, B.; Nussbaum, R.M. Correlation between small-scale rate of heat release and full-scale room flashover for surface linings. *Fire Saf. Sci.* **1989**, *2*, 823–832. [CrossRef]
19. Paul, K.T. Cone calorimeter: Initial experiences of calibration and use. *Fire Saf. J.* **1994**, *22*, 67–87. [CrossRef]
20. An, W.; Jiang, L.; Sun, J.; Liew, K. Correlation analysis of sample thickness, heat flux, and cone calorimetry test data of polystyrene foam. *J. Therm. Anal. Calorim.* **2015**, *119*, 229–238. [CrossRef]
21. El Gazi, M.; Sonnier, R.; Giraud, S.; Batistella, M.; Basak, S.; Dumazert, L.; Hajj, R.; El Hage, R. Fire behavior of thermally thin materials in cone calorimeter. *Polymers* **2021**, *13*, 1297. [CrossRef]
22. Klassen, M.S.; Sutula, J.A.; Holton, M.M.; Roby, R.J.; Izbicki, T. Transmission through and breakage of multi-pane glazing due to radiant exposure. *Fire Technol.* **2006**, *42*, 79–107. [CrossRef]
23. Schartel, B.; Hull, T.R. Development of fire-retarded materials—Interpretation of cone calorimeter data. *Fire Mater. Int. J.* **2007**, *31*, 327–354. [CrossRef]
24. Xu, Q.; Jin, C.; Jiang, Y. Compare the flammability of two extruded polystyrene foams with micro-scale combustion calorimeter and cone calorimeter tests. *J. Therm. Anal. Calorim.* **2017**, *127*, 2359–2366. [CrossRef]
25. Hirschler, M.M. Flame retardants and heat release: Review of data on individual polymers. *Fire Mater.* **2015**, *39*, 232–258. [CrossRef]
26. Sacristán, M.; Hull, T.R.; Stec, A.A.; Ronda, J.C.; Galià, M.; Cádiz, V. Cone calorimetry studies of fire retardant soybean-oil-based copolymers containing silicon or boron: Comparison of additive and reactive approaches. *Polym. Degrad. Stab.* **2010**, *95*, 1269–1274. [CrossRef]
27. Thompson, S.L.; Apostolakis, G.E. A response surface approximation for the bench-scale peak heat release rate from upholstered furniture exposed to a radiant heat source. *Fire Saf. J.* **1994**, *22*, 1–24. [CrossRef]

fire

Review

A Scientometric Research on Applications and Advances of Fire Safety Evacuation in Buildings

Yang Yang, Hongbo Du and Gang Yao *

Key Laboratory of New Technology for Construction of Cities in Mountain Area, School of Civil Engineering, Chongqing University, Chongqing 400045, China
* Correspondence: yaogang@cqu.edu.cn

Abstract: Fire safety evacuation has been used in numerous different kinds of buildings. This research conducts a scientometric review of fire safety evacuation applications and advances in the buildings to clarify the research trends of fire evacuation in the future and provide guidance for relevant research. A total of 3312 journals and conference proceedings were analyzed through different dimensions. The result proves that evacuation environments concentrate mainly on residential building, commercial building, school, and railway station. The characteristics of the evacuee have been gradually refined in recent years, including children, the elderly, patients, and vulnerable groups. The main experimental approaches of fire safety evacuation are evacuation drills, site records, and VR/AR experiments. The crowd behavior models mainly consist of six types: a cellular automata model, a social force model, a lattice gas model, a game-theoretic model, an animal agent-based model, and a computer agent-based model. The analysis results in the theoretical method are becoming gradually closer to the behavioral characteristics and movement data of the crowd during the actual evacuation with improvements of practical considerations. The study of evacuation drills, disaster rescue, emergencies, and other external environmental factors will become the forefront of future research, and subway stations, airports, high-rise building, and other personnel places will be the focus of the study of crowd evacuation.

Keywords: fire safety evacuation; scientometric research; clustering analysis; research method; numerical simulation

Citation: Yang, Y.; Du, H.; Yao, G. A Scientometric Research on Applications and Advances of Fire Safety Evacuation in Buildings. *Fire* **2023**, *6*, 83. https://doi.org/10.3390/fire6030083

Academic Editor: Tiago Miguel Ferreira

Received: 28 January 2023
Revised: 17 February 2023
Accepted: 20 February 2023
Published: 22 February 2023

1. Introduction

Fire safety evacuation is the process by which people are transferred from a dangerous area to a safe area in case of emergency. Studies on safety evacuation have been used in numerous different scientific fields to provide information or data as guides for improving the success rate of escape. Many disasters are predictable to the extent that sufficient warning can be provided to those who need to get out of harm's way, and effective evacuations are critical to reducing disaster-related casualties [1]. When disaster strikes, a well-rehearsed repertoire of responses on safety evacuation can provide fall back with relief. Fire safety evacuation has to address the optimal solutions in dangerous environments. Untimely fire safety evacuation will lead to more casualties when disaster strikes. Many scholars have conducted relevant studies on the evaluation of evacuation achievements, including human behaviors in fire, earthquakes, and other dangerous environments [2–12], and crowd evacuation in buildings and subways [13–18]. It has been found that fire safety evacuation research has diversified with the development of emerging technologies.

Scientometric literature research in a certain field is considered an expedient approach to acquiring a thorough comprehension; it can quantitatively study science including impact, hotspot issues, and the distribution of institutions and journals conducting a certain field. Scientometric research has been used in numerous different scientific fields to provide information or data such as computer vision applications for construction [14], safety science community [15], smart disaster management [16], microalgae-based wastewater treatment [17], computational modelling in built environments [18], dynamics applications in construction management [19], AAP (accident analysis and prevention) [20], CEM (construction engineering and management) [21], or BIM (building information model) [22]. The findings based on scientometric analysis provide excellent research status and development trends for researchers in particular domains of scientific inquiry [23–29]. Researchers can gain quantitative insights in the development direction and efficiently solve complex problems with respect to the theoretical backgrounds, tools, and frameworks on particular domains of scientific inquiry.

There is sustained and fast growth in studies on fire safety evacuation in the buildings research of the past ten years, and the applications and advances in fire safety evacuation have been diverse, with varying degrees of complexity. Nevertheless, few existing publications showcase detailed analyses on fire safety evacuation. Therefore, a research effort is needed to provide the full scope of the applications and advances of fire safety evacuation.

This paper attempts to conduct a scientometric review of the scientific literature relating to fire safety evacuation and to provide an overall description of the applications and advances in fire safety evacuation during the past ten years (from 2010 to 2022). This paper also attempts to clarify the evolution path and development trend of fire safety evacuation in buildings and to provide suggestions for further applications and research directions. The research objectives of this paper are relevant academic publications on fire safety evacuation in the Web of Science database during the past ten years. Data sources and CiteSpace software are introduced to statistically analyze trends in academic publications, subject distribution, journals distribution, research institutions distribution, evacuation objects, evacuation environments, inductive factors, and research methods. This paper aims to review publications that showcase detailed analyses in fire safety evacuation research and to provide research hotspots, development trends, and overall characteristics for researchers on fire safety evacuation domain of scientific inquiry. The achievements in this paper can provide a detailed and comprehensive understanding of the current research state in fire safety evacuation.

The remainder of the paper is organized as follows: Section 2 provides the research methodology in this paper; Section 3 identifies the core journals and conferences, leading source countries and organizations, core authors, and keywords by correlation analysis and processing of academic publication data with WOS and CiteSpace; Section 4 analyzes evacuation models and experiment methods on fire safety evacuation in buildings in the recent ten years; Section 5 derives applications and advances in fire safety evacuation in buildings according to research objects, evacuation environments, and disaster classification; Section 6 summarizes the analysis results.

2. Methodology

Database selection and searching strategy are crucial since they determine academic publications from which conclusions will be drawn. Past studies have considered academic publications from the Web of Science core database (WOS) to be reliable and efficient sources of knowledge with high impact due to rigorous review processes.

A delimitation of the research boundary is necessary since there are many previous academic publications on fire safety evacuation. The time frame is set from January 2010 to December 2022. "Fire evacuation" is used as the key word, and wildcard character is also used to capture variations of one keyword. The keyword search is set as title/abstract/keywords in order to retrieve all the publications containing the selected keywords in the title, abstract, or selected keywords section. In order to use this theme-specific search for quick visibility with an identical construct in terms of research aims and methods, only papers in peer-reviewed English journals and conference proceedings are taken into consideration. Irrelevant journals and conference academic publications are excluded from the analysis data; hence, some academic publications on medical treatment and mechanical engineering are excluded. Duplicated articles are reduced to one record in the analysis. This paper analyzes the retrieved publications with CiteSpace software, and the research route on fire safety evacuation in buildings is given in Figure 1.

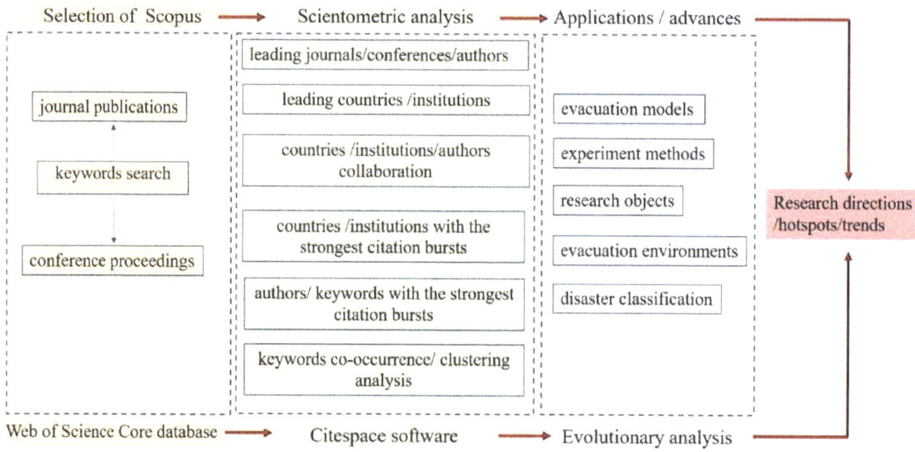

Figure 1. Research route on fire safety evacuation in buildings.

3. Scientometric Analysis

The academic publications on fire safety evacuation are analyzed according to the following methodologies: the number of academic publications, leading journal and conference proceedings, country co-occurrence and burst detection, co-research institution analysis and burst detection, co-author analysis and burst detection, keyword co-occurrence analysis and burst detection, and keyword clustering. Firstly, the co-occurrence analysis and network indicators make an aggregate static representation of fire safety evacuation. Secondly, the strongest citation burst detection sheds further insight on the relative changes of significance over time to identify trends and changes in fire safety evacuation, providing a dynamic representation of fire safety evacuation. Finally, keyword indicators provide evidence for the posterior clustering analysis. Keywords clustering indicates the research patterns within the field in detail, as well as various specific associated research themes to lay out the research conceptual framework. These techniques have been recommended in previous studies of a similar nature.

3.1. Year and Quantitative Analysis of Academic Publications

Exactly 3312 academic publications on fire safety evacuation are evaluated. The number of academic publications in fire safety evacuation from the 2010 to 2022 is given in Figure 2, including journal publications and conference proceedings.

Two main bursts of academic publications are shown in 2012 and 2016, with an increase of 68.7% and 38.9%, respectively. The number of academic publications goes through a steady development stage after each growth. Publications in fire safety evacuation show an overall growth from 2015 to 2022, and the longest period of steady development exists from 2016 to 2022, which shows that the research in this field is developing smoothly and sustainably.

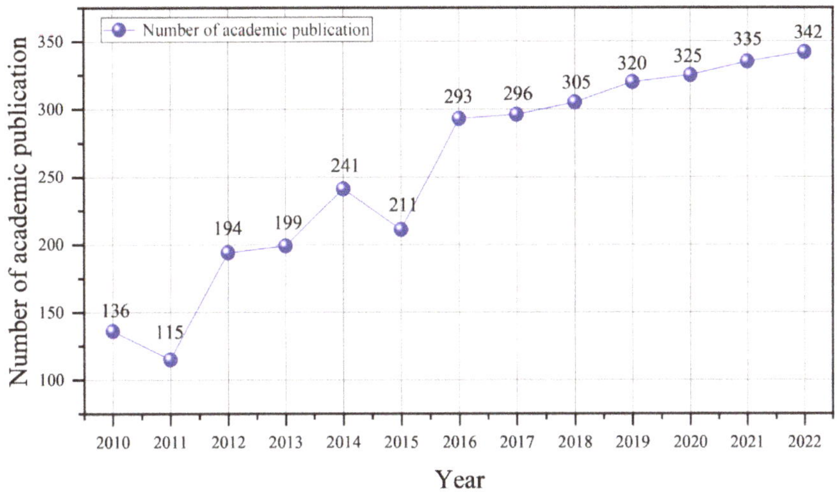

Figure 2. Number of academic publications in fire safety evacuation from 2010 to 2022.

3.2. *Leading Journals and Conference Proceedings*

The selection of an appropriate journal helps the researchers to deliver effective communication in a particular domain. The leading journals and conference proceedings also provide a useful learning channel for the beginner to acquire a quick familiarity and understanding. Leading journals and conference proceedings on fire safety evacuation are concluded in Table 1.

Table 1. Journals and conference proceedings publications in fire safety evacuation from 2010 to 2022: (a) Journals publications. (b) Conference proceedings publications.

(a) Journals Publications		
Journal Title	Number of Articles	% Total Publications
Safety Science	131	3.95%
Physica A: Statistical Mechanics and Its Applications	114	3.42%
Transportation Research Record	92	2.77%
Procedia Engineering	77	2.31%
Fire	70	2.13%
Transportation Research Procedia	42	1.29%
Simulation Modelling Practice and Theory	40	1.18%
Transportation Research Part C: Emerging Technologies	39	1.10%
Journal of Disaster Research	36	1.06%
PLOS ONE	35	1.02%
Applied Mechanics and Materials	33	0.95%
IEEE Access	31	0.83%
Mathematical Problems in Engineering	27	0.83%
Advances in Intelligent Systems Research	27	0.80%
Journal of Advanced Transportation	24	0.72%
Sustainability	24	0.72%
Chinese Physics B	21	0.65%
Disaster Medicine and Public Health Preparedness	21	0.65%
International Journal of Environmental Research and Public Health	21	0.65%
International Journal of Modern Physics C	21	0.65%
Lecture Notes in Artificial Intelligence	21	0.65%

Table 1. *Cont.*

(b) Conference Proceedings Publications		
Conference Title	**Number of Articles**	**% Total Publications**
Conference on Pedestrian and Evacuation Dynamics (PED)	123	3.73%
International Conference on Performance Based Fire and Fire Protection Engineering (ICPBFFPE)	52	1.56%
IOP Conference Series Earth and Environmental Science	31	0.95%
International Conference on Evacuation Modeling and Management	17	0.53%
International IEEE Conference on Intelligent Transportation Systems (ITSC)	16	0.49%
Winter Simulation Conference (WSC)	15	0.46%
International Symposium on Safety Science and Technology (ISSST)	12	0.38%
Traffic and Granular Flow	8	0.27%
IEEE International Conference on Systems Man and Cybernetics (SMC)	7	0.23%
International Conference on Cellular Automata for Research and Industry (CACRI)	6	0.19%
International Conference on Urban Transport and the Environment (ICUTE)	6	0.19%
International Conference on Civil Engineering and Transportation (ICCET)	6	0.19%
IEEE International Conference on Industrial Engineering and Engineering Management (IEEM)	6	0.19%
World Conference of Associated Research Centers for the Urban Underground Space (ACUUS)	4	0.15%
International Symposium on Transportation and Traffic Theory (ISTTT)	4	0.15%
International Conference on Computer Simulation in Risk Analysis and Hazard Mitigation	4	0.15%
Construction Research Congress	4	0.15%
ASIA Conference on Information Systems for Crisis Response and Management (ISCRAM)	4	0.15%

3.3. *Network of Countries*

A network with 92 nodes and 280 links was exhibited in Figure 3 based on the contributions of countries to explore the distribution of research publications on fire safety evacuation, and nodes with high centrality were identified and highlighted with darker outer rings. Research achievements have a positive correlation with research level. Research shows that the more research a country produces, the higher its level of research. As shown in Figure 3, China (959 articles), USA (582 articles), Japan (305 articles), Germany (107 articles), and England (101 articles) have made major contributions to the publications in this field of research. Countries such as the United States of America (centrality = 0.53), France (centrality = 0.21), the United Kingdom (centrality = 0.18), Netherlands (centrality = 0.17), the People's Republic of China (centrality = 0.15), German (centrality = 0.12), or Japan (centrality = 0.12+) have occupied key positions in the network and connected research activities between different countries.

The network of countries makes an aggregate static representation of countries on fire safety evacuation in buildings. The strongest citation burst detection sheds further insight on the relative changes of significance over time to identify trends and changes on countries/organizations/authors/keywords, providing a dynamic representation of these areas. The top 15 countries with the strongest citation bursts are shown in Figure 4.

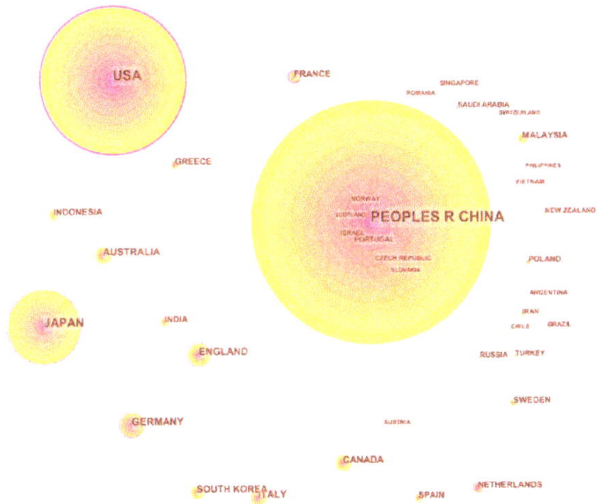

Figure 3. Network of countries.

Top 15 countries with the strongest citation bursts

Countries	Year	Strength	Begin	End	2010–2020
Greece	2010	1.5	2010	2020	
Finland	2010	1.25	2010	2020	
Austria	2010	2.21	2010	2020	
Netherlands	2010	7.72	2010	2020	
Germany	2010	2.5	2011	2020	
Portugal	2010	2.48	2012	2020	
Ecuador	2010	4.54	2014	2020	
Czech Republic	2010	1.56	2015	2020	
Brazil	2010	2.67	2017	2020	
Nepal	2010	1.43	2017	2020	
Indonesia	2010	7.43	2018	2020	
Bangladesh	2010	1.54	2018	2020	
Chile	2010	2.03	2018	2020	
Iran	2010	3.62	2018	2020	
Egypt	2010	1.78	2018	2020	

Figure 4. Top 15 countries with the strongest citation bursts.

The top 15 countries with the strongest citation bursts have a rapid increase in citation frequency in the research time. Articles in these countries tended to affect in great measure the direction of fire safety evacuation research.

3.4. Network of Authorships

The number of research papers could reflect the activity level of scholars in the research domain to a certain degree. The relationship of co-authors could reflect the social interaction in academic communities. The number of research papers and the relationship of co-authors are important indicators by which to evaluate an author. The minimum number of documents and citations of an author are set, respectively, as 5 and 19. Influential authors are given in Figure 5 in the manner of science mapping, and the dot size is proportional to the quantity of research papers.

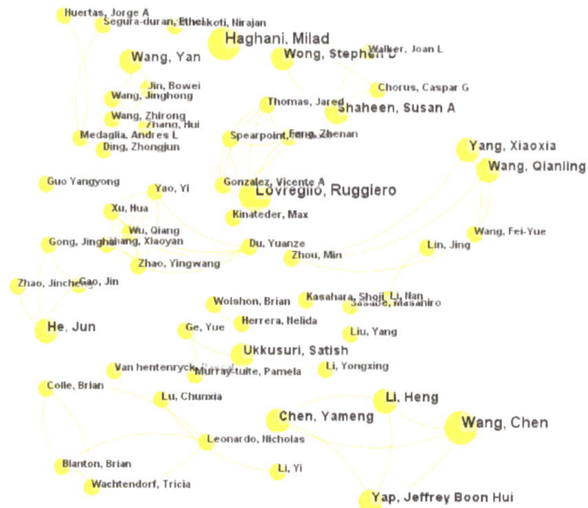

Figure 5. Science mapping of authors.

It can be seen that there are five main research groups in Figure 5. The research group of *Pamela Murray-Tuite* and *Hui Zhang* are located in the center of this mapping and related to the other groups of researchers, showing that they kept a close academic collaboration with leading scholars in the field. *Hui Zhang*, *Tao Chen*, and *Peter B Luh* conduct related research in a team manner. The stair evacuation with considering evacuees' walk preferences is the outstanding contribution in fire safety evacuation of these scholars [28]. Evacuees' walk preference and psychology are introduced into a cellular automata (CA) model, and validation of the model is proved by two fire drills in two high-rise buildings. *Pamela Murray-Tuite*, *Brian Wolshon*, and *Satish V Ukkusuri* have a close cooperative relationship in urban fire safety evacuation. Control strategies of zone phasing and contraflow are proposed to utilize network capacity better and to decrease clearance time [29]. From the analysis results of cooperation academically, it can be summarized that the cooperation among scholars is mainly within the institution, and the core scholars within the institution become the bridge of cooperation with other institutions.

3.5. Keywords Analysis

Keywords can reflect core research content effectively and abstractly. Scholars could identify core research contents and future research trends in a particular domain by keywords. In this scientometric study, keywords are extracted from 3312 publications between 2010 and 2022. The missing keywords in some publications are assigned by the professional indexers from WOS. In this study, the minimum number of occurrences of a keyword is set as 20. Some keywords with the same semantic meanings are combined. The universality and trivial words are ignored. Finally, 43 keywords meet the threshold in total, and the identified keywords are visualized in Figure 6.

Dot size in Figure 6 is proportional to the occurrence frequency of a keyword. The largest point of research directions was "simulation", showing that simulation analysis has been mostly utilized in this area. The distance between the nodes represents the number of co-occurrences of keywords. The higher the number of co-occurrences between two keywords, lesser is the distance between them, and stronger interrelation between those concepts or technologies is depicted. The links are the number of linkages between a given node and others, while the total link strength reflects the total strength linked to a specific item. For instance, the total link strength of simulation is 26, which is at the high level of all the keywords and indicates the strong inter-relatedness between fire safety evacuation and simulation.

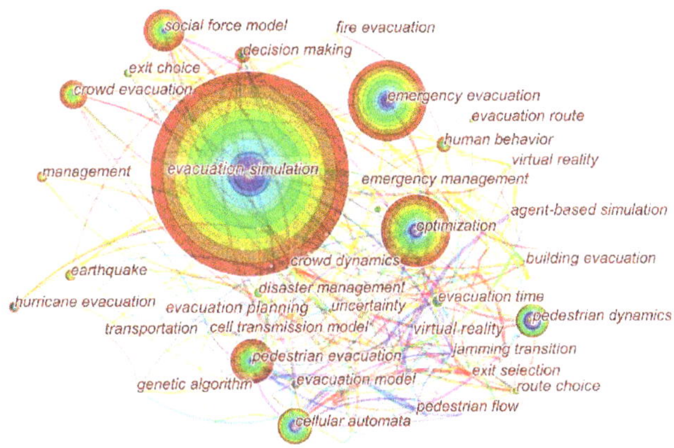

Figure 6. Science mapping of keywords.

According to the results, the keywords with high co-occurrence frequency are "evacuation simulation", "emergency evacuation", "optimization", "pedestrian evacuation", "social force model", "cellular automata", "pedestrian dynamics", and "crowd evacuation". The research hotspots in the field of fire safety evacuation mainly revolve around "evacuation simulation", and it presents a pattern of diversified development. It is also found that the latest research topics relate to "construction safety" and "fire safety evacuation in construction", which indicates a shift in the field of fire safety evacuation. The visualization of the keywords' network could demonstrate the results of the bibliometric analysis of the literature. In order to further understand the research trend in the field of safe evacuation, LLR (log-likelihood rate) algorithm in CiteSpace is used in this paper to carry out cluster analysis on the keyword co-occurrence network map. The time range of cluster analysis is from 2010 to 2022, and the time slice is 3.0. Different colors in the cluster analysis are clustering topics, and it is shown in Figure 7.

Clustering analysis is used to study the keyword about fire safety evacuation. The cluster structure is taken as significant when modularity is beyond 0.3. The cluster mean contour value is considered to be reasonable when the silhouette is beyond 0.3. In Figure 7, modularity is 0.392, and the silhouette is 0.7126. Hence, cluster analysis has high reliability. It also can be found that keywords are clustered into seven categories from #0 to #6, and they are emergency preparedness, evacuation planning, evacuation method, FDS plus evacuation, disaster simulation, ICT (information communication technology), and restrictive evacuation. The number of keywords decreases in turn from clustering #0 to clustering #6, which indicates that emergency preparedness covers a wider range of research topics.

The basic knowledge Is the co-cited document, and the research frontier is the cited document of the co-cited document. The cluster naming of the knowledge base in CiteSpace is determined by the nominal terms extracted from the cited document, and this naming can be considered as the research frontier. The research frontier is embodied by the clustering of emergent words in the literature that forms the co-citation matrix and in the cited literature. Therefore, the emergent clustering of research keywords is applied to determine the research frontier in the field of fire safety evacuation. In order to better identify and predict the latest evolution and development trend of fire safety evacuation research, keywords with the strongest citation bursts are taken to be analyzed. Compared with high-frequency keywords in fire safety evacuation, keywords with the strongest citation bursts are more suitable for detecting emerging trends and sudden changes in the development of fire safety evacuation. The burst detection algorithm is used in obtaining keywords with the strongest citation bursts, and the threshold is set as Top 30 in this analysis. The top 30 keywords with the strongest citation bursts are exhibited in Figure 8.

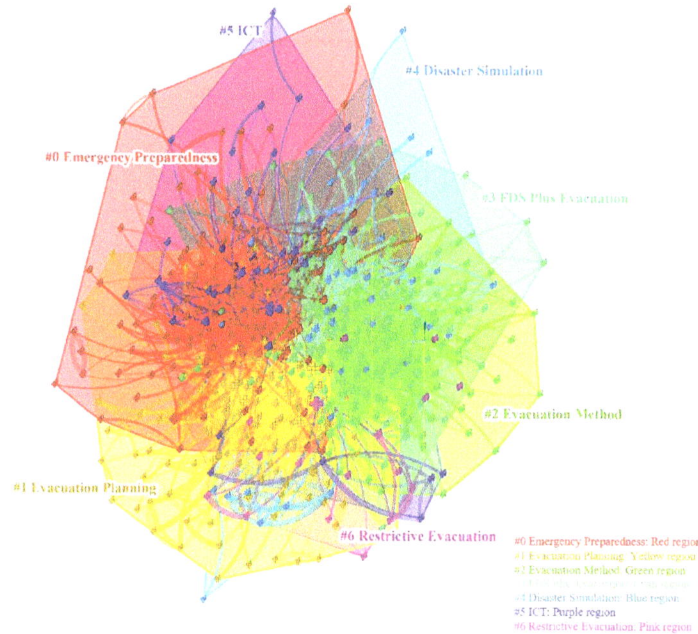

Figure 7. Clustering analysis of keywords.

Keywords	Year	Strength	Begin	End	2010–2022
Evacuation Model	2010	5.23	2010	2022	
Pedestrian Dynamics	2010	4.58	2010	2022	
Subway Station	2010	4.2	2010	2022	
Occupant Evacuation	2010	3.48	2010	2022	
Fire	2010	3.42	2010	2022	
Cell Transmission Model	2010	3.29	2010	2022	
Physics	2010	3.17	2010	2022	
Evacuation	2010	2.82	2010	2022	
Experiment	2010	2.67	2010	2022	
Traffic Flow	2010	2.52	2010	2022	
Transition	2010	4.41	2011	2022	
Evacuation Simulation	2010	4.43	2011	2022	
Jamming Transition	2010	3.92	2011	2022	
Evacuation Planning	2010	2.78	2011	2022	
Wireless Sensor Network	2010	4.3	2012	2022	
Pedestrian Flow	2010	4.13	2012	2022	
Routing	2010	3.52	2012	2022	
Decision Support System	2010	3.18	2012	2022	
Cellular Automata Model	2010	2.72	2012	2022	
ICT	2010	3.36	2013	2022	
Assignment	2010	3.22	2013	2022	
Congestion	2010	2.66	2013	2022	
Road Tunnel	2010	2.65	2013	2022	
Traffic Simulation	2010	2.53	2014	2022	
Floor Field Model	2010	2.5	2014	2022	
Network flow	2010	2.77	2016	2022	
Crowd	2010	4.67	2017	2022	
Earthquake	2010	3.29	2017	2022	
Emergency Response	2010	2.69	2017	2022	
Selection	2010	2.84	2018	2022	

Figure 8. Top 30 keywords with the strongest citation bursts.

The evacuation model has the highest strength and the longest persistent period in the Top 30 keywords, which indicates that the evacuation model is the most interesting research topic. Keywords with citation bursts in the last five years are network flow, crowd, earthquake, emergency response, and selection, which indicates research trends in fire safety evacuation. In 2013, ICT has become the representative of the emerging hot words, and scholars pay more attention to high-rise buildings and building fire evacuation with the help of wireless sensor networks and mobile devices. From the keywords with the strongest citation bursts and clustering analysis of keywords, both ICT and building evacuation are research trends. ICT technology is more closely combined with crowd evacuation, more integrated, more widely applied, and more systematic. It benefits from the mature manufacturing technology of hardware facilities such as navigation equipment, sensor devices, and computers in recent years, as well as the development and application of algorithms and related software such as path optimization, simulation modeling and cloud computing. The rapid development of building evacuation is because the study of crowd evacuation depends on the external environment. Therefore, the study of evacuation drills, disaster rescues, emergencies, and other external environmental factors should also become the forefront of future research, and subway stations, airports, high-rise buildings, and other personnel places will be the focus of the study of crowd evacuation.

4. Research Approaches of Fire Safety Evacuation

Many studies have concentrated on fire safety evacuation that aims to guide evacuees out of hazardous areas safely and efficiently. Research approaches regarding fire safety evacuation are mainly from evacuation models and experiment methods. Evacuation models are established to predict the egress time for all evacuees, provide evacuation paths for evacuees, and optimize the design of crowded sites. Experiment methods could validate and improve the proposed evacuation models. Records from past incidents, evacuation drills, controlled experiments, questionnaire surveys, and VR/AR experiments are the main experimenting methods.

4.1. Evacuation Models

It is not realistic to carry out the research of evacuation behavior during disasters. Therefore, computational tools are widely accepted as the best approach for simulating evacuation behavior during disasters. Parametric numerical models of evacuation could be employed as behavioral comparison tools for various aspects of evacuees' decision-making. Numerical simulation tools are considered valuable in most fire safety evacuation studies with easier alternatives to experiments. The simulation results can be used to predict safety performance and to make an evacuation plan to reduce casualties and unfold the saving work smoothly.

The evacuation model concentrates on guiding evacuees out of emergency safely and efficiently from the wayfinding algorithm. Crowd behaviors including clogging, pushing, and trampling could lead to serious fatalities in the evacuation process. The crowd behavior models could be divided into the cellular automata model [30–36], social force model [37–43], lattice gas model [44–51], game-theoretic model [52–54], animal agent-based model [55–62], and computer agent-based model [63–72]. The functions of evacuation models have been analyzed and are shown in Table 2.

There are many examples of evacuation simulation software and they are classified according to the approach of physical space simulation, including fine grid models, rough grid models, and continuous models. Evacnet [73] and CFAST [74] are example of simulation software with fine grid models. Building Exodus [75] and Simulex [76] are examples of simulation software with rough grid models. Legion [77] and Pathfinder [78] are examples of simulation software with continuous models. The results of evacuation simulation software with fine grid models are well in agreement with the actual situation in detail. The computation speed of the adaptive grid roughening algorithm increases greatly compared with that of the fine grid method.

Table 2. Functional analysis of evacuation models.

Model Type	Evacuee Characteristic			Evacuation Behavior				
	Motion	Variable	Type	Avoidance	Reentry	Following	Nearby	Guideline
Cellular automata	UD (direction)	US (0/1)	SD	UI	UI	UI	UD	UD
Social force	UD (speed)	UD	UD	UI	UI	UD	UD	UD
Lattice gas	UD (direction)	US (0/1)	SD	UI	UI	UD	UD	UD
Game-theoretic	UD	SD	UD	UD	UD	UD	UD	UD
Animal agent-based	UD	SD	US	SD	SD	SD	SD	SD
Computer agent-based	UD	UD	UD	SD	SD	SD	SD	SD

UD—could define the variable according to demand. UD (direction) represents that the direction is the only defined variable. **US**—users could select the variable from the original model system. US (0/1) represents that there are only 0 and 1 options in the function. **UI**—users need to improve the original model system to realize the function. **SD**—the original model system has set options separately; users can realize the function by turning on and off.

ICT in fire safety evacuation integrate fire science, traffic science, psychological science, and other disciplines. The Internet of Things, radio frequency technology, the geographic information system, and other technologies have been applied in fire safety evacuation. Big data, virtual reality crowdsourcing, and other technologies are used to quantify and analyze the human psychology and social behavior of groups under different evacuation situations. By integrating the Internet of Things, BIM, and fire dynamics simulation (FDS) technologies, real-time fire monitoring and intelligent simulation and formulation of fire safety evacuation routes are realized [79–90]. The emergence of ICT [91,92] enables people to begin to quantify and analyze human psychology and social behavior in evacuation situations on a large scale and accurately. The quantitative model of human evacuation behavior is of great practical significance for updating emergency evacuation coordination systems, optimizing intelligent decision-making methods, and improving core evacuation capabilities, such as psychological and behavioral intervention, before and after safe evacuation.

The exploration of various correlations from the available data can support the scientometric analysis. Therefore, various data visualization tools including CiteSpace [93], VOSviewer [94], Gephi [95], and CiteNetExplorer [96] are used to analyze information such as year, author, journal, affiliation, country, document-type, and domain distribution.

4.2. Experiment Methods

The validity and transferability of theory and numerical analysis models put forward by researchers need to be proved by conducting experiments. Safety evacuation can hardly be completely conducted because of danger and risk. Both evacuation data records from actual emergencies and data from evacuation experiments without danger and risk are valid sources of safety evacuation data. Even based on incomplete evacuation data, the behaviors from accidents and emergencies can reveal safety awareness, true reflection, and the details of the escaping process of evacuees. Conclusions regarding safety evacuation could be drawn and summarized. There are four primary sources of real safety evacuation data, including earthquake/fire evacuation record, terror attack incident evacuation record, crush incident evacuation record, and trajectories record of pedestrians in a natural setting [97–118]. Evacuation time depends on many factors, including the behavior of evacuees and characteristics of evacuated surroundings. An evacuation drill could assess the evacuation procedure, observe the behavior of evacuees, and give a prediction of the relationship between evacuation time and the number of evacuees in a specific environment. Evacuees also could gain training experience in evacuating surroundings from participation. Evacuation drills in different evacuated surroundings have been researched for egress design, selection, and optimizing of the evacuation route in case of emergency. A series of controlled experiments were conducted to explore the commonality behavior of evacuees in evacuation procedures: exit and route choice behavior, stepping and conflict/collision avoidance behavior of pedestrians in different circumstances, including limited visibility conditions, decision-making behavior of social groups in evacuations, fundamental factors (speed, step, pace, etc.) of different human crowds (male/female, young/elder/child, unidirectional/bidirectional/multi-directional) [119–128].

A questionnaire survey is a conventional form of hypothetical choice survey. Investigators could obtain the expected data accurately by the design of questionnaire, including choice surveys under emergency scenarios in different evacuate surroundings [116,125,129,130]. A questionnaire survey is often used in combination with other experiment methods by investigators.

VR and AR experiments are emerging technologies which, initially, were mainly applied in evacuation education. VR/AR experiments, including immersive and non-immersive experiments, could record and analyze test results of emotional, psychological, and physiological responses, wayfinding performances, exit and route choice behaviors, and pre-evacuation time of evacuees [97,131–151]. It has been confirmed that VR/AR experiment is a reasonable proxy of evacuee's performance under emergency scenarios.

Through the analysis of recent studies, it can be found that evacuation experiments have gradually changed from traditional evacuation exercises to VR and AR experiments. This is mainly because the traditional evacuation exercises have certain security risks and inauthenticity, and the participants do not have a sense of urgency in the event of an actual emergency. In contrast, the test results of VR and AR experiments will be more accurate, because the exercisers will be substituted into the crisis by visual and auditory stimuli. Therefore, the use of VR and AR experimental methods will be the trend of future research on fire evacuation. The outline for the implementation of fire safety evacuation simulation are shown in Figure 9.

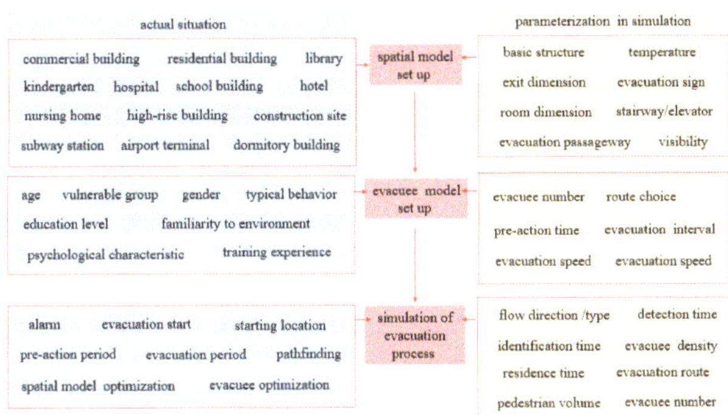

Figure 9. Outline for the implementation of fire safety evacuation simulation.

5. Research Contents of Fire Safety Evacuation

5.1. Research Objects

Physical property, subjective consciousness, and psychological features of evacuees are research objects in safety evacuation. The subjective consciousness of evacuee includes inclusive cognition under emergency circumstances [18,54,69,140], familiarity with evacuation environment [55,129,137,152,153], and individual sense of direction [154]. The psychological features of evacuees include panic, conformity, despair, and impulse [155–160]. The physical properties of evacuees include personnel type, human characteristic, crowd density, walking speed, and crowd flow [77,79,123,141,142,152–154,161]. Evacuee behaviors show great differences under the influence of personality, age, gender, and other factors inhibiting the implementation of evacuation [6,32,71,109]. It poses a great challenge to the study of the crowd dynamics for vulnerable populations. Safety evacuation has a trend of delicacy management. It has been found that the relationship of velocity-step width, velocity-step length, velocity-stepping time, and step length-step frequency for different groups have considerable differences affected by evacuee gender, age, height, and state of health. Children, older people, disability people, patient, and building workers are studied on classification to identify the factors that inhibit the implementation of evacuation in each group [78,79,81,85,99,102,121,123,162–166]. The habitual behavior, herd behavior, and avoidance behavior are observed when people escaped from emergency circumstances. According to the different directions of pedestrians in the space, the pedestrian flow is divided into the unidirectional pedestrian flow, bidirectional pedestrian flow, and multi-directional pedestrian flow [113–115,167–169]. There are different categories of research objects and influencing factors in safety evacuation; nevertheless, the time line in safety evacuation has universal stages in different research objects, and it is shown in Figure 10.

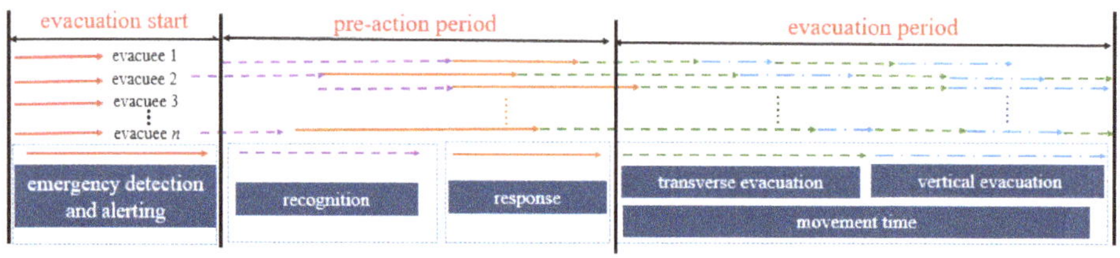

Figure 10. Time line in fire safety evacuation.

The physical properties of evacuees are the most studied objects at present because its appearance is the most intuitive and easy to be digitized. However, the subjective consciousness and psychological features of evacuees also have a great impact on evacuation behavior, which is often ignored in previous studies.

5.2. Evacuation Environments

Evacuation environments have a number of unique features in crowd density and physical characteristics. Hence, effective evacuation planning should be primarily put forward according to environmental features in order to minimize the number of casualties and property loss. Most attention has been paid to safety evacuation in buildings, and a number of environmental characteristics of buildings have been studied. Influences of spatial type, quantity and location of stairs, quantity, and location of exits, quantity and location of facilities, and state of exits have been points of focus. By studying the influence factor of environments, safety evacuation models in different buildings are set up, and targeted safety evacuation strategies are proposed.

Mass audience venues with high-density pedestrian flows, rapid oxygen consumption, limited space for movement, and fixed exits are always the research focus in safety evacuation, including railway stations [170], subway stations [15,171], airports [31], hospitals [99,102], schools [108,120], commercial buildings [65], residential buildings [107,128], and public places of entertainment [172–174]. Safety evacuation of offshore platforms are one of the frontiers of research and a new branch in offshore engineering to which increasing attention has been paid [122].

Safety evacuation in high-rise buildings is the emphasis in this domain, and it becomes difficult as interior structure and architectural form of high-rise building [110,112,136,155]. People in high-rise buildings are highly concentrated, and some areas have poor mobility for various reasons. If there is an emergency, a large number of people rush to the security exit and go downstairs to evacuate. However, the stairs of high-rise buildings are too limited to evacuate pedestrian flow which is over the design condition. Hence, it is impossible to evacuate all people to the ground safely in a short time. When there is an emergency, the crowd becomes panicked and disorder. In an unorganized situation, it is easy to lead to congestion and stampede, which further affects the evacuation efficiency. Research flow of fire safety evacuation is summarized and shown in Figure 11. There are mainly six categories of factors affecting safety evacuation performance, including evacuation object, psychological characteristic, typical behavior, evacuation environment, evacuation route and evacuation time.

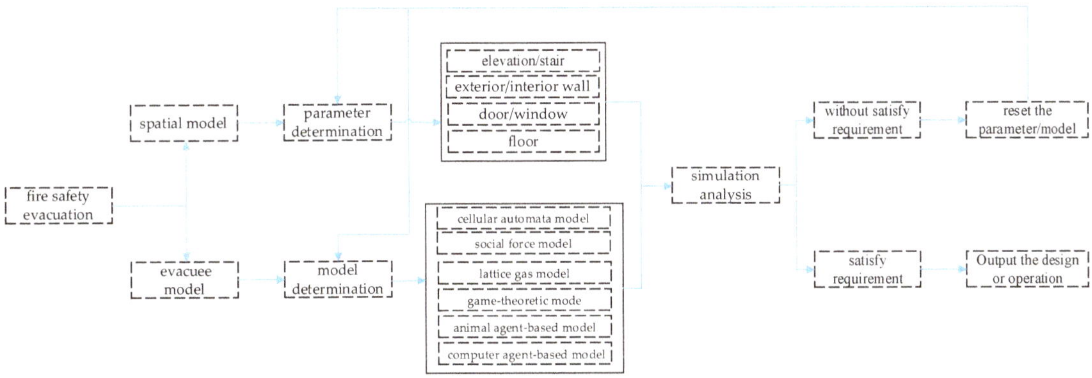

Figure 11. Research flow of fire safety evacuation.

The research on the fire evacuation environment has gradually shifted from the relatively simple mass venues to marine engineering and high-rise buildings with complex structures and difficult evacuation. With the development of the construction industry, the research on fire evacuation to ensure personal safety should also keep pace with the times.

With the emergence of high-rise buildings, it is necessary to study the appropriate use of elevators for fire safety evacuation. After analyzing the structural characteristics of high-rise buildings, fire spread law, evacuation behavior, stair-elevator hybrid evacuation strategy, elevator operation control mode, evacuation system operability, and other factors, the elevator evacuation system is summarized for reference, and it is shown in Figure 12.

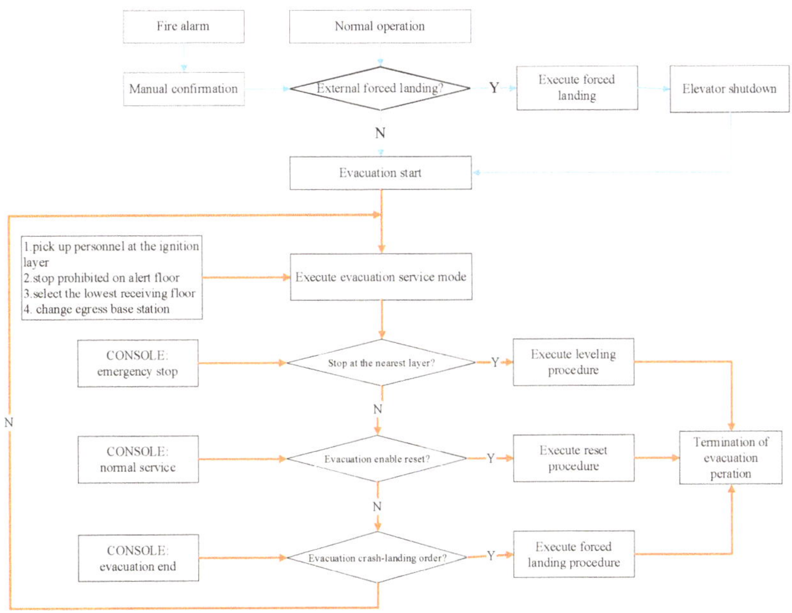

Figure 12. Elevator fire evacuation.

5.3. Disaster Classification

Development and implementation of more efficient targeted safety evacuation could significantly reduce lives and property losses in disasters. Targeted safety evacuation studies have been conducted in natural disasters and man-made/humanistic disasters, including fires [3], earthquakes [8,100,104], landslides [175], tsunamis [61], floods [63], hurricanes [176], typhoons [177], tornados [178], transportation accidents [18,29,42,67,69, 108,145,152], and terrorist attacks [179]. In the scientific research on disaster risk reduction, disasters could be classified into three types: natural disasters, man-made/humanistic disasters, and industrial/technological disasters. Fire safety evacuation received the most attention in recent years.

6. Conclusions

Fire safety evacuation research is of great significance to whether a person can reach the safe area smoothly, and thus to preventing the occurrence of casualties. Fire safety evacuation research plays a very positive role in the development of social health and stability. There is a bias in the research literature towards traditional fire safety evacuation in buildings and a concerning lack of fire safety evacuation innovation; conducting similar research at future crucial junctures will continue to address the evolving nature of fire safety evacuation in buildings and help monitor its development. With the help of CiteSpace and WOS, this paper analyzes the research frontier, application trends, and knowledge basis of fire safety evacuation through correlation analysis and the processing of literature data and information. Based on co-citation theory and the pathfinding network algorithm, the trends and hotspots in fire safety evacuation research are revealed using reference burst-detection analysis. Through visualization of the research literature, the key evolution path of the discipline's frontier and the key nodes of the discipline are clearly and intuitively analyzed. This paper points out that the future research trends of fire evacuation are intelligence, visualization, and interaction, which have a certain guiding role for the development of related research.

(1) New fire safety evacuation research themes have emerged or have been visibly strengthened, including fire safety evacuation of elderly, children, people with mobility difficulties, and vulnerable groups. While the hottest traditional topics have been explored primarily through laboratory crowd experiments, VR/AR approaches have proved more helpful in studying the hottest emerging topics with applications that involve fire safety evacuation design, decision-making, and forecasting;

(2) In the fire safety evacuation research, the hotspots of the study mainly include the emergency events caused by different factors (natural disaster and man-made disaster), evacuation environment, evacuation objects, psychological characteristics, behavioral characteristics, pathfinding characteristics of evacuees, and overall evacuation time. The core technologies and methods include questionnaire surveys, evacuation drills, actual data collection through an information system, and numerical simulations;

(3) ICT and restrictive evacuation are the most active research directions by clustering analysis of keywords. The outline for the implementation of fire safety evacuation is the establishment of a spatial model, the establishment of and evacuee model, and the simulation of the evacuation process;

(4) The main experimental approaches of fire safety evacuation are evacuation drills, site records, and VR/AR experiments. The experimental approaches are developing in the direction of intelligence. The crowd behavior models could be divided into a cellular automata model, a social force model, a lattice gas model, a game-theoretic model, an animal agent-based model, and a computer agent-based model. The factors considered in the theoretical analysis process are gradually complete and closer to the behavioral characteristics and movement data of the crowd during the actual evacuation. Keywords mainly focus on "evacuation simulation", "emergency evacuation", and "optimization". Intelligent fire safety evacuation, visual fire safety evacuation, and multi-dimensional dynamic collaborative interactive fire safety evacuation are the future research trends of fire safety evacuation;

(5) In the papers related to fire evacuation, "simulation" is the most frequent keyword, showing that simulation analysis has been mostly utilized in this area. Meanwhile,

"evacuation" is the most frequent word, indicating the research hotspots in the field of fire safety evacuation mainly revolve around "evacuation simulation", and this presents a pattern of diversified development. It is also found that the latest research topics relate to "construction safety" and "fire safety evacuation in construction", which indicates a shift in the field of fire safety evacuation in the buildings;

(6) The rapid development of building evacuation is because the study of crowd evacuation depends on the external environment. Therefore, the study of evacuation drills, disaster rescues, emergencies, and other external environmental factors should also become the forefront of future research, and subway stations, airports, high-rise buildings, and other personnel places will be the focus of the study of crowd evacuation.

Author Contributions: Conceptualization, Y.Y. and G.Y.; methodology, Y.Y. and G.Y.; software, H.D.; formal analysis, Y.Y. and H.D.; writing—original draft preparation, Y.Y. and H.D.; writing—review and editing, Y.Y. and G.Y. All authors have read and agreed to the published version of the manuscript.

Funding: This work was funded by the 111 project of the Ministry of Education and the Bureau of Foreign Experts of China (No. B18062).

Institutional Review Board Statement: Not applicable.

Informed Consent Statement: Not applicable.

Data Availability Statement: Not applicable.

Conflicts of Interest: The authors declare no conflict of interest.

References

1. Thompson, R.R.; Garfin, D.R.; Silver, R.C. Evacuation from natural disasters: A systematic review of the literature. *Risk Anal.* **2017**, *37*, 812–839. [CrossRef]
2. Kobes, M.; Helsloot, I.; de Vries, B.; Post, J.G. Building safety and human behaviour in fire: A literature review. *Fire Saf. J.* **2010**, *45*, 1–11. [CrossRef]
3. Thompson, O.F.; Galea, E.R.; Hulse, L.M. A review of the literature on human behaviour in dwelling fires. *Saf. Sci.* **2018**, *109*, 303–312. [CrossRef]
4. Kinateder, M.T.; Kuligowski, E.D.; Reneke, P.A.; Peacock, R.D. Risk perception in fire evacuation behavior revisited: Definitions, related concepts, and empirical evidence. *Fire Sci. Rev.* **2015**, *4*, 1. [CrossRef]
5. Ding, N.; Chen, T.; Zhu, Y.; Lu, Y. State-of-the-art high-rise building emergency evacuation behavior. *Phys. A Stat. Mech. Its Appl.* **2021**, *561*, 125168. [CrossRef]
6. Glauberman, G. Scoping review of fire safety behaviors among high-rise occupants: Implications for public health nursing. *Public Health Nurs.* **2020**, *37*, 371–379. [CrossRef]
7. Bernardini, G.; D'Orazio, M.; Quagliarini, E. Towards a "behavioural design" approach for seismic risk reduction strategies of buildings and their environment. *Saf. Sci.* **2016**, *86*, 273–294. [CrossRef]
8. Kaveh, A.; Javadi, S.; Moghanmi, R.M. Emergency management systems after disastrous earthquakes using optimization methods: A comprehensive review. *Adv. Eng. Softw.* **2020**, *149*, 102885. [CrossRef]
9. Kako, M.; Steenkamp, M.; Ryan, B.; Arbon, P.; Takada, Y. Best practice for evacuation centres accommodating vulnerable populations: A literature review. *Int. J. Disaster Risk Reduct.* **2020**, *46*, 101497. [CrossRef]
10. Luna, S.; Pennock, M.J. Social media applications and emergency management: A literature review and research agenda. *Int. J. Disaster Risk Reduct.* **2018**, *28*, 565–577. [CrossRef]
11. Omori, H.; Kuligowski, E.D.; Gwynne, S.M.V.; Butler, K.M. Human response to emergency communication: A review of guidance on alerts and warning messages for emergencies in buildings. *Fire Technol.* **2017**, *53*, 1641–1668. [CrossRef]
12. Zhu, R.; Lin, J.; Becerik-Gerber, B.; Li, N. Human-building-emergency interactions and their impact on emergency response performance: A review of the state of the art. *Saf. Sci.* **2020**, *127*, 104691. [CrossRef]
13. Zhou, M.; Dong, H.; Ioannou, P.A.; Zhao, Y.; Wang, F.-Y. Guided crowd evacuation: Approaches and challenges. *J. Autom. Sin.* **2019**, *6*, 1081–1094. [CrossRef]
14. Li, Y.; Chen, M.; Dou, Z.; Zheng, X.; Cheng, Y.; Mebarki, A. A review of cellular automata models for crowd evacuation. *Phys. A Stat. Mech. Its Appl.* **2019**, *526*, 120752. [CrossRef]
15. Nouri, F.; Khorasani-Zavareh, D.; Kavousi, A.; Mohammadi, R. A system approach on safe emergency evacuation in subways: A systematic literature review. *Arch. Trauma Res.* **2019**, *8*, 119–143. [CrossRef]
16. Gilchrist, N.; Simpson, J.N. Pediatric disaster preparedness: Identifying challenges and opportunities for emergency department planning. *Curr. Opin. Pediatr.* **2019**, *31*, 306–311. [CrossRef]

17. Lin, J.; Zhu, R.; Li, N.; Becerik-Gerber, B. How occupants respond to building emergencies: A systematic review of behavioral characteristics and behavioral theories. *Saf. Sci.* **2020**, *122*, 104540. [CrossRef]
18. Iliopoulou, C.; Konstantinidou, M.; Kepaptsoglou, K.; Stathopoulos, A. Its technologies for decision making during evacuation operations: A review. *J. Transp. Eng. Part A-Syst.* **2020**, *146*, 04020010. [CrossRef]
19. Martinez, P.; Al-Hussein, M.; Ahmad, R. A scientometric analysis and critical review of computer vision applications for construction. *Autom. Constr.* **2019**, *107*, 102947. [CrossRef]
20. Li, J.; Goerlandt, F.; Reniers, G. An overview of scientometric mapping for the safety science community: Methods, tools, and framework. *Saf. Sci.* **2021**, *134*, 105093. [CrossRef]
21. Neelam, S.; Sood, S.K. A scientometric review of global research on smart disaster management. *IEEE Trans. Eng. Manag.* **2021**, *68*, 317–329. [CrossRef]
22. Li, Z.; Zhu, L.D. The scientometric analysis of the research on microalgae-based wastewater treatment. *Environ. Sci. Pollut. Res.* **2021**, *28*, 25339–25348. [CrossRef]
23. Aghimien, E.I.; Aghimien, L.M.; Petinrin, O.O.; Aghimien, D.O. High-performance computing for computational modelling in built environment-related studies- a scientometric review. *J. Eng. Des. Technol.* **2020**, *19*, 1138–1157. [CrossRef]
24. Wu, Z.; Yang, K.; Lai, X.; Antwi-Afari, M. A scientometric review of system dynamics applications in construction management research. *Sustainability* **2020**, *12*, 7474. [CrossRef]
25. Zou, X.; Vu, H.L.; Huang, H.L. Fifty years of accident analysis& prevention: A bibliometric and scientometric overview. *Accid. Anal. Prev.* **2020**, *144*, 105568. [CrossRef]
26. Jin, R.Y.; Zou, Y.; Gidado, K.; Ashton, P.; Painting, N. Scientometric analysis of bim based research in construction engineering and management. *Eng. Constr. Archit. Manag.* **2018**, *26*, 1750–1776. [CrossRef]
27. Zhao, X.B. A scientometric review of global bim research: Analysis and visualization. *Autom. Constr.* **2017**, *80*, 37–47. [CrossRef]
28. Deng, K.; Zhang, Q.; Zhang, H.; Xiao, P.; Chen, J. Optimal emergency evacuation route planning model based on fire prediction data. *Mathematics* **2022**, *10*, 3146. [CrossRef]
29. Renne, J.; Wolshon, B.; Murray-Tuite, P.; Pande, A. Emergence of resilience as a framework for state departments of transportation (dots) in the United States. *Transp. Res. Part D* **2019**, *82*, 102178. [CrossRef]
30. Spartalis, E.; Georgoudas, I.G.; Sirakoulis, G.C. Ca crowd modeling for a retirement house evacuation with guidance. *Lect. Notes Comput. Sci.* **2014**, *8751*, 481–491. [CrossRef]
31. Giitsidis, T.; Dourvas, N.I.; Sirakoulis, G.C. Parallel implementation of aircraft disembarking and emergency evacuation based on cellular automata. *Int. J. High Perform. Comput. Appl.* **2015**, *18*, 1094342015584533. [CrossRef]
32. de Carvalho, P.V.R.; Ranauro, D.O.; Mol, A.C.D.A.; Jatoba, A.; de Siqueira, A.P.L. Using Serious Game in Public Schools for Training Fire Evacuation Procedures. *Int. J. Serious Games* **2022**, *9*, 125–139. [CrossRef]
33. Wąs, J. Egress Modeling through Cellular Automata Based Multi-Agent Systems. *Lect. Notes Comput. Sci.* **2012**, *7270*, 222–235. [CrossRef]
34. Xu, M.C.; Hijazi, I.; Mebarki, A.; Meouche, R.E.; Abune'Meh, M. Indoor guided evacuation: Tin for graph generation and crowd evacuation. *Geomat. Nat. Hazards Risk* **2016**, *7*, 47–56. [CrossRef]
35. Wu, W.; Li, J.; Yi, W.; Zheng, X. Modeling crowd evacuation via behavioral heterogeneity-based social force model. *IEEE Trans. Intell. Transp. Syst.* **2022**, *23*, 15476–15486. [CrossRef]
36. Yi, J.X.; Pan, S.L.; Chen, Q. Simulation of pedestrian evacuation in stampedes based on a cellular automaton model. *Simul. Model. Pract. Theory* **2020**, *104*, 102147. [CrossRef]
37. Abu Bakar, N.A.; Majid, M.A.; Adam, K.; Allegra, M. Social force as a microscopic simulation model for pedestrian behavior in crowd evacuation. *Adv. Sci. Lett.* **2018**, *24*, 7611–7616. [CrossRef]
38. Li, J.; Zhang, H.X.; Ni, Z.R. Improved social force model based on navigation points for crowd emergent evacuation. *J. Inf. Process. Syst.* **2020**, *16*, 1309–1323. [CrossRef]
39. Cao, N.; Zhao, L.; Chen, M.; Luo, R. Fuzzy social force model for pedestrian evacuation under view-limited condition. *Math. Probl. Eng.* **2020**, *2020*, 2879802. [CrossRef]
40. Makmul, J. A social force model for pedestrians' movements affected by smoke spreading. *Model. Simul. Eng.* **2020**, *2020*, 8819076. [CrossRef]
41. Kolivand, H.; Rahim, M.S.; Sunar, M.S.; Fata, A.Z.A.; Wren, C. An integration of enhanced social force and crowd control models for high-density crowd simulation. *Neural Comput. Appl.* **2020**, *33*, 6095–6117. [CrossRef]
42. Wei, J.; Fan, W.; Li, Z.; Guo, Y.; Fang, Y.; Wang, J. Simulating crowd evacuation in a social force model with iterative extended state observer. *J. Adv. Transp.* **2020**, *2020*, 4604187. [CrossRef]
43. Tian, X.Y.; Cui, H.J.; Zhu, M.Q. Improved social force model for rescue action during evacuation. *Mod. Phys. Lett. B* **2020**, *34*, 2050273. [CrossRef]
44. Chen, T.; Wang, W.; Tu, Y.; Hua, X. Modelling unidirectional crowd motion in a corridor with statistical characteristics of pedestrian movements. *Math. Probl. Eng.* **2020**, *2020*, 7483210. [CrossRef]
45. Tao, Y.Z.; Dong, L.Y. A floor field real-coded lattice gas model for crowd evacuation. *Europhys. Lett.* **2017**, *119*, 10003. [CrossRef]
46. Zhu, K.; Yang, Y.; Niu, Y.; Fu, Z.; Shi, Q. Modeling pedestrian flow on multi-storey stairs considering turning behavior. *Int. J. Mod. Phys. C* **2017**, *28*, 1750034. [CrossRef]

47. Zou, Y.; Xie, J.; Wang, B. Evacuation of pedestrians with two motion modes for panic system. *PLoS ONE* **2016**, *11*, e0153388. [CrossRef]
48. Shang, H.Y.; Huang, H.J.; Zhang, Y.M. An extended mobile lattice gas model allowing pedestrian step size variable. *Phys. A Stat. Mech. Its Appl.* **2015**, *424*, 283–293. [CrossRef]
49. Bouzat, S.; Kuperman, M.N. Game theory in models of pedestrian room evacuation. *Phys. Rev. E* **2014**, *89*, 032806. [CrossRef]
50. Viswanathan, V.; Lee, C.E.; Lees, M.H.; Cheong, S.A.; Sloot, P. Quantitative comparison between crowd models for evacuation planning and evaluation. *Eur. Phys. J. B* **2014**, *87*, 27. [CrossRef]
51. Guo, X.; Chen, J.; You, S.; Wei, J. Modeling of pedestrian evacuation under fire emergency based on an extended heterogeneous lattice gas model. *Phys. A Stat. Mech. Its Appl.* **2013**, *392*, 1994–2006. [CrossRef]
52. Rigos, A.; Mohlin, E.; Ronchi, E. The cry wolf effect in evacuation: A game-theoretic approach. *Phys. A Stat. Mech. Its Appl.* **2019**, *526*, 120890. [CrossRef]
53. Ibrahim, A.M.; Venkat, V.I.; De Wilde, P. The impact of potential crowd behaviours on emergency evacuation: An evolutionary game-theoretic approach. *J. Artif. Soc. Soc. Simul.* **2019**, *22*, 3. [CrossRef]
54. Turnwald, A.; Wollherr, D. Human-like motion planning based on game theoretic decision making. *Int. J. Soc. Robot.* **2019**, *11*, 151–170. [CrossRef]
55. Morten, G.; Granmo, O.-C.; Radianti, J. Escape planning in realistic fire scenarios with Ant Colony Optimisation. *Appl. Intell.* **2015**, *42*, 24–35. [CrossRef]
56. Liu, H.; Xu, B.; Lu, D.; Zhang, G. A path planning approach for crowd evacuation in buildings based on improved artificial bee colony algorithm. *Appl. Soft Comput.* **2018**, *68*, 360–376. [CrossRef]
57. Zhang, G.J.; Lu, D.J.; Liu, H. Strategies to utilize the positive emotional contagion optimally in crowd evacuation. *IEEE Trans. Affect. Comput.* **2018**, *11*, 708–721. [CrossRef]
58. Wang, P.L.; Zhang, T.; Xiao, Y.J. Application research of ant colony cellular optimization algorithm in population evacuation path planning. *Acta Phys. Sin.* **2020**, *69*, 080504. [CrossRef]
59. Niyomubyeyi, O.; Pilesjö, P.; Mansourian, A. Evacuation planning optimization based on a multi-objective artificial bee colony algorithm. *ISPRS Int. J. Geo-Inf.* **2019**, *8*, 110. [CrossRef]
60. Liu, M.; Zhang, F.; Ma, Y.; Pota, H.R.; Shen, W. Evacuation path optimization based on quantum ant colony algorithm. *Adv. Eng. Inform.* **2016**, *30*, 259–267. [CrossRef]
61. Forcael, E.; Gonzalez, V.; Orozco, F.; Vargas, S.; Pantoja, A.; Moscoso, P. Ant colony optimization model for tsunamis evacuation routes. *Comput.-Aided Civ. Infrastruct. Eng.* **2014**, *29*, 723–737. [CrossRef]
62. Khamis, N.; Selamat, H.; Ismail, F.S.; Lutfy, O.F.; Haniff, M.F.; Nordin, I.N.A.M. Optimized exit door locations for a safer emergency evacuation using crowd evacuation model and artificial bee colony optimization. *Chaos Solitons Fractals* **2019**, *131*, 109505. [CrossRef]
63. Wang, Z.; Huang, J.; Wang, H.; Kang, J.; Cao, W. Analysis of flood evacuation process in vulnerable community with mutual aid mechanism: An agent-based simulation framework. *Int. J. Environ. Res. Public Health* **2020**, *17*, 560. [CrossRef]
64. Yang, P.Z.; Wang, X.; Liu, T. Agent-based simulation of fire emergency evacuation with fire and human interaction model (Retraction of vol 49, pg1130, 2011). *Saf. Sci.* **2012**, *50*, 1171. [CrossRef]
65. Azar, E.; Menassa, C.C. Agent-based modeling of occupants and their impact on energy use in commercial buildings. *J. Comput. Civ. Eng.* **2012**, *26*, 506–518. [CrossRef]
66. Heliövaara, S.; Korhonen, T.; Hostikka, S.; Ehtamo, H. Counterflow model for agent based simulation of crowd dynamics. *Build. Environ.* **2012**, *48*, 89–100. [CrossRef]
67. Bernardini, G.; D'Orazio, M.; Quagliarini, E.; Spalazzi, L. An agent-based model for earthquake pedestrians' evacuation simulation in urban scenarios. *Transp. Res. Procedia* **2014**, *2*, 255–263. [CrossRef]
68. Cimellaro, G.P.; Ozzello, F.; Vallero, A.; Mahin, S.; Shao, B. Simulating earthquake evacuation using human behavior models. *Earthq. Eng. Struct. Dyn.* **2017**, *46*, 985–1002. [CrossRef]
69. Wang, H.; Mostafizi, A.; Cramer, L.A.; Cox, D.; Park, H. An agent-based model of a multimodal near-field tsunami evacuation: Decision-making and life safety. *Transp. Res. Part C Emerg. Technol.* **2016**, *64*, 86–100. [CrossRef]
70. Mostafizi, A.; Wang, H.; Cox, D.; Dong, S. An agent-based vertical evacuation model for a near-field tsunami: Choice behavior, logical shelter locations, and life safety. *Int. J. Disaster Risk* **2019**, *34*, 467–479. [CrossRef]
71. Joo, J.; Kim, N.; Wysk, R.A.; Rothrock, L.; Son, Y.-J.; Oh, Y.-G.; Lee, S. Agent-based simulation of affordance-based human behaviors in emergency evacuation. *Simul. Model. Pract. Theory* **2013**, *32*, 99–115. [CrossRef]
72. Kasereka, S.; Kasoro, N.; Kyamakya, K.; Goufo, E.-F.D.; Chokki, A.P.; Yengo, M.V. Agent-based modelling and simulation for evacuation of people from a building in case of fire. *Procedia Comput. Sci.* **2018**, *130*, 10–17. [CrossRef]
73. Min, Y.; Yu, Y. Calculation of mixed evacuation of stair and elevator using evacnet4. *Procedia Eng.* **2013**, *62*, 478–482. [CrossRef]
74. Yemelyanenko, S.; Ivanusa, A.; Yakovchuk, R.; Kuzyk, A. Fire risks of public buildings. *Ser. Geol. Technol. Sci.* **2020**, *6*, 75–82. [CrossRef]
75. Wang, J.; Guo, J.; Wu, X.; Guo, X. Study on intelligent algorithm of guide partition for emergency evacuation of a subway station. *IET Intell. Transp. Syst.* **2020**, *14*, 1440–1446. [CrossRef]
76. Shao, C.H.; Shao, P.C.; Kuo, F.M. Stampede events and strategies for crowd management. *J. Disaster Res.* **2019**, *14*, 949–958. [CrossRef]

77. Alginahi, Y.M.; Mudassar, M.; Kabir, M.N.; Tayan, O. Analyzing the crowd evacuation pattern of a large densely populated building. *Arab. J. Sci. Eng.* **2019**, *44*, 3289–3304. [CrossRef]
78. Yao, Y.S.; Lu, W. Research on kindergarten children evacuation: Analysis of characteristics of the movement behaviours on stairway. *Int. J. Disaster Risk Reduct.* **2020**, *50*, 101718. [CrossRef]
79. Sun, Q.; Turkan, Y. A bim-based simulation framework for fire safety management and investigation of the critical factors affecting human evacuation performance. *Adv. Eng. Inform.* **2020**, *44*, 101093. [CrossRef]
80. Wang, T.; Du, M.H.; Tang, Y.F.; Zhang, Q. An analysis on the fire model and the safety evacuation based on bim. *Adv. Mater. Res.* **2015**, *1065–1069*, 2386–2389. [CrossRef]
81. Chen, X.S.; Liu, C.C.; Wu, I. A bim-based visualization and warning system for fire rescue. *Adv. Eng. Inform.* **2018**, *37*, 42–53. [CrossRef]
82. Xu, Z.; Zhang, Z.; Lu, X.; Zeng, X.; Guan, H. Post-earthquake fire simulation considering overall seismic damage of sprinkler systems based on bim and femap-58. *Autom. Constr.* **2018**, *90*, 9–22. [CrossRef]
83. Yue, H.; Zhang, B.-Y.; Shao, C.-F.; Xing, Y. Exit selection strategy in pedestrian evacuation simulation with multi-exits. *Chin. Phys. B* **2014**, *23*, 050512. [CrossRef]
84. Dimakis, N.; Filippoupolitis, A.; Gelenbe, E. Distributed building evacuation simulator for smart emergency management. *Comput. J.* **2010**, *53*, 1384–1400. [CrossRef]
85. Marzouk, M.; Daour, I.A. Planning labor evacuation for construction sites using bim and agent-based simulation. *Saf. Sci.* **2018**, *109*, 174–185. [CrossRef]
86. Wehbe, R.; Shahrour, I. A bim-based smart system for fire evacuation. *Future Internet* **2021**, *13*, 221. [CrossRef]
87. Yamamoto, K.; Sawaguchi, Y.; Nishiki, S. Simulation of tunnel fire for evacuation safety assessment. *Safety* **2018**, *4*, 12. [CrossRef]
88. Zhou, Y.; Pang, Y.; Chen, F.; Zhang, Y. Three-dimensional indoor fire evacuation routing. *ISPRS Int. J. Geo-Inf.* **2020**, *9*, 558. [CrossRef]
89. Deng, H.; Ou, Z.; Zhang, G.; Deng, Y.; Tian, M. BIM and computer vision-based framework for fire emergency evacuation considering local safety performance. *Sensors* **2021**, *21*, 3851. [CrossRef]
90. Zhang, N.; Liang, Y.; Zhou, C.; Niu, M.; Wan, F. Study on Fire Smoke Distribution and Safety Evacuation of Subway Station Based on BIM. *Appl. Sci.* **2022**, *12*, 12808. [CrossRef]
91. Sarker, M.N.I.; Peng, Y.; Yiran, C.; Shouse, R.C. Disaster resilience through big data: Way to environmental sustainability. *Int. J. Disaster Risk Reduct.* **2020**, *51*, 101769. [CrossRef]
92. Mañas, E.L.; Plá, J.; Herrero, G.M.; Gervás, P. Augmented reality and indoors Wi-Fi positioning for conducting fire evacuation drills using mobile phones. In Proceedings of the 4th Symposium of Ubiquitous Computing and Ambient Intelligence UCAmI, Toledo, Spain, 1 January 2010.
93. Song, J.; Zhang, H.; Dong, W. A review of emerging trends in global PPP research: Analysis and visualization. *Scientometrics* **2016**, *107*, 1111–1147. [CrossRef]
94. Van Eck, N.J.; Waltman, L. Software survey: VOSviewer, a computer program for bibliometric mapping. *Scientometrics* **2009**, *84*, 523–538. [CrossRef]
95. Rajagopal, V.; Venkatesan, S.P.; Goh, M. Decision-making models for supply chain risk mitigation: A review. *Comput. Ind. Eng.* **2017**, *113*, 646–682. [CrossRef]
96. Eck, V.; Jan, N.; Waltman, L. CitNetExplorer: A new software tool for analyzing and visualizing citation networks. *J. Informetr.* **2014**, *8*, 802–823. [CrossRef]
97. Tsiftsis, A.; Georgoudas, I.G.; Sirakoulis, G.C. Real data evaluation of a crowd supervising system for stadium evacuation and its hardware implementation. *IEEE Syst. J.* **2016**, *10*, 649–660. [CrossRef]
98. Bernardini, G.; Lovreglio, R.; Quagliarini, E. Proposing behavior-oriented strategies for earthquake emergency evacuation:a behavioral data analysis from New Zealand, Italy and Japan. *Saf. Sci.* **2019**, *116*, 295–309. [CrossRef]
99. Lambie, E.S.; Wilson, T.M.; Brogt, E.; Johnston, D.M.; Ardagh, M.; Deely, J.; Jensen, S.; Feldmann-Jensen, S. Closed circuit television(cctv) earthquake behaviour coding methodology: Analysis of christchurch public hospital video data from the 22 february christchurch earthquake event. *Nat. Hazards* **2017**, *86*, 1175–1192. [CrossRef]
100. Gu, Z.; Liu, Z.; Shiwakoti, N.; Yang, M. Video-based analysis of school students' emergency evacuation behavior in earthquakes. *Int. J. Disaster Risk Reduct.* **2016**, *18*, 1–11. [CrossRef]
101. Lambie, E.; Wilson, T.M.; Johnston, D.M.; Jensen, S.; Brogt, E.; Doyle, E.E.H.; Lindell, M.K.; Helton, W.S. Human behaviour during and immediately following earthquake shaking: Developing a methodological approach for analysing video footage. *Nat. Hazards* **2016**, *80*, 249–283. [CrossRef]
102. Lovreglio, R.; Gonzalez, V.; Feng, Z.; Amor, R.; Spearpoint, M.; Thomas, J.; Trotter, M.; Sacks, R. Prototyping virtual reality serious games for building earthquake preparedness: The auckland city hospital case study. *Adv. Eng. Inform.* **2018**, *38*, 670–682. [CrossRef]
103. Gabriele, P.; Valeria, C.; Luca, P. Emotional and behavioural reactions to tremors of the umbria-marche earthquake. *Disasters* **2012**, *36*, 439–451. [CrossRef]
104. Mora, K.; Chang, J.; Beatson, A.; Morahan, C. Public perceptions of building seismic safety following the canterbury earthquakes:a qualitative analysis using twitter and focus groups. *Int. J. Disaster Risk Reduct.* **2015**, *13*, 1–9. [CrossRef]

105. Chen, J.; Wang, J.; Wang, B.; Liu, R.; Wang, Q. An experimental study of visibility effect on evacuation speed on stairs. *Fire Saf. J.* **2018**, *96*, 189–202. [CrossRef]
106. Huo, F.; Song, W.; Liu, X.; Jiang, Z.; Liew, K. Investigation of human behavior in emergent evacuation from an underground retail store. *Procedia Eng.* **2014**, *71*, 350–356. [CrossRef]
107. Kobes, M.; Helsloot, I.; de Vries, B.; Post, J. Exit choice, (pre-) movement time and (pre-) evacuation behavior in hotel fire evacuation—Behavioural analysis and validation of the use of serious gaming in experimental research. *Procedia Eng.* **2010**, *3*, 37–51. [CrossRef]
108. Poulos, A.; Tocornal, F.; de la Llera, J.C.; Mitrani-Reiser, J. Validation of an agent-based building evacuation model with a school drill. *Transp. Res. Part C Emerg. Technol.* **2018**, *97*, 82–95. [CrossRef]
109. Fang, Z.-M.; Song, W.-G.; Li, Z.-J.; Tian, W.; Lv, W.; Ma, J.; Xiao, X. Experimental study on evacuation process in a stairwell of a high-rise building. *Build. Environ.* **2012**, *47*, 316–321. [CrossRef]
110. Fang, H.; Qiu, H.; Lin, P.; Lo, S.M.; Lo, J.T.Y. Towards a smart elevator-aided fire evacuation scheme in high-rise apartment buildings for elderly. *IEEE Access* **2022**, *10*, 90690–90705. [CrossRef]
111. Liao, Y.J.; Lo, S.M.; Ma, J.; Liu, S.B.; Liao, G.X. A study on people's attitude to the use of elevators for fire escape. *Fire Technol.* **2014**, *50*, 363–378. [CrossRef]
112. Ma, J.; Song, W.; Tian, W.; Lo, S.M.; Liao, G. Experimental study on an ultra high-rise building evacuation in china. *Saf. Sci.* **2012**, *50*, 1665–1674. [CrossRef]
113. Zhang, J.; Klingsch, W.; Schadschneider, A.; Seyfried, A. Experimental study of pedestrian flow through a T-junction. *Traffic Granul. Flow* **2013**, *11*, 241–249. [CrossRef]
114. Zhang, J.; Klingsch, W.; Schadschneider, A.; Seyfried, A. Ordering in bidirectional pedestrian flows and its influence on the fundamental diagram. *J. Stat. Mech. Theory Exp.* **2012**, *02*, P02002. [CrossRef]
115. Zhang, J.; Seyfried, A. Comparison of intersecting pedestrian flows based on experiments. *Phys. A Stat. Mech. Its Appl.* **2014**, *405*, 316–325. [CrossRef]
116. Boltes, M.; Zhang, J.; Tordeux, A.; Schadschneider, A.; Seyfried, A. Empirical results of pedestrian and evacuation dynamics. In *Encyclopedia of Complexity and Systems Science*; Springer: Berlin/Heidelberg, Germany, 2018; pp. 1–29. [CrossRef]
117. Seto, D.; Jones, C.; Trugman, A.T.; Varga, K.; Plantinga, A.J.; Carvalho, L.M.V.; Thompson, C.; Gellman, J.; Daum, K. Simulating potential impacts of fuel treatments on fire behavior and evacuation time of the 2018 Camp Fire in northern California. *Fire* **2022**, *5*, 37. [CrossRef]
118. Nauslar, N.J.; Abatzoglou, J.T.; Marsh, P.T. The 2017 North Bay and Southern California fires: A case study. *Fire* **2018**, *1*, 18. [CrossRef]
119. Peacock, R.D.; Hoskins, B.L.; Kuligowski, E.D. Overall and local movement speeds during fire drill evacuations in buildings up to 31 stories. In *Pedestrian and Evacuation Dynamics*; Springer: Berlin/Heidelberg, Germany, 2012; pp. 25–35. [CrossRef]
120. Nakano, G.; Yamori, K.; Miyashita, T.; Urra, L.; Mas, E.; Koshimura, S. Combination of school evacuation drill with tsunami inundation simulation: Consensus-making between disaster experts and citizens on an evacuation strategy. *Int. J. Disaster Risk Reduct.* **2020**, *51*, 101803. [CrossRef]
121. Hamilton, G.N.; Lennon, P.F.; O'Raw, J. Toward fire safe schools: Analysis of modelling speed and specific flow of children during evacuation drills. *Fire Technol.* **2020**, *56*, 605–638. [CrossRef]
122. Zhang, J.; Zhao, J.; Song, Z.; Gao, J. Evacuation performance of participants in an offshore platform under smoke situations. *Ocean Eng.* **2020**, *216*, 107739. [CrossRef]
123. Fang, Z.M.; Jiang, L.X.; Li, X.L.; Qi, W.; Chen, L.Z. Experimental study on the movement characteristics of 5–6 years old chinese children when egressing from a pre-school building. *Saf. Sci.* **2019**, *113*, 264–275. [CrossRef]
124. Sugiyama, T.; Yamori, K. Consideration of evacuation drills utilizing the capabilities of people with special needs. *J. Disaster Res.* **2020**, *15*, 794–801. [CrossRef]
125. Rahouti, A.; Lovreglio, R.; Gwynne, S.; Jackson, P.; Datoussaïd, S.; Hunt, A. Human behaviour during a healthcare facility evacuation drills: Investigation of pre-evacuation and travel phases. *Saf. Sci.* **2020**, *129*, 104754. [CrossRef]
126. Jaime, S. Factors motivating mexico city residents to earthquake mass evacuation drills. *Int. J. Disaster Risk Reduct.* **2020**, *49*, 101661. [CrossRef]
127. Kawai, J.; Mitsuhara, H.; Shishibori, M. Game-based evacuation drill using augmented reality and head-mounted display. *Emerald Insight* **2016**, *13*, 186–201. [CrossRef]
128. Kobes, M.; Helsloot, I.; de Vries, B.; Post, J.G.; Oberijé, N.; Groenewegen, K. Way finding during fire evacuation; an analysis of unannounced fire drills in a hotel at night. *Build. Environ.* **2010**, *45*, 537–548. [CrossRef]
129. Abolghasemzadeh, P. A comprehensive method for environmentally sensitive and behavioral microscopic egress analysis in case of fire in buildings. *Saf. Sci.* **2013**, *59*, 1–9. [CrossRef]
130. Chen, Y.; Wang, C.; Yap, J.B.H.; Li, H.; Zhang, S. Emergency evacuation simulation at starting connection of cross-sea bridge: Case study on haicang avenue subway station in xiamen rail transit Line. *J. Build. Eng.* **2020**, *29*, 101163. [CrossRef]
131. Kinateder, M.; Ronchi, E.; Nilsson, D.; Kobes, M.; Mülberger, A. Virtual Reality for Fire Evacuation Research. In Proceedings of the 2014 Federated Conference on Computer Science and Information Systems, Warsaw, Poland, 7–10 September 2014; IEEE: Piscataway, NJ, USA, 2015; pp. 313–321. [CrossRef]

132. Lu, X.; Luh, P.B.; Marsh, K.L.; Gifford, T.; Tucker, A. Guidance optimization of building evacuation considering psychological features in route choice. In Proceedings of the 11th World Congress on Intelligent Control and Automation, Shenyang, China, 29 June–4 July 2014; IEEE: Piscataway, NJ, USA, 2015; pp. 2669–2674. [CrossRef]
133. Marsh, K.L.; Wilkie, C.T.; Luh, P.B.; Zhang, Z.; Gifford, T.; Olderman, N. Crowd guidance in building emergencies: Using virtual reality experiments to confirm macroscopic mathematical modeling of psychological variables. In *Pedestrian and Evacuation Dynamics*; Springer: Berlin/Heidelberg, Germany, 2012; pp. 197–212. [CrossRef]
134. Ronchi, E.; Nilsson, D.; Kojić, S.; Eriksson, J.; Lovreglio, R.; Modig, H.; Walter, A.L. Virtual reality experiment on flashing lights at emergency exit portals for road tunnel evacuation. *Fire Technol.* **2016**, *52*, 623–647. [CrossRef]
135. Zou, H.; Li, N.; Cao, L.J. Emotional response- based approach for assessing the sense of presence of subjects in virtual building evacuation studies. *J. Comput. Civ. Eng.* **2017**, *31*, 04017028. [CrossRef]
136. Andrée, K.; Nilsson, D.; Eriksson, J. Evacuation experiments in a virtual reality high-rise building: Exit choice and waiting time for evacuation elevators. *Fire Mater.* **2016**, *40*, 554–567. [CrossRef]
137. Bourhim, E.M.; Cherkaoui, A. Simulating pre-evacuation behavior in a virtual fire environment. In Proceedings of the IEEE 2018 9th International Conference on Computing, Communication and Networking Technologies, Bengaluru, India, 10–12 July 2018. [CrossRef]
138. Cao, L.J.; Lin, J.; Li, N. A virtual reality based udy of indoor fire evacuation after active or passive spatial exploration. *Comput. Hum. Behav.* **2018**, *90*, 37–45. [CrossRef]
139. Gu, T.; Wang, C.B.; He, G.Q. A vr-based, hybrid modeling approach to fire evacuation simulation. *ASSOC Comput. Mach.* **2018**, *19*, 1–8. [CrossRef]
140. He, G.-Q.; Chen, Q.; Yang, Y.; Jiang, Z.-Q.; Pan, Z.-G. Efficient scene playback and evacuation decision in the configurable 3d virtual emergency scenes. In Proceedings of the International Conference on Virtual Reality and Visualization, Hangzhou, China, 24–26 September 2016; pp. 331–334. [CrossRef]
141. Kinateder, M.; Wirth, T.D.; Warren, W.H. Crowd dynamics in virtual reality. In *Crowd Dynamics, Volume I Modeling and Simulation in Science, Engineering and Technology*; Springer: Berlin/Heidelberg, Germany, 2019; pp. 15–36. [CrossRef]
142. Schaffer, D.; Boeira, C.; Rockenbach, G.; Maurer, G.; Antonitsch, A.; Musse, S.R. Simulating virtual humans crowds in facilities. In Proceedings of the 2018 17th Brazilian Symposium on Computer Games and Digital Entertainment (SBGames), Foz do Iguacu, Brazil, 29 October–1 November 2018; IEEE: Piscataway, NJ, USA, 2018; pp. 343–344. [CrossRef]
143. Liu, Y.; Ren, D.-W.; Liu, Z.-G.; Lou, Y.-R. *The Development and Evaluation of Virtual Reality Platform for Emergency Evacuation in Coal Mines*; Springer: Berlin/Heidelberg, Germany, 2016; pp. 381–387. [CrossRef]
144. Mühlberger, A.; Kinateder, M.; Brütting, J.; Eder, S.; Müller, M.; Gromer, D.; Pauli, P. Influence of information and instructions on human behavior in tunnel accidents: A virtual reality study. *J. Virtual Real. Broadcast.* **2015**, *12*. [CrossRef]
145. Li, H.L.; Zhang, J.; Xia, L.; Song, W.; Bode, N.W. Comparing the route-choice behavior of pedestrians around obstacles in a virtual experiment and a field study. *Transp. Res. Part C Emerg. Technol.* **2019**, *107*, 120–136. [CrossRef]
146. Iguchi, K.; Mitsuhara, H.; Shishibori, M. Evacuation instruction training system using augmented reality and a smartphone-based head mounted display. In Proceedings of the 2016 3rd International Conference on Information and Communication Technologies for Disaster Management, Vienna, Austria, 13–15 December 2016; pp. 158–163. [CrossRef]
147. Mitsuhara, H.; Iguchi, K.; Kawai, J. Using digital game, augmented reality, and head mounted displays for immediate-action commander training. *Int. J. Emerg. Technol. Learn.* **2017**, *12*, 101–117. [CrossRef]
148. Ahn, J.; Han, R. Rescueme: An indoor mobile augmented-reality evacuation system by personalized pedometry. In Proceedings of the 2011 IEEE Asia-Pacific Services Computing Conference, Jeju, Republic of Korea, 12–15 December 2011; pp. 70–77.
149. Lochhead, I.; Hedley, N. Mixed reality emergency management: Bringing virtual evacuation simulations into real-world built environments. *Int. J. Digit. Earth* **2018**, *12*, 190–208. [CrossRef]
150. Arias, S.; La Mendola, S.; Wahlqvist, J.; Rios, O.; Nilsson, D.; Ronchi, E. Virtual reality evacuation experiments on way-finding systems for the future circular collider. *Fire Technol.* **2019**, *55*, 2319–2340. [CrossRef]
151. Wang, J.H.; Lo, S.M.; Sun, J.H.; Wang, Q.-S.; Mu, H.-L. Qualitative simulation of the panic spread in large scale evacuation. *Simulation* **2012**, *88*, 1465–1474. [CrossRef]
152. Haghani, M.; Sarvi, M. Stated and revealed exit choices of pedestrian crowd evacuees. *Transp. Res. Part B Methodol.* **2017**, *95*, 238–259. [CrossRef]
153. Najmanová, H.; Kuklík, L.; Pešková, V.; Bukáček, M.; Hrabák, P.; Vašata, D. Evacuation trials from a double-deck electric train unit: Experimental data and sensitivity analysis. *Saf. Sci.* **2022**, *146*, 105523. [CrossRef]
154. Chu, M.L.; Parigi, P.; Law, K.H.; Latombe, J.-C. Simulating individual, group, and crowd behaviors in building egress. *Simulation* **2015**, *91*, 825–845. [CrossRef]
155. Kodur, V.K.R.; Venkatachari, S.; Naser, M.Z. Egress parameters influencing emergency evacuation in high-rise buildings. *Fire Technol.* **2019**, *56*, 2035–2057. [CrossRef]
156. Kurdi, H.A.; Al-Megren, S.; Althunyan, R.; Almulifi, A. Effect of exit placement on evacuation plans. *Eur. J. Oper. Res.* **2018**, *269*, 749–759. [CrossRef]
157. Boguslawski, P.; Mahdjoubi, L.; Zverovich, V.; Fadli, F. A dynamic approach for evacuees distribution and optimal routing in hazardous environments. *Autom. Constr.* **2018**, *94*, 11–21. [CrossRef]

158. Shaikh, N.; Kakosimos, K.E.; Adia, N.; Véchot, L. Concept and demonstration of a fully coupled and dynamic exposure-response methodology for crowd evacuation numerical modelling in airborne-toxic environments. *J. Hazard. Mater.* **2020**, *399*, 123093. [CrossRef]
159. Bernardini, G.; Quagliarini, E.; D'Orazio, M. Towards creating a combined database for earthquake pedestrians' evacuation models. *Saf. Sci.* **2016**, *82*, 77–94. [CrossRef]
160. Wang, J.-H.; Yan, W.-Y.; Zhi, Y.-R.; Jiang, J.-C. Investigation of the panic psychology and behaviors of evacuation crowds in subway emergencies. *Procedia Eng.* **2016**, *135*, 128–137. [CrossRef]
161. Ye, J.H.; Chen, X.H.; Jian, N.J. Impact analysis of human factors on pedestrian traffic characteristics. *Fire Saf. J.* **2012**, *52*, 46–54. [CrossRef]
162. Kim, H.J.; Zhao, Y.C.; Kim, N.H.; Ahn, Y.H. Home modifications for older people with cognitive impairments: Mediation analysis of caregivers' information needs and perceptions of fall risks. *Int. J. Older People Nurs.* **2019**, *14*, e12240. [CrossRef]
163. Li, W.; Li, Y.; Yu, P.; Gong, J.; Shen, S.; Huang, L.; Liang, J. Modelling simulation and analysis of the evacuation process on stairs in a multi-floor classroom building of a primary school. *Phys. A Stat. Mech. Its Appl.* **2017**, *469*, 157–172. [CrossRef]
164. Li, H.; Zhang, J.; Yang, L.; Song, W.; Yuen, K.K.R. A comparative study on the bottleneck flow between preschool children and adults under different movement motivations. *Saf. Sci.* **2020**, *121*, 30–41. [CrossRef]
165. Hu, J.J.; Wu, H.Y.; Chou, C.C. Evacuation simulation in a cultural asset fire: Impact of fire emergency evacuation facilities for people with disabilities on evacuation time. *Fire* **2022**, *6*, 10. [CrossRef]
166. Choi, M.; Lee, S.; Hwang, S.; Park, M.; Lee, H.-S. Comparison of emergency response abilities and evacuation performance involving vulnerable occupants in building fire situations. *Sustainability* **2019**, *12*, 87. [CrossRef]
167. Hu, J.; Li, Z.; Zhang, H.; Wei, J.; You, L.; Chen, P. Experiment and simulation of the bidirectional pedestrian flow model with overtaking and herding behavior. *Int. J. Mordern Phys. C* **2015**, *26*, 1550131. [CrossRef]
168. Sun, L.S.; Yang, Z.F.; Rong, J.; Liu, X. Study on the weaving behavior of high density bidirectional pedestrian flow. *Math. Probl. Eng.* **2014**, *2014*, 765659. [CrossRef]
169. Bernardini, G.; Santarelli, S.; Quagliarini, E.; D'Orazio, M. Dynamic guidance tool for a safer earthquake pedestrian evacuation in urban systems. *Comput. Environ. Urban Syst.* **2017**, *65*, 150–161. [CrossRef]
170. Fang, Z.M.; Lv, W.; Jiang, L.-X.; Xu, Q.-F.; Song, W.-G. Observation, simulation and optimization of the movement of passengers with baggage in railway station. *Int. J. Mod. Phys. C* **2015**, *26*, 1550124. [CrossRef]
171. Wang, C.; Song, Y. Fire evacuation in metro stations: Modeling research on the effects of two key parameters. *Sustainability* **2020**, *12*, 684. [CrossRef]
172. Kirik, E.; Bogdanov, A.; Sushkova, O.; Gravit, M.; Shabunina, D.; Rozov, A.; Vitova, T.; Lazarev, Y. Fire Safety in Museums: Simulation of Fire Scenarios for Development of Control Evacuation Schemes from the Winter Palace of the Hermitage. *Buildings* **2022**, *12*, 1546. [CrossRef]
173. Gravit, M.; Kirik, E.; Savchenko, E.; Vitova, T.; Shabunina, D. Simulation of evacuation from stadiums and entertainment arenas of different epochs on the example of the Roman Colosseum and the Gazprom Arena. *Fire* **2022**, *5*, 20. [CrossRef]
174. Kurdi, H.; Alzuhair, A.; Alotaibi, D.; Alsweed, H.; Almoqayyad, N.; Albaqami, R.; Althnian, A.; Alnabhan, N.; Al Islam, A.B.M.A. Crowd evacuation in Hajj stoning area: Planning through modeling and simulation. *Sustainability* **2022**, *14*, 2278. [CrossRef]
175. Hu, T.Y.; Ho, W.M. A modified entropy-based dynamic gravity model for the evacuation trip distribution problem during typhoons. *J. Chin. Inst. Eng.* **2016**, *39*, 548–555. [CrossRef]
176. Durage, S.W.; Kattan, L.; Wirasinghe, S.C.; Ruwanpura, J.Y. Evacuation behaviour of households and drivers during a tornado. *Nat. Hazards* **2014**, *71*, 1495–1517. [CrossRef]
177. Ma, L.; Chen, B.; Qiu, S.; Li, Z.; Qiu, X. Agent-based modeling of emergency evacuation in a railway station square under sarin terrorist attack. *Int. J. Model. Simul. Sci. Comput.* **2017**, *8*, 1750022. [CrossRef]
178. Kuo, T.-W.; Lin, C.-Y.; Chuang, Y.-J.; Hsiao, G.L.-K. Using smartphones for indoor fire evacuation. *Int. J. Environ. Res. Public Health* **2022**, *19*, 6061. [CrossRef]
179. Lorusso, P.; De Iuliis, M.; Marasco, S.; Domaneschi, M.; Cimellaro, G.P.; Villa, V. Fire emergency evacuation from a school building using an evolutionary virtual reality platform. *Buildings* **2022**, *12*, 223. [CrossRef]

fire

MDPI

Article

Critical Factors Affecting Fire Safety in High-Rise Buildings in the Emirate of Sharjah, UAE

Musab Omar *, Abdelgadir Mahmoud and Sa'ardin Bin Abdul Aziz

Razak Faculty of Technology and Informatics, Universiti Teknologi Malaysia, Kuala Lumpur 54100, Malaysia
* Correspondence: alkhaldy70@hotmail.com; Tel.: +97-1527771351

Abstract: The purpose of this paper is to identify the critical factors affecting fire accidents in high-rise residential buildings in the Emirate of Sharjah in order to find solutions that contribute to reducing injuries and deaths from fire accidents. A large urban expansion of the Emirate of Sharjah has taken place in the form of high-rise buildings, and the Emirate is now third in the UAE in terms of the number of high-rise buildings and is home to 19% of the population. As a consequence, an increase in the rate of fire accidents has also been observed. As such, there is a need to conduct research on enhancing fire safety in high-rise buildings by conducting a literature review, in which nineteen factors affecting fire globally were identified. Because the fire characteristic is unique in every country, to identify the nature of fire in the Emirate of Sharjah, we consulted sixteen subject matter experts in the field of fire in the Emirate of Sharjah to identify the factors applicable to the Emirate. We used the failure mode, effect, and criticality analysis methodology to accomplish this goal. The outcome of the consultations resulted in the three main factor categories, which are management factors, human factors, and technical factors, and the critical factors affecting the high-rise buildings in the Emirate of Sharjah were identified, which are: fire regulations, fire enforcement regulations, accident investigation, rescue speed, human behavior, lack of proper maintenance, fire training, building design, fire knowledge, combustible materials, fire culture of society, and urbanization. Using the Analytical Hierarchy Process (AHP) tools implemented to measure the effect level of the sub-critical fire factors, 45 effects were identified, and the most common effects were: the building is fully covered by cladding, the effect of stopping activities in HRBs that are non-compliant with fire regulations, the residents practice activities related to fire knowledge, fire regulations efficiency, the training of new employees by their employers, and the residents have fire-related knowledge.

Keywords: fire factor; residential buildings; fire accidents

check for updates

Citation: Omar, M.; Mahmoud, A.; Abdul Aziz, S.B. Critical Factors Affecting Fire Safety in High-Rise Buildings in the Emirate of Sharjah, UAE. *Fire* **2023**, *6*, 68. https://doi.org/10.3390/fire6020068

Academic Editor: Tiago Miguel Ferreira

Received: 26 December 2022
Revised: 20 January 2023
Accepted: 30 January 2023
Published: 15 February 2023

1. Introduction

With the development of urbanization and the growth of complex industries in the Emirate of Sharjah, the fire accident rate increased in the Emirate of Sharjah compared to those in other Emirates inside the UAE for the period from 2013 to 2018, according to data published in the UAE Ministry of Interior report of 2019. Fire accidents are a real problem that should be addressed to avoid them affecting society in the Emirate of Sharjah. Fire accidents will affect the economy of the Emirate of Sharjah because they have a direct effect on the real estate market, industrial activities, and business reputation, eventually leading to a decrease in the competitiveness of the Emirate of Sharjah in the region.

High-rise buildings and very tall buildings have dramatically increased in number, and consequently, the number of factors that affect the cause and/or development of fire has also increased. It is difficult to quantify the factors, and they are not independent from each other. A degree of ambiguity exists, so fire-related problems in high-rise building have become a worldwide concern [1]. Fire safety studies are of great significance in improving our understanding of the nature of fire phenomena and how fires develop. As such, it is necessary that we carry out fire prevention and control measures [2]. Firefighters can accurately predict the places, types, and regional distribution of potential fire hazards, and they can focus on the seasons and populations prone to fire disasters [3]. The definition of a fire disaster here is a fire that is burning out of control in a space over time. Fire disasters have become one of the most destructive disasters in modern society due to their high frequency and serious destructiveness [4]. Buildings are major sources of urban fires; thus, fire prevention training programs should be provided, particularly to those in densely populated urban areas [5]. The development of fire safety strategies should be a continuous process such that fire safety systems are regularly reviewed and maintained [6]. The fire safety framework involves the enhancement of fire safety in four key areas: fire protection features in buildings; regulation and enforcement; consumer awareness; technology and resource advancements [7]. Effective fire safety management is a critical task in the planning, design, and operation of a building; furthermore, the occupants/users of a building should be familiar with the escape routes in case of a fire, and maintenance staff must be provided with the relevant information about responsible staff, key locations, and fire safety equipment to ensure that the equipment is in good working order [8]. The failure of all of the alarm and extinguishing systems can accelerate the progress of the fire and hinders people's awareness of the accident and their timely response. Therefore, it is necessary to develop a safe environment that allows more time for people to leave a risky place [9]; the influencing factors on the high-rise building fires are related to people, objects, environment, technology, and management [10]. A study in Nigeria indicated that the most common causes of fire incidents in high-rise buildings are electrical faults [11]. Crowd evacuation in high-rise buildings in case of a fire becomes a major safety issue. In a fire environment, personnel evacuation behavior in high-rise buildings shows complex multi-directional characteristics [12]. Fire load and heat release rate are important considerations during a fire. In order to assess the fire risk of high-rise residential buildings, possible fire scenarios should be identified. There is an urgent need to collect data on the fire load and identify the heat release rate for this type of building [13]. High-rise building fires have many characteristics, such as the diversity of the blazes, factors affecting them, various ways of the fires spreading, and the difficulty of evacuation [14].

The meaning of high-rise buildings (HRBs) varies from one country to another. According to the NFPA, a high-rise building is "A building where the floor of an occupiable story is greater than 75 ft (23 m) above the lowest level of fire department vehicle access". In China, residential buildings with seven stories or more are defined as high-rise buildings. They could be further subdivided into middle-high-rise, high-rise, and super high-rise buildings according to their number of stories and height. The term "high-rise building" in Korea is defined as a reinforced concrete structure with 30 stories or more [15].

In the Emirate of Sharjah, UAE, the definition of a high-rise building is "The occupancies or Multiple and Mixed occupancies, facilities, buildings and structures having total height of the building (excluding roof parapets) is between 23 m to 90 m from the lowest grade or lowest level of Fire Service access into that occupancy", and the definition of a super high-rise building is "The occupancies or Multiple and Mixed occupancies, facilities, buildings and structures having total height of the building (excluding roof parapets) is more than 90 m from the lowest grade or lowest level of Fire Service access into that occupancy" [16].

The Emirate of Sharjah is third in the UAE in terms of the number of high-rise buildings it has, and it is classified as one of the Emirates with rapid urban growth. The UAE ranks in third place in the world in terms of the number of skyscrapers it has, and it is home to 251 buildings that are taller than 150 m. As the safety of high-rise buildings is a global concern, in this study, we review the factors that affect the fire prevention systems of HRBs and super high-rise buildings in the Emirate of Sharjah.

The region of UAE has dry, subtropical weather with year-round sunny days and rare, shallow rainfall. The weather is extremely hot and humid along the shoreline. The summers from the months of June to September are extremely hot and humid, with temperatures reaching 48 °C (118 °F) and the humidity being as high as 80–90% [17]. The Emirate of

Sharjah is considered to be the third largest Emirate in the UAE in terms of area, which covers 2600 km^2; 19% of the UAE population live in Sharjah, and the Emirate is home to people of 200 different nationalities. Moreover, 1.5 million tourists visit the Emirate of Sharjah annually.

The contribution of this paper is to identify the critical factors that affect the current fire management system implemented in high-rise buildings in the Emirate of Sharjah, the first step in correction, and the diagnostic procedure and to determine the area of failure, which are provided in detail in this paper through the fire effect weight listed in the sub-critical factors effecting the fire management in HRBs in the Emirate of Sharjah.

2. Methods

By reviewing the literature related to fire in HRBs, 15 research papers were reviewed and the factors affecting the safety of HRBs from fire risks were monitored, and based on the frequency of factors in the reviewed papers, 20 factors affecting the fire system in HRBs were identified. These factors must have been applicable to the Emirate of Sharjah, and therefore, 16 experts in the field of fire in the Emirate of Sharjah were consulted; their experience spans more than 10 years, and they were identified as stakeholders in Civil Defense, the Prevention and Safety Authority, maintenance and installation companies, and distribution companies. The Failure Mode Effect and Criticality Analysis (FMECA) tool was used to analyze the pattern affecting the identified factors directly related to the Emirate of Sharjah, including fire equipment, fire equipment factories, and fire systems designed for offices. After determining the factors, an analysis was performed for each factor separately to determine the possible failure patterns by calculating their severity, occurrence, and means of detection, and the factors that were evaluated as being very important were considered as critical factors that affect HRBs in the Emirate of Sharjah. The data of critical factors were used to develop a fire factor effect index for high-rise buildings based on the subject matter experts through the application of the Analytical Hierarchy Process (AHP) tools to measure the level of fire sub factor effect on the high-rise buildings, and the method was implemented as shown in Figure 1.

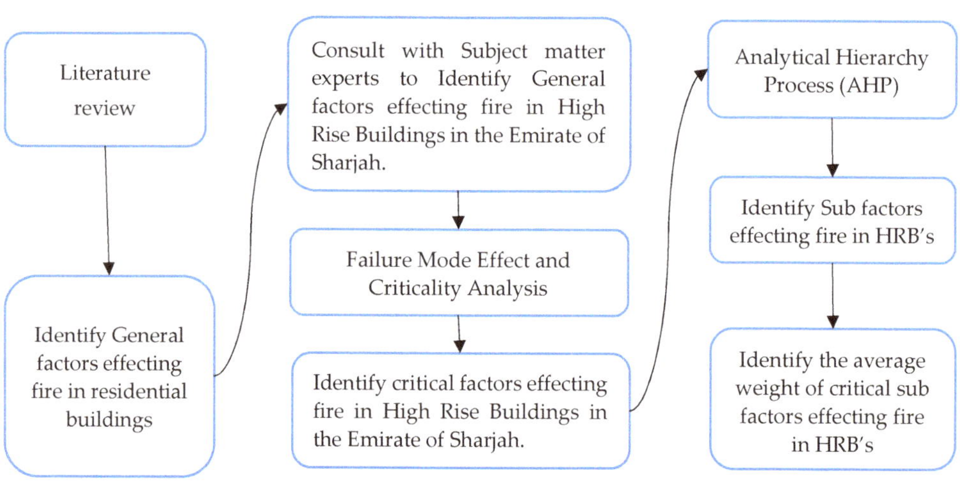

Figure 1. Method of Research.

3. Results

3.1. Fire Factors Effecting Fire Management System in the Residential Buildings

After reviewing research and studies that dealt with the fire factors in residential buildings, and after reviewing 15 journals related to fire as shown in Table 1, 19 factors affecting fire in residential buildings were identified, as shown in Table 2. It is not necessarily the case that all of these factors affect the Emirate of Sharjah. Each region has a different kind of fire, which depends on the weather, laws, materials used in construction, fire culture, and the compliance of residential buildings with fire regulations, so we deemed it necessary to present the factors identified from published research to experts in the Emirate of Sharjah in order to determine the general factors affecting high-rise residential buildings in the Emirate. Specifically, and based on the review results, four main categories of factors were identified: administrative factors, technical factors, human factors, and other factors, as specified in Table 3.

Table 1. Reviewed papers.

[18]	[19]	[20]	[4]	[21]
[7]	[22]	[23]	[24]	[8]
[5]	[25]	[26]	[6]	[27]

Table 2. Factors effecting fire management system in the residential buildings.

No.	Factors	Weight
1.	Building design	5%
2.	Fire regulations	6%
3.	Facilities management and policies	5%
4.	Rescue speed	6%
5.	Fire knowledge	6%
6.	Fire equipment	6%
7.	Human behavior	5%
8.	Firefighting maintenance	5%
9.	Fire culture of society	5%
10.	Fire training	6%
11.	Combustible materials	6%
12.	Fire enforcement regulations	6%
13.	Fire data analysis/availability	4%
14.	Accident investigation	6%
15.	Fire R&D	4%
16.	Fire technology	5%
17.	Public/contractor attitude	5%
18.	Staff assignment	5%
19.	Climate change	4%

Administration factors such as fire regulation, rescue speed, fire regulation enforcement, and accident investigation management have the highest frequency in the literature review, while human factors such as fire training and fire knowledge have the highest frequency, and technical factors such as fire equipment and combustible material counted in are at the top of the list.

The general factors affecting HRB fire management systems that were identified from the literature review were sent to experts in the field of fire protection in the Emirate of Sharjah who are stakeholders in the fire management system such as: government authorities, firefighting installation and maintenance contractors, fire consultant offices, or fire agents and distributers, and the results are shown below. The three factors identified by the subject matter experts are the government structure factors, as well as the urban planning and urbanization factors affecting the efficiency of fire prevention management systems in the Emirate of Sharjah. As shown in Table 3, based on the Delphi technique method, the third round of factors achieving 75% of the subject matter expert consensus are identified as factors affecting the Emirate of Sharjah.

Table 3. General factors effecting the fire management system in the Emirate of Sharjah.

No.	Factors	Weight
1.	Building design	6%
2.	Fire regulations	7%
3.	Facilities management and policies	2%
4.	Rescue speed	7%
5.	Fire knowledge	9%
6.	Fire equipment	2%
7.	Human behavior	5%
8.	Firefighting maintenance	5%
9.	Fire culture of society	8%
10.	Fire training	5%
11.	Combustible materials	9%
12.	Fire enforcement regulations	9%
13.	Accident investigation	9%
14.	Fire technology	2%
15.	Public/contractor attitude	8%
16.	Urbanization	4%
17.	Government structure	3%
18.	Urban planning	1%
19.	Resource allocation	6%

The top factors ranking ones are: fire knowledge, fire enforcement regulations, combustible materials, accident investigation, public/contractor attitude, fire culture of society, fire regulations, and rescue speed, which reflect the general fire factors affecting the high-rise buildings in the Emirate of Sharjah, and the general factors need to be evaluated to identify the critical factors affecting the HRBs fire safety by using the failure mode, effect, and criticality analysis (FMECA).

3.2. Failure Mode, Effect, and Criticality Analysis

To determine the critical factors affecting fires in high-rise residential buildings in the Emirate of Sharjah, FMECA was used. Failure mode, effect, and criticality analysis (FMECA) is one of the most robust and widely implemented engineering risk management tools. To enhance its applicability of addressing the different aspects of engineering problems, FMECA is often integrated with other techniques related to multicriteria decision-making (MCDM) processes [28]. The main factors affecting the HRB fire prevention management systems in the Emirate of Sharjah were examined with the help of subject matter experts, and the possible failures in each factor were identified.

The criteria for dealing with failure are classified in Table 4. FMECA descriptions according to the fault type, the degree, and the number of impacts, and they are assessed in terms of severity, occurrence, and detection. In terms of severity and occurrence, one represents the least impacted one, and ten represents the most impacted one. In terms of detection, one represents a defect being detectable, and ten represents a defect being non-detectable. The criteria were validated by the subject matter experts.

Table 4. FMECA descriptions.

Degree	Number	Severity
low	1	The defect is limited and cannot affect the effectiveness of the fire prevention management system
	2	
	3	
Medium	4	It can cause controllable failure
	5	
	6	
High	7	It can weaken the fire protection system
	8	
	9	
	10	
Degree	Number	Occurrence
low	1	The defect applies to only a few parts of the system
	2	
	3	
Medium	4	The defect applies to 50% or more of the system components
	5	
	6	
High	7	The defect applies to more than 75% of the components of the fire fighting system
	8	
	9	
	10	
Degree	Number	Detectability
High	1	There is a possibility of identifying the defect
	2	
	3	
Medium	4	There is a possibility of us not being able to identify the defect
	5	
	6	
Low	7	There is a high probability of us not being able to identify the defect
	8	
	9	
	10	

The risks involved in a fire prevention system are completely dependent on the defect severity, but severity is not the only influencing factor that determines the critical situation of a failure. The possibility of a fault occurring is an important factor, but the possibility of detecting the fault is the most important and influential factor, as the possibility of detection determines the possibility of controlling the malfunction. In Table 5. FMECA rules, samples of the basis of the risk assessment are given, and the relevant criteria are specified, which involves a combination of impact severity, the possibility of defect occurrence, and the possibility of defect detection. The faults classified by the experts into the medium- and high-severity groups with the possibility of medium- and high-severity occurrences when the possibility of detection is low are treated as very important. Moreover, if the possibility of detection is medium, it is addressed on the basis that it is important. The basis for the risk evaluation was validated by the subject matter experts.

Table 5. FMECA rules.

Severity	Occurrence	Detection	Risk
Medium	Medium	Low	Very Important
Medium	Medium	Medium	Important
High	Medium	Low	Very Important
High	Medium	Medium	Important
High	High	Low	Very Important
High	High	Medium	Important

3.3. Critical Success Factors Affecting the Fire Prevention Management Systems in the Emirate of Sharjah

Based on the assessment of subject matter experts in the field of HRB fire prevention systems in the Emirate of Sharjah, the severity of the impact of a potential defect was evaluated for each of the factors that were previously identified as those affecting HRBs in the Emirate of Sharjah. The probability of occurrence and the possibility of detection were evaluated, and the majority of the experts agreed on the evaluation according to Table 6, in which the types of defects are listed in order of importance, ranging from very important to important to non-important. Accordingly, 91 very important faults, 62 important faults, and 6 non-important faults were identified, as shown in Figure 2.

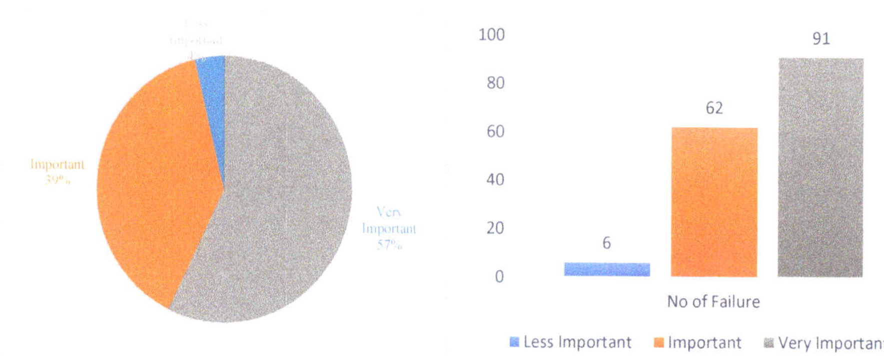

Figure 2. FMECA outcome.

The failure in each factor was identified, as shown in Table 6, and after the analysis according to the sample of the basis for risk evaluation, the degree of importance of each failure was determined based on the risks it contains. Some of the factors for which the failures were evaluated were of a high risk, so they were rated as being very important. These factors were classified as critical in the HRB fire prevention systems in the Emirate of Sharjah.

Based on the number of failures rated as very important in each factor, 12 factors were observed to have a high frequency of very important ratings, as shown in Table 7. Critical factors effecting the fire management in the HRBs in the Emirate of Sharjah. According to the very important failure mode frequency, the critical factors affecting HRB fire management systems have been identified.

Table 6. Failure mode effect criticality analysis.

Factor: Fire Regulations	S	O	D	Risk	Fire Training	S	O	D	Risk
It does not include all types of residential buildings.	M	M	L	V.Imp	Workers not receiving fire fighting training.	H	M	L	V.Imp
Do not support continuous improvement in the fire prevention management system.	M	M	M	Important	Ineffective training.	H	M	L	V.Imp
Do not comply with the applicable procedures.	M	M	H	Not Imp	Training of employees is not mandatory.	H	M	L	V.Imp
Does not support research and development.	H	M	L	V.Imp	There are no mechanisms to achieve mandatory training	H	M	L	V.Imp
The investigation procedures in fire accidents are not required or specified.	H	H	L	V.Imp	Resetting the system and stopping the alarm by the building guard without verifying the fire.	H	H	M	Important
It did not specify the mechanisms and procedures for response and rescue.	H	M	L	V.Imp	Shutdown the system completely by disconnecting the electrical power source in the event of a frequent bell.	H	M	M	Important
It does not support optimum utilization of resources.	M	H	L	V.Imp	Fire pumps are in the off position.	H	M	L	V.Imp
The requirements of the administrative structure did not specify the fire protection system at the level of the authority and stakeholders.	H	M	L	V.Imp	Fire pumps are isolated from power.	H	M	M	Important
It did not specify the procedures required to educate people about the fire.	H	L	L	Not Imp	Diesel pump without fuel.	H	M	L	V.Imp
Do not request reporting of near misses or fire incidents that did not require the intervention of firefighters.	H	H	L	V.Imp	The starting batteries in the diesel pump are disconnected or not working.	H	M	M	Important
There are no procedures in the legislation regulating the investigation of fire accidents.	H	H	L	V.Imp	The water tank is not full according to the design capacity of the fire extinguishing system.	H	M	L	V.Imp
Factor: fire enforcement regulations	**S**	**O**	**D**	**Risk**	Use the pump room as a material store room.	M	M	M	Important
The process of inspection of residential building is not carried out in a regular periodic manner.	H	M	L	V.Imp	Lack of knowledge of dealing with fire alarm panel and other extinguishing equipment.	H	M	L	V.Imp
Audit for licensed entities not carried out periodically.	H	M	L	V.Imp	The use of fire hoses to wash the corridors.	H	M	M	Important
The process of issuing certificates of compliance is not based on precise, specific, and strict criteria.	H	M	L	V.Imp	**Fire Knowledge**	**S**	**O**	**D**	**Risk**
The process of issuing certificates of completion is not based on precise, specific, and strict criteria.	H	M	L	V.Imp	Residents lack knowledge of fire hazards.	H	H	L	V.Imp
Licensing processes for companies are not based on precise, specific, and strict criteria.	H	H	L	V.Imp	Residents lack knowledge of fire behavior and its causes.	H	H	L	V.Imp
Failure to take the necessary measures against the procedures related to residential establishments that violate.	H	M	L	V.Imp	Residents lack knowledge of the procedures required in the event of a fire.	H	H	L	V.Imp
Not to take punitive measures for licensed companies that violate legislation.	H	H	L	V.Imp	Residents lack knowledge of when to use fire equipment.	H	H	L	V.Imp
Failure to check the qualification requirements of employees of licensed companies.	H	H	L	V.Imp	**Fire Society culture**	**S**	**O**	**D**	**Risk**
Lack of focus on obligating high-rise buildings.	H	M	L	V.Imp	The community's religious culture of predestination does not support the possibility of avoiding fire accidents.	M	M	L	V.Imp
Facilities Management and policies	**S**	**O**	**D**	**Risk**	The culture of the community about the causes of fire varies among the residents of the same residential establishment.	M	M	L	V.Imp
Lack of safety management policies or procedures.	H	H	M	Important	The culture of the community does not support taking preventive measures against fire hazards.	M	M	L	V.Imp

Table 6. *Cont.*

Factor: Fire Regulations	S	O	D	Risk	Fire Training	S	O	D	Risk
					The nature of the community in residential establishments				
No fire risk assessment is done.	H	H	L	V.Imp	Awareness of the consequences that can occur if a fire occurs in a residential building with large groups of residents.	M	M	M	Important
There is no record of fire hazards.	H	H	L	V.Imp	Negative culture about fires, which results from religious or ideological beliefs or racial behaviors.	H	M	L	V.Imp
There is no internal inspection or audit.	H	H	M	Important	The culture of the population when hearing the sound of the fire alarm, as it is dealt with on the basis that it is a recurring technical failure, and the response is not performed.	H	H	M	Important
There is no emergency management plan for high-risk building.	H	H	M	Important	**Fire Technology**	**S**	**O**	**D**	**Risk**
There is no qualified employee who supervises the management of the fire system and risk management.	H	H	M	Important	It is not suitable for the nature of residential establishments in the Emirate of Sharjah.	H	M	L	V.Imp
The absence of mandatory requirements for the sustain serviceability of fire protection systems in residential building.	M	M	M	Important	Inefficent.	H	M	L	V.Imp
The owners of residential building consider that the resources provided for the management of the fire protection system are a waste of money and time.	H	M	M	Important	They are not certified based on reliable reliability procedures.	M	M	M	Important
Real estate companies that manage building do not put fire protection systems among their priorities.	H	H	M	Important	Not related to the latest technologies in the field of fire fighting.	M	M	M	Important
HVAC not included in the required preventive maintenance.	H	H	M	Important	Early detection of fires works poorly.	H	M	M	Important
Electrical connections are not included in the required preventive maintenance.	H	H	M	Important	The competent authority does not rely on early fire detection data.	H	H	M	Important
The elevator system is not connected to the alarm and fire fighting system.	M	M	M	Important	The early fire detection system is not approved by the Federal Fire Authority.	M	H	M	Important
Owners of residential building do not care about preventive maintenance.	H	H	L	V.Imp	Early detection of fires does not comply with the requirements and specifications of the Federal Authority.	H	M	M	Important
Real estate companies do not care about preventive maintenance.	H	M	M	Important	The technology used in fire detection is unsuitable and has frequent breakdowns.	H	M	M	Important
Accident investigation	**S**	**O**	**D**	**Risk**	The technology used is not compared to other similar areas that apply good practices.	M	M	M	Important
Accidents are not investigated by the relevant fire authority.	H	H	L	V.Imp	The technology used has not been evaluated and tested.	H	M	M	Important
The investigation of fire accidents is carried out by the Public Prosecution Office for the purposes of providing evidence to the court, for the purpose of compensation procedures related to insurance, or for lawsuits only. The data of the investigation are confidential.	H	M	L	V.Imp	It does not support Fourth Industrial Revolution technologies such as artificial intelligence, big data, robotics, and the Internet of Things.	M	M	M	Important
There is no specialized department in the structure of the competent authority to investigate fire accidents.	H	H	L	V.Imp	**Lack/improper Maintenance**	**S**	**O**	**D**	**Risk**
There are no qualified personnel to investigate fire accidents at the competent authority.	H	H	L	V.Imp	Inefficiency of installation and maintenance contractors.	H	H	M	Important
Fire accident investigation data are not seen as important data that need to be obtained.	H	H	L	V.Imp	Lack of clarity about the mechanism for reporting alarm and fire fighting system malfunctions to the responsible installation and maintenance contractors.	H	H	M	Important

Table 6. *Cont.*

Factor: Fire Regulations	S	O	D	Risk	Fire Training	S	O	D	Risk
Some fire accidents are repeated periodically because fire accidents are not investigated.	H	H	L	V.Imp	The equipment used are of poor reliability.	H	M	L	V.Imp
The root causes of fire accidents are unknown.	H	H	L	V.Imp	Lack of continuous supply of spare parts for devices and equipment.	H	M	M	Important
Fire accident data are confidential and may not be viewed or available for research and scientific studies.	H	H	L	V.Imp	Workers in fire fighting installation and maintenance contractors are not competent.	H	M	L	V.Imp
There is not enough staff to carry out the investigation of fire accidents.	M	M	L	V.Imp	Contracts regulating the relationship between the management of the residential building and the maintenance companies of fire extinguishing systems have defects.	H	M	L	V.Imp
The cause unknown in the fire accident investigations is acceptable to close the investigation.	H	M	L	V.Imp	Competent authority oversight is ineffective.	H	H	M	Important
The prevailing culture is that the task of the competent authority is to fight fires only.	M	M	L	V.Imp	Contractual procedures with residents restrict entry to residential apartments to remove faults.	H	M	L	V.Imp
Contractor Attitude	**S**	**O**	**D**	**Risk**	Manufacturing and design defects of the fire detection system.	H	M	L	V.Imp
Use the cheapest products to make the most profit.	H	M	M	Important	Absence of a maintenance record for the alarm and fire fighting system.	H	H	L	V.Imp
The general view of fire requirements as a governmental requirement, rather than as a means to save lives.	H	M	M	Important	Preventive maintenance history labels for fire equipment can be tampered with.	H	M	M	Important
The lack of adequate control over the implementation of the requirements by the consulting firms.	H	H	M	Important	**Fire equipment**	**S**	**O**	**D**	**Risk**
Rescue speed	**S**	**O**	**D**	**Risk**	Fire extinguishing equipment is not in line with the development of fire hazards.	H	M	M	Important
The type of vehicles and equipment used by the competent authority.	H	M	H	Not Imp	Fire alarm systems in buildings use the Conventional type.	M	M	M	Important
Inadequate training of firefighters.	H	M	M	Important	Fire extinguishers rely on training residents to be able to use them.	H	H	M	Important
Traffic congestion to reach residential areas.	H	H	L	V.Imp	Not compatible with the technologies of the fourth industrial revolution	M	M	L	V.Imp
Geographical distribution of fire stations in the Emirate.	H	M	L	V.Imp	**Building Design**	**S**	**O**	**D**	**Risk**
Distribution of firefighters to fire stations.	M	M	L	V.Imp	Failure to give sufficient priority to fire fighting at the design stage.	H	M	L	V.Imp
Incident-reporting mechanism.	H	M	L	V.Imp	Focusing on the areas of the apartments without taking into account the times and sufficient escape exits.	H	H	L	V.Imp
Fire trucks are not given priority on the road.	H	L	L	Not Imp	Not allocating storage rooms in the apartments, forcing residents to use escape corridors as storages.	H	H	L	V.Imp
Procedures followed during the accident.	H	M	L	V.Imp	Not focusing on the use of fire-insulating materials in the design stages of residential building.	H	H	L	V.Imp
Resource allocation	**S**	**O**	**D**	**Risk**	Escape routes do not correspond to the population of the building.	H	M	L	V.Imp
Unequal distribution of workers in the centers of the competent fire authority.	M	M	L	V.Imp	The pumps of the fire fighting system do not correspond to the height of the residential building.	H	M	M	Important

Table 6. *Cont.*

Factor: Fire Regulations	S	O	D	Risk	Fire Training	S	O	D	Risk
There is no equality between workers in rescue centers and workers in preventive maintenance.	M	M	M	Important	Focus on reducing prices in order to reduce the safety of the population.	H	M	L	V.Imp
The distribution of workers between fire fighting and fire prevention centers is not based on studies, research and scientific methodologies.	M	M	M	Important	Evacuation of residents from the upper floors of high-rise residential towers is not effective.	H	M	L	V.Imp
Lack of workers in the centers of the competent authority.	M	M	M	Important	Failure to take into account the design, evacuating the elderly and then other people.	H	H	L	V.Imp
Economic measures at the level of the Government of Sharjah.	M	M	M	Important	Comparisons with successful and similar experiences in the field of designing residential building.	H	M	L	V.Imp
Establishing a competent local authority that affected the distribution of workers in fire fighting tasks.	M	L	M	Not Imp	Not including a control room specialized in monitoring alarm systems and surveillance cameras in residential building.	H	M	M	Important
Lack of information and comparisons about previous accidents	H	M	M	Important	Smoke detectors are not distributed over the entire area of the apartments.	H	M	M	Important
The prediction of accidents is inaccurate.	M	M	M	Important	Malfunctions in the sprinklers used in the fire extinguishing system.	H	M	L	V.Imp
Government structure	**S**	**O**	**D**	**Risk**	The gas sensor alarm is not connected to the fire alarm system.	H	H	L	V.Imp
The structure of fire fighting at the level of the government of Sharjah, with overlapping roles, responsibilities, and authorities.	M	M	M	Important	Failure to link the status of the fire pumps to the main alarm panel.	H	H	M	Important
The structure of the Sharjah government does not support the flexibility of coordination between government agencies regarding the plan of fire fighting measures.	H	M	L	V.Imp	The absence of an alarm system or fire fighting in the old residential building.	H	M	M	Important
The position of the Sharjah Civil Defense Authority as a local authority and its compliance with federal and local requirements may hinder efficiency and impact.	M	M	M	Important	The absence of a pressure test mechanism in the fire fighting system in the entire residential building.	H	M	M	Important
Fire fighting training which is supervised by another body in the government structure.	M	H	M	Important	**Combustible materials**	**S**	**O**	**D**	**Risk**
The early warning system, which is not directly supervised by the Sharjah Civil Defense.	H	H	L	V.Imp	The use of flammable materials in the exterior cladding of residential building.	H	H	L	V.Imp
The management of the early warning system by a semi-governmental company, which hinders its accountability.	H	H	L	V.Imp	The use of combustible materials in different areas of residential building when carrying out construction.	H	H	L	V.Imp
Absence of a national strategy to combat fire in the Emirate of Sharjah.	H	M	M	Important	The use of flammable materials by residents of residential building.	H	H	L	V.Imp
Urban planning	**S**	**O**	**D**	**Risk**	Flammable materials are not precisely defined and precautions are not taken to reduce their risks.	H	M	L	V.Imp
Overcrowded residential areas do not allow fire fighting vehicles to reach the accident at the required speed.	H	H	L	V.Imp	**Urbanization**	**S**	**O**	**D**	**Risk**
Planning parking lots around residential building hinders the work of emergency and fire fighting teams.	H	H	L	V.Imp	Residents from outside the country are not prepared to deal with the dangers of fire in residential building.	H	H	L	V.Imp
Concentration of high-rise buildings in specific areas.	H	H	M	Important	Immigrants from non-urban areas are causing fires because they are not aware of its dangers.	H	H	L	V.Imp
The closeness of the towers to each other, which threatens the possibility of the transmission of fire from one tower to another.	H	H	M	Important	Lack of knowledge of the correct procedures for dealing at the time of fire for the expatriate population from non-urban areas.	H	H	L	V.Imp

Table 6. *Cont.*

Factor: Fire Regulations	**S**	**O**	**D**	**Risk**	**Fire Training**	**S**	**O**	**D**	**Risk**
The narrow distance between residential buildings and the main road, which increases the risks to the residents in the event of evacuation and hinders emergency operations.	H	H	M	Important	Failure to conduct studies of fire risks resulting from the residence of expatriates from non-urban areas in high-rise buildings.	H	H	L	V.Imp
The lack of planning for fire stations among the public building in the city.	H	M	M	Important	Not specifying the maximum height in residential areas.	H	M	M	Important
Narrow roads, which impede the arrival of ambulances and fire fighting on time.	H	M	M	Important	Accommodation of state and non-urban migrant workers in multi-floored housing building.	H	M	L	V.Imp
Human behavior	**S**	**O**	**D**	**Risk**					
Some religious beliefs.	H	L	L	Not Imp					
Smoking addiction.	H	M	M	Important					
Improper use of electrical appliances and equipment.	H	M	L	V.Imp					
Cooking and grilling.	H	M	L	V.Imp					
Deliberately closing smoke detectors	H	M	M	Important					
Handling of cooking gas.	H	M	L	V.Imp					
Dealing with HVAC equipment.	H	M	M	Important					
Children's behavior	M	M	L	V.Imp					
The use of incense.	H	H	L	V.Imp					

Table 7. Critical factors effecting the fire management in the HRBs in the Emirate of Sharjah.

No.	Factors	Weight
1.	Building design	7%
2.	Fire regulations	8%
3.	Rescue speed	8%
4.	Fire knowledge	11%
5.	Human behavior	6%
6.	Firefighting maintenance	6%
7.	Fire culture of society	9%
8.	Fire training	6%
9.	Combustible materials	11%
10.	Fire enforcement regulations	11%
11.	Accident investigation	11%
12.	Urbanization	9%

The failure mode, effect, and criticality analysis provide a clear and deep evaluation of the etch factor to measure the effect and possible failure mode. It is calculated based on severity, occurrence, and detection. Ten experts in the fire management system in the Emirate of Sharjah who are different stakeholders were involved in the analysis. The data collected and the result will be of added value to the Sharjah Civil Defense Authority and other stakeholders, who will use the analysis as guidance to predict failures and take the necessary preventive action to avoid fire accidents and increase the level of prevention in residential high-rise buildings in the Emirate of Sharjah.

To increase the level of possible protection, the factors classified as important were considered as critical ones, but for the purpose of this paper, only the factors classified as very important were considered to be critical factors affecting the fire management system in the high-rise building in the Emirate of Sharjah.

More analyses were carried out for the critical factors identified by failure mode, effect, and criticality analysis to determine the weight of each sub-factor through the Analytical Hierarchy Process (AHP) and use the outcome as an index to measure the effect of fire factors on the fire management system in high-rise buildings in the Emirate of Sharjah.

According to FMECA tools, the critical factors were identified, as shown in Table 7, and the top critical factors are: combustible materials, fire knowledge, fire enforcement regulations, accident investigation, urbanization, and the fire culture of the society.

The Analytic Hierarchy Process (AHP), since its invention, has been a tool at the hands of decision makers and researchers, and it is one of the most widely used multiple criteria decision-making tools [29] for the general factors affecting the Emirate of Sharjah, according to the outcomes of subject matter experts. They were analyzed using the AHP tool to identify the weight and the priority to measure the HRB's fire effect.

The AHP produces an index for the critical factors affecting high-rise buildings in the Emirate of Sharjah, as shown in Table 8.

Through the Analytical Hierarchy Process (AHP), a further analysis was carried out for the critical factors to identify the sub-critical factors and the weight of each of them to use them to build an index to measure the effect, as shown in Table 8.

The top sub-factors effecting high-rise building in the Emirate of Sharjah are: the building is fully covered with cladding, the effect of stopping activities in HRBs that are non-compliant with the fire regulations, the residents practice activities related to fire knowledge, fire regulations efficiency, the training of new employees by their employers, and the residents has fire-related knowledge, as shown in Table 9.

Table 8. Sub-factors affecting the fire management system in the HRBs.

Factors	Weight	Sub-Factor	Sub-weight	Total Weight
Building design	0.069	Building in design phase	0.209	0.014
		Building in construction phase	0.248	0.017
		Building in use phase	0.226	0.016
		Building in change phase	0.184	0.013
		Building in demolition phase	0.134	0.009
Fire regulations	0.077	Legislation breakdown	0.193	0.015
		Fire regulations scope	0.268	0.021
		Fire regulations efficiency	0.539	0.042
Rescue speed	0.079	Compliance to the fire regulations	0.316	0.025
		Distance from the fire station	0.208	0.016
		Building height	0.214	0.017
		Knowledge, ability, training, and experience	0.262	0.021
Fire knowledge	0.106	Beliefs related to fire knowledge	0.362	0.038
		Practices related to fire knowledge	0.411	0.043
		Philosophies related to fire knowledge	0.228	0.024
Human behavior	0.059	Human behavior: proactive	0.572	0.034
		Human behavior: reactive	0.257	0.015
		Human behavior: neutral	0.171	0.010
Firefighting maintenance	0.058	Firefighting maintenance: training	0.106	0.006
		Firefighting maintenance: resources	0.138	0.008
		Firefighting maintenance: integration	0.135	0.008
		Firefighting system: reactive maintenance	0.174	0.010
		Firefighting system: proactive maintenance	0.257	0.015
		Firefighting system: predictive maintenance	0.190	0.011
Fire culture of society	0.088	Values related to fire culture	0.200	0.017
		Conditions related to fire culture	0.179	0.016
		Procedures related to fire culture	0.282	0.025
		Behaviors related to fire culture	0.339	0.030
Fire training	0.060	Fire training: theory	0.188	0.011
		Fire training: practical	0.459	0.028
		Fire training: methodology	0.353	0.021
Combustible materials	0.106	The building is fully covered with cladding	0.457	0.048
		The building is partial covered with cladding	0.293	0.031
		The building is without cladding	0.250	0.026
Fire enforcement regulations	0.106	Fire enforcement regulations depend on fines for non-compliant facilities	0.270	0.029
		Non-compliant facilities may be prohibited by fire enforcement regulations	0.412	0.043
		Building owners who violate fire regulations are brought to trial	0.318	0.034
Accident investigation	0.106	Report major/minor/near-miss fire accident	0.188	0.020
		Investigation major/minor/near-miss fire accident	0.231	0.024
		Analysis major/minor/near-miss fire accident	0.189	0.020
		Corrective actions to the cause of fire accident	0.218	0.023
		Preventative actions to the cause of fire accident	0.175	0.018
Urbanization	0.088	Awareness of newcomer by the real estate companies	0.329	0.029
		Training of new employees by their employers	0.460	0.040
		Awareness of newcomer with visa procedures	0.210	0.018

Table 9. Sub-critical factors effecting the fire management in the HRBs in the Emirate of Sharjah.

No.	Sub-Factors	Weight
1.	The building is fully covered with cladding	5%
2.	The effect of stopping activities in the HRBs that are non-compliant with fire regulations	4%
3.	Residents practices related to fire knowledge	4%
4.	Fire regulations efficiency	4%
5.	Training of new employees by their employers	4%
6.	Residents believes related to fire knowledge	4%
7.	The effect of proactive resident behavior during the fire accident	3%
8.	The effect of brought to trial HRBs owners violate fire regulations	3%
9.	The building is partially covered with cladding	3%
10.	Resident behaviors related to fire culture	3%
11.	Fire awareness of newcomer by the real estate companies	3%
12.	The effect of fines for non-compliant HRBs	3%
13.	Practical fire training	3%
14.	The effect of the building being without cladding	3%
15.	HRBs in full compliance with the fire regulations	3%
16.	Procedures implemented in the HRB by the residents related to fire culture	2%
17.	Investigation major/minor/near-miss fire accident	2%
18.	Philosophies of residents related to fire knowledge	2%
19.	Corrective actions to the cause of fire accident	2%
20.	Fire training: methodology	2%
21.	Firefighters' knowledge, ability, training and experience	2%
22.	Scope of fire regulations	2%
23.	Analysis of major/minor/near-miss fire accident	2%
24.	Report major/minor/near-miss fire accident	2%
25.	Preventative actions to the cause of fire accident	2%
26.	Awareness of newcomer with visa procedures	2%
27.	Values related to fire culture	2%
28.	Fire arrangements for the building during construction phase	2%
29.	High-rise building height	2%
30.	Building distance from the fire station	2%
31.	Conditions of the HRB effected the fire culture	2%
32.	The fire arranegements for building in use phase	2%
33.	The effect of reactive residents behavior during fire accident	2%
34.	Firefightin system proactive maintenance	1%
35.	Legislation to be breakdown (laws, regulations, standards, and guidelines)	1%
36.	Fire to be considered from building design phase	1%
37.	Fire arranegements for the building in case of change of purpose	1%
38.	Fire training: theory	1%
39.	Firefighting system predictive maintenance	1%
40.	Firefighting system reactive maintenance	1%
41.	Residents behavior to be nutral during fire accident	1%
42.	Fire arrangements for the building in demolition phase	1%
43.	Firefighting maintenance: resources	1%
44.	Firefighting maintenance: integration	1%
45.	Firefighting maintenance: training	1%

4. Discussion

4.1. Fire Fighting Legislation

It is necessary to develop fire fighting legislation that is in line with the development of strategies to identify the causes of fires. The causes of fires vary largely with the development of equipment used in HRB residential buildings. The Sharjah Civil Defense Authority was established as a local authority, and it needs to develop legislation that is compatible with the nature of fires in the Emirate of Sharjah, especially in relation to fires occurring in HRBs. Stakeholders must be updated on the applicable fire legislation to ensure their opinions are well informed in a changing world. The stakeholders' perspective is that the most important form of legislation to be that which effectively provides the maximum degree of protection from fire to real-estate developers, financing agencies, insurance companies, fire companies, and residents.

4.2. Compulsory Fire Legislation

The role of the Sharjah Civil Defense Authority is vital to ensure that all HRBs comply with fire legislation. Periodic supervision and inspection visits are carried out, requiring all HRBs to issue annual certificates of completion and penalizing HRB that have not obtained the annual certificate of compliance or that do not comply with fire fighting legislation. Converting compliance monitoring to digital monitoring contributes to the effectiveness of compliance monitoring by taking advantage of the early warning system, Aman, which links 7000 residential and commercial buildings with a unified control system, according to the statistics published by the Sharjah Prevention and Safety Authority for the year 2022. Using the techniques of the Fourth Industrial Revolution can contribute to increasing the percentage of compliance with fire legislation.

4.3. Management of HRBs

The safety management of HRBs is an important step to ensure the protection of residents from fire risks. Legislation requires the appointment of a fire risk officer in each facility. A fire risk management system helps the owners and the Sharjah Civil Defense Authority to reduce and control fires when they arise. The poor management of fire risks endangers the safety of the residents. Moreover, it contributes to an increase in the number of fire accidents, thus providing guidelines that help real-estate developers and owners to manage fire safety in high-rise residential buildings, which can contribute to improving the efficiency of the applied procedures.

4.4. Fire Research and Development

The Emirate of Sharjah is home to one of the largest universities in the UAE. The cooperation between the Sharjah Civil Defense Authority and the existing universities in the Emirate of Sharjah in the field of research and development contributes to our knowledge of the nature of fires, as well as the development of scientific approaches to fire fighting, benefiting from the latest international studies and research in the field.

4.5. Accident Investigation

The investigation of fire accidents provides important information to understand the nature of fires and their causes. The current investigation process in the Emirate of Sharjah is carried out at the request of the Public Prosecution for the purpose of determining the compensation procedures that will be paid by insurance companies. The Public Prosecution undertakes investigations into fire accidents, and it assigns specialized technical processes to competent authorities in relation to fire protection. To ensure access to the root cause of a fire is obtained, and to ensure that another fire does not arise again as a result of the same cause, the investigation of accidents needs to be included in the organizational structure of the Sharjah Authority for Civil Defense, and relevant employees should conduct accident investigations.

4.6. Contractor Attitude

Simpson describes perception as a way of seeing or understanding, attitude as a way of thinking or behaving, and behavior as a way of acting or functioning [30], and the process of installing fire fighting systems in facilities requires consulting offices to be effective in order to ensure the quality of the materials used and the accuracy and effectiveness of designs. The contractors' lack of awareness of the risks of a system failure during operations may affect the quality of the implementation of fire and alarm systems in the facilities in the Emirate of Sharjah.

4.7. Speed of Response and Rescue

The residents in residential facilities are in need of a response from the Sharjah Civil Defense Authority when their fire system fails to control a fire, the efforts of the staff in the residential facility fail, and the efforts of the residents also fail. In this case, the situation becomes completely out of control. A slow response from the Sharjah Civil Defense increases the human and asset losses, thus, it is essential that the speed of the response to fire disasters is considered to be one of the main performance indicators for the effectiveness of the work of the Sharjah Civil Defense Authority.

4.8. Optimization of Fire Resources

Some European countries, such as the United Kingdom and those in Scandinavia, have worked to implement stronger preventive measures that contribute to reducing fire accidents. The Emirate of Sharjah should similarly focus on preventive measures to reduce the occurrence of fires, such as employing more firefighters to control fires and ensuring the optimal distribution of resources.

4.9. Human Behavior

The fire risk in informal settlements is a function of complex interactions between the built environment, the natural environment, and people [31]. One of the most important factors in the fire fighting process is the behavior of the residents in residential facilities. Negative behaviors cause fires. Some residents have possibly fire-causing habits, such as the use of incense, which may contribute to the occurrence of fires. Negative behaviors need to be addressed via the continuous spread of awareness.

4.10. Fire Training

Training, in general, aims to convey knowledge and situational skills relevant to a specific context to a trainee [32], and training workers in the residential facilities on the dangers of fire contributes to removing the causes of fire, improves the procedures for dealing with fire in the event of an outbreak, and improves the emergency response operations.

4.11. Knowledge of Fire Hazards

Fire-related knowledge, beliefs, and practices that have been developed and applied on specific landscapes for specific purposes by long-time inhabitants [33]. It is necessary to increase the knowledge of the population about fires and their causes. Increasing the population's knowledge of the dangers of fires contributes to a reduction of the possibility of a fire being caused.

4.12. Culture of Society

"Safety culture" is defined as a set of values, conditions, procedures, and behaviors recognized both individually and collectively in the organization that is under consideration, regarding the organization of a management system to prevent and protect against errors, incidents, breakdowns, cyber-attacks, system integration, and accidents, and to promote safety-oriented behaviors between cooperating organizations in normal and emergency situations [34], and the culture surrounding fire in the community needs continuous improvement in order to remove the negative attitude toward the dangers of fire. The process of educating the community is a continuous process that starts with educating school students, workers in the facilities, and the population in order to establish a positive culture that prevents the occurrence of fires.

4.13. Fire Fighting Technology

The technology used in fire fighting determines the effectiveness of carrying out the task of fire fighting. Using the latest and most advanced technology will contribute to building a strong fire fighting culture and reducing the rate of fires in the Emirate of Sharjah.

4.14. Absence or Poor Preventive Maintenance of Fire Fighting Systems

While these newer maintenance strategies require increased commitments to training, resources, and integration, there are three basic types of maintenance programs, including reactive, preventive, and predictive maintenance [35]. Maintenance is necessary to maintain the serviceability of fire fighting systems. The absence or poor quality of preventive maintenance makes the first firewall weak, and the fire fighting system may be unable to deal with fires. Continuously operating fire pumps in an automatic mode and ensuring the serviceability of the bare minimum of the systems, e.g., the water in the water tank, contribute to fighting fires that might break out in a facility. The serviceability of alarms is also important, as they ensure that residents are alerted in the event of a fire, especially in high-rise buildings.

4.15. Fire Fighting Equipment

The presence of complete and advanced equipment assists in fighting and controlling fires. The selection of equipment is an important element in the fire fighting process, as is the identification and testing of advanced specifications and ensuring their suitability to the environment and the nature of the Emirate of Sharjah.

4.16. Residential Building Design

An appropriate fire safety design should ensure occupant safety first when a building fire occurs [36]; the fire fighting process starts from the design stage of the facility using fire-resistant materials in all materials used in the construction process, providing adequate and appropriate escape exits, designing an extinguishing system that covers all parts of the facility, and reducing the possibility of a fire and reducing its effects in the event of its occurrence.

4.17. Flammable Materials

External cladding has been identified as a more critical component in buildings than it has been before due to many catastrophic fire incidents that have occurred in recent decades [36], but the use of external cladding in high-rise building needs more testing processes to improve the materials that are used for it, as the currently used materials are flammable when they are exposed to high temperatures. Flammable materials may be used by residents or during the construction of a residential facility. As such, the monitoring of flammable materials and prevention of their use could contribute to reducing the rate of fires in the Emirate of Sharjah.

5. Conclusions

According to the literature review, 19 factors affecting fire prevention management systems were identified, and they were then further classified into four categories: management factors, human factors, technical factors, and other factors. After consulting experts in the field of fire prevention management systems in the Emirate of Sharjah, 17 factors were identified, and an extra 3 factors related to the Emirate of Sharjah were added: government structure, urban planning, and urbanization. By using the FMECA tools, 12 factors were identified as critical success factors affecting the fire management systems in the Emirate of Sharjah.

The critical factors identified were analyzed by using Analytical Hierarchy Process (AHP) tools to identify the weight and priority of the sub-factors; the outcome is a list of 45 fire sub-factors affecting high-rise buildings in the Emirate of Sharjah that were identified. This is considered to be an index for Sharjah government authorities to increase the level of fire protection in high-rise buildings through the correction of the factors and sub-factors identified in this paper.

Author Contributions: Conceptualization, A.M.; methodology, A.M.; validation, S.B.A.A.; investigation, M.O.; data curation, M.O.; writing—original draft preparation, M.O.; writing—review and editing, S.B.A.A.; visualization, M.O.; supervision, A.M. All authors have read and agreed to the published version of the manuscript.

Funding: This research received no external funding.

Informed Consent Statement: Study did not involve humans.

Data Availability Statement: The data presented in this study are available on reasonable request from the corresponding author.

Conflicts of Interest: The authors declare no conflict of interest.

References

1. Li, S.-Y.; Tao, G.; Zhang, L.-J. Fire Risk Assessment of High-rise Buildings Based on Gray-FAHP Mathematical Model. *Procedia Eng.* **2018**, *211*, 395–402. [CrossRef]
2. Lang, Z.; Liu, H.; Meng, N.; Wang, H.; Wang, H.; Kong, F. Mapping the knowledge domains of research on fire safety—An informetrics analysis. *Tunn. Undergr. Space Technol.* **2020**, *108*, 103676. [CrossRef]
3. Wang, B.; Wang, Y. Big data in safety management: An overview. *Saf. Sci.* **2021**, *143*, 105414. [CrossRef]
4. Wang, K.; Yuan, Y.; Chen, M.; Wang, D. A POIs based method for determining spatial distribution of urban fire risk. *Process. Saf. Environ. Prot.* **2021**, *154*, 447–457. [CrossRef]
5. Zhang, Y.; Shen, L.; Ren, Y.; Wang, J.; Liu, Z.; Yan, H. How fire safety management attended during the urbanization process in China? *J. Clean. Prod.* **2019**, *236*, 117686. [CrossRef]
6. Ebenehi, I.Y.; Mohamed, S.; Sarpin, N.; Masrom, M.A.N.; Zainal, R.; Azmi, M.A.M. The management of building fire safety towards the sustainability of Malaysian public universities. *IOP Conf. Ser. Mater. Sci. Eng.* **2017**, *271*, 012034. [CrossRef]
7. Kodur, V.; Kumar, P.; Rafi, M.M. Fire hazard in buildings: Review, assessment and strategies for improving fire safety. *PSU Res. Rev.* **2019**, *4*, 1–23. [CrossRef]
8. Wang, S.-H.; Wang, W.-C.; Wang, K.-C.; Shih, S.-Y. Applying building information modeling to support fire safety management. *Autom. Constr.* **2015**, *59*, 158–167. [CrossRef]
9. Sun, Q.; Turkan, Y. A BIM-based simulation framework for fire safety management and investigation of the critical factors affecting human evacuation performance. *Adv. Eng. Inform.* **2020**, *44*, 101093. [CrossRef]
10. Haitao, C.; Leilei, L.; Jiuzi, Q. Accident Cause Analysis and Evacuation Countermeasures on the High-Rise Building Fires. *Procedia Eng.* **2012**, *43*, 23–27. [CrossRef]
11. Nimlyat, P.S.; Audu, A.U.; Ola-Adisa, E.O.; Gwatau, D. An evaluation of fire safety measures in high-rise buildings in Nigeria. *Sustain. Cities Soc.* **2017**, *35*, 774–785. [CrossRef]
12. Sun, J.; Hu, L.; Zhang, Y. A review on research of fire dynamics in high-rise buildings. *Theor. Appl. Mech. Lett.* **2013**, *3*, 042001. [CrossRef]
13. Liu, J.; Chow, W. Determination of Fire Load and Heat Release Rate for High-rise Residential Buildings. *Procedia Eng.* **2014**, *84*, 491–497. [CrossRef]
14. Liu, X.; Zhang, H.; Zhu, Q. Factor Analysis of High-Rise Building Fires Reasons and Fire Protection Measures. *Procedia Eng.* **2012**, *45*, 643–648. [CrossRef]
15. Wang, Y.; Mauree, D.; Sun, Q.; Lin, H.; Scartezzini, J.; Wennersten, R. A review of approaches to low-carbon transition of high-rise residential buildings in China. *Renew. Sustain. Energy Rev.* **2020**, *131*, 109990. [CrossRef]
16. Chhabra, A. Thomas Bell-Wright International Consultants The Cladding Problem: Establishing and Assessing Safe Building Envelopes. *Int. J. Conform. Assess.* **2022**, *1*, 9–12. [CrossRef]
17. Salam, A. Climate change: The challenges for public health and environmental effects in UAE. *WIT Trans. Ecol. Environ.* **2015**, *193*, 457–466.
18. Wang, Y.; Ni, X.; Wang, J.; Hu, Z.; Lu, K. A Comprehensive Investigation on the Fire Hazards and Environmental Risks in a Commercial Complex Based on Fault Tree Analysis and the Analytic Hierarchy Process. *Int. J. Environ. Res. Public Health* **2020**, *17*, 7347. [CrossRef]
19. McNamee, M.; Meacham, B.; van Hees, P.; Bisby, L.; Chow, W.; Coppalle, A.; Dobashi, R.; Dlugogorski, B.; Fahy, R.; Fleischmann, C.; et al. IAFSS agenda 2030 for a fire safe world. *Fire Saf. J.* **2019**, *110*, 102889. [CrossRef]

20. Luo, Y.-X.; Li, Q.; Jiang, L.-R.; Zhou, Y.-H. Analysis of Chinese fire statistics during the period 1997–2017. *Fire Saf. J.* **2021**, *125*, 103400. [CrossRef]
21. Rathnayake, R.; Sridarran, P.; Abeynayake, M. Factors contributing to building fire incidents: A review. In Proceedings of the International Conference on Industrial Engineering and Operations Management, Dubai, United Arab Emirates, 10–12 March 2020.
22. Yeung, J.F.; Chan, D.W. Developing a holistic fire risk assessment framework for building construction sites in Hong Kong. *J. Construct. Res.* **2020**, *1*, 43–58. [CrossRef]
23. Gautami, C.M.; Prajapati, M.; Khurana, R. Analysis of Factors Affecting Fire Safety Management of Residential Building: A Case Study. *Int. Res. J. Eng. Technol.* **2020**, *7*, 81–87.
24. Chen, C.; Reniers, G. Chemical industry in China: The current status, safety problems, and pathways for future sustainable development. *Saf. Sci.* **2020**, *128*, 104741. [CrossRef]
25. Rahardjo, H.A.; Prihanton, M. The most critical issues and challenges of fire safety for building sustainability in Jakarta. *J. Build. Eng.* **2020**, *29*, 101133. [CrossRef]
26. Wahed, T.A. Impact of Facility Management on Fire Safety Crisis in Bangladesh's AEC Industry. Master's Thesis, Metropolia University of Applied Science, Helsinki, Finland, 2018.
27. Akhimien, N.; Isiwele, A.J.; Adamolekun, M.O. Fire Safety in Buildings. *J. Civil Environ. Eng.* **2017**, *4*, 63–79.
28. Dabous, S.A.; Ibrahim, F.; Feroz, S.; Alsyouf, I. Integration of failure mode, effects, and criticality analysis with multi-criteria decision-making in engineering applications: Part I–Manufacturing industry. *Eng. Failure Ana.* **2021**, *122*, 105264. [CrossRef]
29. Vaidya, O.S.; Kumar, S. Analytic hierarchy process: An overview of applications. *Eur. J. Oper. Res.* **2006**, *169*, 1–29. [CrossRef]
30. McEwan, S.L.; De Man, A.F.; Simpson-Housley, P. Ego-Identity Achievement and Perception of Risk in Intimacy in Survivors of Stranger and Acquaintance Rape. *Sex Roles* **2002**, *47*, 281–287. [CrossRef]
31. Arce, S.G.; Jeanneret, C.; Gales, J.; Antonellis, D.; Vaiciulyte, S. Human behaviour in informal settlement fires in Costa Rica. *Saf. Sci.* **2021**, *142*, 105384. [CrossRef]
32. Menzemer, L.W.; Ronchi, E.; Karsten, M.M.V.; Gwynne, S.; Frederiksen, J. A scoping review and bibliometric analysis of methods for fire evacuation training in buildings. *Fire Saf. J.* **2023**, *136*, 103742. [CrossRef]
33. Huffman, M.R. The Many Elements of Traditional Fire Knowledge: Synthesis, Classification, and Aids to Cross-cultural Problem Solving in Fire-dependent Systems Around the World. *Ecol. Soc.* **2013**, *18*, 3. [CrossRef]
34. Jiang, W.; Zhou, J.; Su, H.; Wu, Z. The design of experimental courses in safety culture. *Heliyon* **2022**, *8*, e11915. [CrossRef] [PubMed]
35. Swanson, L. Linking maintenance strategies to performance. *Int. J. Prod. Econ.* **2001**, *70*, 237–244. [CrossRef]
36. Thevega, T.; Jayasinghe, J.; Robert, D.; Bandara, C.; Kandare, E.; Setunge, S. Fire compliance of construction materials for building claddings: A critical review. *Constr. Build. Mater.* **2022**, *361*, 129582. [CrossRef]

fire

Article

Development of a Novel Quantitative Risk Assessment Tool for UK Road Tunnels

Razieh Khaksari Haddad [1] and Zambri Harun [2,*]

1 London Bridge Associates Ltd., London SE1 1TY, UK
2 Faculty of Engineering and Built Environment, National University of Malaysia, Bangi 43600, Selangor, Malaysia
* Correspondence: zambri@ukm.edu.my; Tel.: +60-3-8911-8390

Abstract: Some of the most critical transportation infrastructures are road tunnels. Underground passageways for motorists are provided through this cost-effective engineering solution, which allows for high traffic volumes. A crucial aspect of the operation of road tunnels is fire safety. Risk assessments have been established to ensure the level of safety in tunnels. As the existing quantitative risk analysis (QRA) models are inapplicable to assess the fire risk in UK road tunnels, this paper presents a novel QRA model, named LBAQRAMo, for UK road tunnels. This model consists of two main sections: quantitative frequency analysis, to estimate the frequency of fire incidents via an event tree; and quantitative consequences analysis, to model the consequences of fire incidents. LBAQRAMo covers the risk to tunnel users. The result of the risk analysis is the expected value of the societal risk of the investigated tunnel, presented via F/N curve. Another major result of this model is the estimation of the number of fatalities for each scenario based on the comparison between required safe egress time (RSET) and available safe egress time (ASET). Risk evaluation was carried out by comparison of the tunnel under study with the UK ALARP limit. The operation of the model is demonstrated by its application to the Gibraltar Airport Tunnel as a case study. Simulation of 34 different possible scenarios show that the tunnel is safe for use. The sensitivity of the model to HGV fire incident frequency and basic pre-movement times was studied as well.

Keywords: quantitative risk analysis; road tunnel; fire safety; LBAQRAMo; F/N curve

check for updates

Citation: Haddad, R.K.; Harun, Z. Development of a Novel Quantitative Risk Assessment Tool for UK Road Tunnels. *Fire* **2023**, *6*, 65. https://doi.org/10.3390/fire6020065

Academic Editor: Tiago Miguel Ferreira

Received: 19 September 2022
Revised: 6 October 2022
Accepted: 11 October 2022
Published: 10 February 2023
Corrected: 14 April 2023

1. Introduction

High population density is one of the main specifications of urban areas, resulting in high population in cities and steadily growing traffic volumes. Consequently, traffic congestion, and its adverse effects such as noise and air pollution, increase. To reduce traffic congestion, a road network's infrastructure should be built more comprehensively. An influential section of road infrastructure is road tunnels. Road tunnels provide effective engineering solutions and increase traffic capacity and accessibility, thus saving traveling time. They are also a good alternative to travel through physical barriers such as mountains.

Despite the benefits mentioned above, the significant weakness of road tunnels is the severity of the accidents that may occur in them. Accidents which may result in catastrophic consequences. Although the accident rate is lower in road tunnels, those accidents have a greater severity than those on open roads. For instance, a study in Italy has shown that road tunnels have a severe accident rate between 9.13 and 20.45 crashes/108 vehicles per km, while on the associated motorways the rate was between 8.62 and 10.14 crashes/108 vehicles per km [1].

Fire is the most disastrous hazard in road tunnels [2]. It can cause catastrophic consequences, such as human loss and structural damage. The current severity of tunnel fires is indicated by previous fire accidents, such as the Trans-Alpine accidents that occurred in St. Gotthard, Switzerland in 2001; Wuxi Lihu, China in 2010; and Viamala, Switzerland in 2006, accidents which cost the life of 11, 24, and nine people, respectively, and also caused extensive destruction of facilities and significant economic losses that by far exceeded the rehabilitation of the infrastructure [3]. The common causes of tunnel fires are collisions, mechanical or electrical defects in vehicles, and driver behavior [4–7]. Zhang et al. [8] studied the causes of vehicle fire through statistical analysis of defective vehicle recall data caused by fires in China and the United States. Based on their study, 37.24% of the total fires happened because of electrical system defects, 41.03% happened due to defects in the fuel system, and 15.17% from the flammable liquid transportation system.

In addition to fires resulting from a collision, an electrical fault in passenger cars and overheating brakes of heavy goods vehicles (HGVs) on a long downward slope may also result in fires in tunnels [9]. A summarized list of fires in road and rail tunnels is presented in Table 1. A more comprehensive list is given in [10].

Table 1. A summarized list of fires in road tunnels.

Location	Country	Year	Fatalities
Mont-Blanc	France/Italy	1999	39
Gotthard	Switzerland	2001	11
Frejus	France/Italy	2004	-
Gudvanga	Norway	2013	-
Oslofjord	Norway	2017	-
Eiksund	Norway	2005	5

As mentioned, researchers have been inspired to learn more about tunnel fire dynamics because of the increasing number of large-scale incidents in urban tunnels and the catastrophic influences of tunnel fire. Fire flame, smoke, and toxic combustion products travel to the tunnel ceiling and then spread in the left and right directions when a fire happens and threatens the drivers' and firefighters' lives. The structure of the tunnel is built up of concrete with steel bars that lose their strength and structural stability in contact with hot gases. Another consequence of fire is its spread from one vehicle to another. A lot of research about tunnel fire dynamics and the parameters that they influence has been undertaken within the last 20 years. Ref. [11] offers a summary of the main research in that field.

One of the possible protections of tunnel structures in case of fire is by investigating the maximum smoke temperature under the tunnel ceiling so that the building owner understands the limits of the structure. Other than the stability of the tunnel structure, sprinklers or detectors installed in tunnels are activated by the smoke traveling through the ceiling or when the latter detects high temperature. A series of experimental tests and theoretical analyses have been conducted to investigate the maximum smoke temperature beneath the tunnel ceiling [12]. Tunnel fire experiments have been carried out by Haddad et al. [13] to examine the temperature distribution along the tunnel ceiling. This study includes both experimental and theoretical research with the results of the theoretical study compared with experimental data to present an acceptable prediction.

The aforementioned studies considered the maximum smoke temperature regardless of the blockage effect of vehicles inside the tunnel. The obstruction impact of vehicles influences the local velocity around the fire, the smoke flow pattern, and the temperature. Tang et al. [14] proposed a global model including the effect of blockage ratio and blockage fire distance on the maximum smoke temperature. Scaled-down experiments have also been performed to study the effect of blockages on critical velocity, backlayering length [15] and maximum smoke temperature [16]. In [16] the influence of blockage percentages on the maximum smoke temperature beneath the ceiling was investigated and the model of maximum smoke temperature published earlier by Li [12] using local velocity near the fire source was improved.

In recent years, various computer models, such as computational fluid dynamic (CFD) models and zone models, have been used to simulate specific characteristics of tunnel fire, such as its heat release rate (HRR) and smoke and their effects, and the evacuation of people in the tunnel. A numerical simulation of a tunnel equipped with jet fans was carried out by [17] to assess the safe evacuation. Based on their study under controlled ventilation, the evacuation time was prolonged as the flow of smoke could be controlled. The evacuation process in a tunnel under the contraflow condition was studied by [18]. Traffic and passenger conditions are based on real data from a tunnel in the UK. Simulation results indicate that all evacuees can survive before the combustion gases and heat influence their survivability when a realistic worst-case fire scenario is modeled without longitudinal ventilation in the tunnel under study.

The severe consequences of accidents and fire incidents in road tunnels have also highlighted the importance of evaluating the safety level and effectiveness of safety measures to reduce these consequences. Risk analysis is an effective tool to improve and optimize the level of safety in road tunnels. Thus, safety analysts and policymakers have employed risk assessments for more than ten years to study fire risks related to road tunnels [4]. There are two main types of risk analysis: qualitative and quantitative. Quantitative risk analysis (QRA) is divided into deterministic and probabilistic categories.

The deterministic type requires exact input parameters to derive accurate results. On the other hand, the probabilistic method is a better way to consider the uncertainty related with the risk analysis process and assess the long-term risk. The probabilistic method includes the identification of hazards, the estimations of probability and the consequences of each hazard, and quantifies the risk as the sum of probabilities multiplied by consequences. One of the outputs of QRA is societal risk, such as the expected number of fatalities in the tunnel per year, and is presented via the F/N curve, where F is the cumulative frequency that the number of fatalities is equal to or greater than a given number N. The F/N curve is required to be compared with the area between tolerable and intolerable thresholds, otherwise known as 'As Low as Reasonably Practicable' (so-called ALARP).

QRA has received more attention in past years, and has been demonstrated as an effective and efficient method for quantitatively evaluating the risks of tunnels in many countries such as the TuRisMo model of Austria, the TUNPRIM model of the Netherlands, the Italian risk analysis model, the OECD/PIARC model [19], and the quantitative risk assessment for road tunnels (QRAFT) model of Singapore [2,20,21]. PIARC 1995 sponsored a project on the transport of dangerous goods through road tunnels and the PIARC/OECD/EU QRA model (QRAM) was developed. This model has 13 hazardous scenarios and was computerized by spreadsheet-based software [22]. The validation of the QRAM model was studied in Austria, France, Netherlands, Norway, Sweden, and Switzerland and various risk reduction measures were examined employing the QRAM software [23].

All these QRA models choose societal risk as the risk indice to evaluate the safety level of a road tunnel. The risk analysis of a road tunnel is determined by different input parameters such as traffic flow, tunnel geometries, tunnel safety provisions, etc. It has been generally acknowledged that uncertainty is an unavoidable component in risk analysis.

A novel methodology to perform a quantitative fire risk assessment of road tunnels is explained in this paper. This model has two main sections: quantitative consequence analysis and quantitative frequency analysis. The quantitative consequence analysis section of this model was inspired by [24]. Based on the results of this risk assessment, safety management strategies and safe evacuation policies can be developed.

This model was applied to the Gibraltar Airport Tunnel to show how it is operated and how the fire risk is analyzed via the application of this model. Temperature and toxic fume concentration were measured at human height to estimate the fatality rate of each fire scenario.

2. LBAQRAMo—Tunnel Risk Analysis Model for UK Road Tunnels

At a statistical level, tunnels are marked by a lower accident rate than the open-air sections because the user is more careful when driving and the weather and visibility conditions are constant. Although the possible consequences of a relevant event in a tunnel can be more significant than open-air events, especially if fatalities and damage to the infrastructure are considered. The recent history of tragedies in tunnels has resulted in the topic of tunnel safety being the center of attention and more resources are now being spent to improve the safety levels in tunnels. Whilst much has been done to improve tunnel safety since the early 2000s, there are still improvements to be made to harmonize and optimize the approach to tunnel safety. From the literature survey, it can be concluded that there is no generic QRA model for UK road tunnels and the lack of a QRA model has been felt. Therefore, it was decided to develop the most appropriate quantitative risk analysis model for safety in UK road tunnels, leading to improved and consistent decisions on the grounds of safety in the design and operation of road tunnels, the LBA quantitative risk analysis model (LBAQRAMo).

This article aims to propose a new QRA model for UK road tunnels by considering the unique characteristics of the tunnel under study. This type of analysis is usually the most detailed method and therefore requires more time and resources. It should be kept in mind that, while quantitative analysis is more objective, it is still an estimate.

Subsidiary risk assessments shall be made to determine the probability and likely impact of hazards with a range of potential impacts [25]. Risks must be evaluated considering the tunnel geometry and its infrastructures, fire safety measures, as well as equipment and management procedures. In addition, they require the analysis of many complex factors and processes related to human behavior, such as pre-evacuation times, e.g., reluctance to leave the vehicle, interactions between occupants, interactions between occupants and smoke, etc. In this model, the risk assessment analysis is divided into quantitative consequence analysis and quantitative frequency analysis.

The quantitative consequence analysis model comprises three parts, a queue formation model which estimates the number of potential tunnel users, the distribution model which calculates the evacuation distance, and the egress model which calculates the required egress time.

In each scenario, a set of variables, including the influence of the different combinations of fire safety equipment, pre-movement time, movement speed, fire source location, type and number of vehicles involved, and tunnel characteristics, including its dimensions and lanes are considered. The possibility for safe egress is estimated by 3D fluid dynamics simulation via measuring tenability thresholds through the evacuation path.

The quantitative frequency analysis is based on an event tree analysis to calculate the frequency of a specific fire incident, including UK road tunnels' basic fire rate, the time of fire incidents, traffic conditions, accident type, vehicle type, and fire source locations.

F/N curves of societal risk are provided with the results of quantitative consequence analysis and quantitative frequency analysis. Risk acceptance is obtained using the ALARP criterion in the UK.

Five different vehicle types were considered: passenger car, small van or 2–3 cars, bus or empty HGV, a truck with combustible load, and HGV. Five accident locations inside the tunnel were investigated.

Two different ventilation regimes can be selected: natural and longitudinal. The design of the model allows a detailed investigation of the performance of the ventilation system regarding the number of activated jet fans and the time of activation.

This model was developed with the aid of Microsoft Excel. This model enables the safety analyst to create many scenarios, predict potential losses among trapped users in the event of a fire, and illustrate the effectiveness of fire safety systems.

Although this model provides a depiction of the current level of knowledge, it needs to update the relevant information as necessary. This information requires continuous updates with new data from new academic studies, and accidents and fire incidents data that may change the frequency.

2.1. Quantitative Consequence Analysis

For each scenario, the number of fatalities is calculated through the quantitative consequence analysis. Estimation of the number of fatalities due to accidents involving fire includes three main sections:

- Queue formation model
- Distribution model
- Egress model

The key input parameters required for the LBAQRAMo are of three main categories: traffic parameters, tunnel safety systems, and tunnel characteristics. Tunnel characteristics include its length, width, height, the number of lanes, the number of emergency exits and the distance between them, and gradient. Traffic parameters include traffic volume and vehicle composition.

Functional parameters of a tunnel's safety systems, including their specific characteristics and their reliabilities, are in its safety systems category.

Tunnel safety systems are divided into three categories in this model, primary, intermediate, and secondary. Tunnel safety systems that the activation of intermediate and secondary measures rely on are primary measures. Intermediate measures are a connection between the primary measures and the secondary measures, such as the tunnel control center. Secondary measures, such as the suppression system, are those whose applications depend on the intermediate and primary systems.

2.1.1. Queue Formation Model

The main output of the queue formation model of the LBAQRAMo is the estimation of the number of vehicles queueing in each lane. Traffic is divided into three categories; free flow, congested, and stoppage. Since the average peak traffic density, percentage, and type of vehicles are not uniform between various lanes in a multi-lane tunnel, in the first stage, the total traffic flow of the tunnel under study is divided between the left and right lanes in two-lane tunnels and between shoulder, middle, and median lanes in three-lane tunnels. Saad Yousif and his colleagues [26] studied the distribution of traffic flow among the available number of lanes and modelled lane utilization. They studied a relatively large dataset which was more representative of current lane utilization on United Kingdom motorways. Therefore, their model was utilized in this model in the absence of tunnel-related information to estimate the lane utilization factor for each lane based on the total traffic flow of vehicles (veh/h) in two-lane and three-lane tunnels. The composition of different types of vehicles must also be determined at this stage. Then, the closure time of the tunnel (t_{cl}) was calculated based on the basic value of the closure time, which is 120 s in this model, and all fire safety measures influencing it.

In this model, the stopping distance between vehicles in the queue can be estimated based on the type of traffic [27]. If it is assumed that the average length of light vehicles (passenger cars and vans) is 5 m and 15 m for large vehicles, the stopping distance is calculated as below:

$$d_{stop} = [1000 - D \times (\%SV \times 5 + \%LV \times 15)]/D, \quad (1)$$

where D is the average peak traffic density (veh/km), %SV is the average of %PC and %VAN, and %LV is the average of %BUS, %TRUCK, and %HGV for each lane. Considering the stopping distance between vehicles, and the percentage and the average length of each type of vehicle in the traffic, the density of stopped vehicles was estimated. The density of stopped vehicles, D_q, traffic flow, Q (veh/h), and traffic density, D (veh/km) were used to calculate queue formation speed for congested and stoppage traffic as below for each lane:

$$u_q = \frac{-Q}{D_q - D} \quad (2)$$

The queue formation speed for the free flow traffic is equal to the traffic speed. The saturation time of each lane depends on the fire source location and queue formation speed. By comparing the closure time and the saturation time (for each lane in a multi-lane tunnel), the queue length, L_q is calculated. If the saturation time is smaller than the closure time, L_q is the same as the saturation length. otherwise:

$$L_q = u_q \times t_{cl} \quad (3)$$

The number of vehicles in each lane is determined as follows:

$$N_{veh} = L_q \times D_q \quad (4)$$

The total number of occupants in a specified tunnel is the sum of the individuals who are present in various lanes because traffic composition is different from one lane to the other. The composition of different types of vehicles in under-studied lanes and the average occupation coefficient of each specific type of vehicle must be known to estimate the number of occupants in the i-th lane. Figure 1 shows the process of calculating the number of people in the queue.

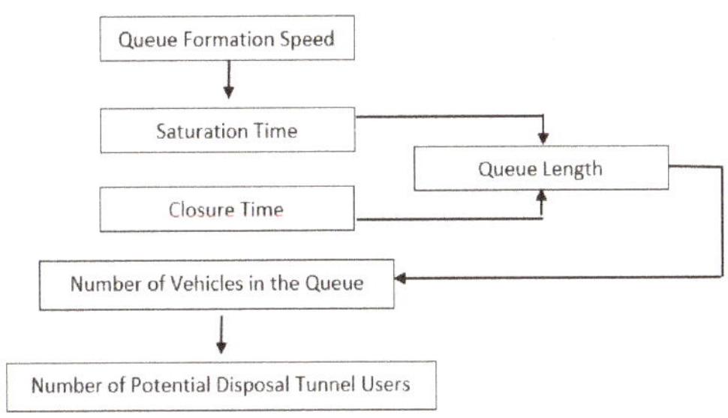

Figure 1. The process sequence of calculating the number of people in the queue.

2.1.2. Distribution Model

After calculating the number of potentially exposed people, the tunnel users were distributed into the queue via the distribution model. It was assumed that the users are distributed homogeneously along the queue. The queue length was divided into sections, named cells. All cells have the same length, and they extend to the end of the queue. Figure 2 shows the schematic view of the distribution of users along the lane.

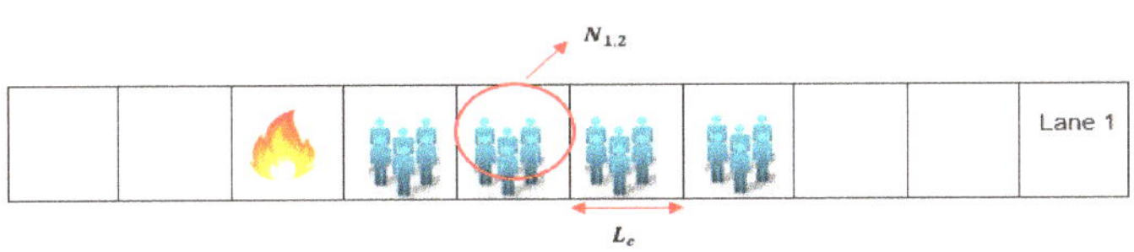

Figure 2. Distribution of users in cells adopted from [24].

The evacuation starts from the evacuees' initial position, and they travel cell by cell towards a place of safety i.e., an emergency exit. A two-dimensional evacuation simulation was conducted, as trapped users' evacuation route is predominantly limited to two dimensions. The total distance of the evacuation path is calculated by considering both the horizontal travel distance and transversal travel distance:

$$DE_{(i,j)} = L_{x,i} + L_{y,i}, \quad (5)$$

where:

$L_{x,i}$ is the horizontal distance related to the center of c-th cell of the *i*-th lane
$L_{y,i}$ is the vertical distance related to the *c*-th cell center of the *i*-th lane.

The horizontal travel distance depends on the distance between each group of users (each cell) and the nearest emergency exit to them. The vertical travel distance depends on the tunnel width and the lane which is studied. Figure 3 shows an illustration of the $DE_{(i,j)}$ from the centre of the first cell to the closest emergency exit, where the longitudinal $L_{x,i}$ and lateral $L_{y,i}$ contributions are shown.

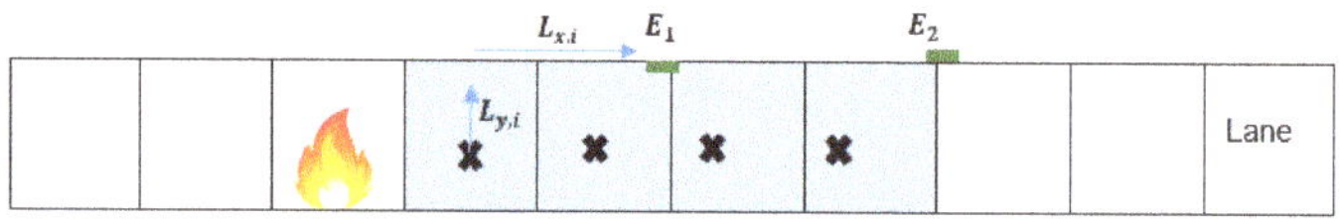

Figure 3. $C_{1,1}$ escape path based on the initial position adopted from [24].

2.1.3. Egress Model

A timeline model is used for the egress model, which describes the sequence of events as a list of continuous phases. A four-stage evacuation process was considered. The first stage is detection which depends on the safety equipment in the tunnel. The second stage is the alarm stage which is the time between detection and the time when the alarm system is activated. The third stage is the pre-movement stage including recognition time, response time, and the time to exit the vehicle. The last stage is traveling which depends on the movement speed and the distance to the emergency exit. Figure 4 illustrates the stages of the evacuation process. The total time of these stages defines the required safe escape time (RSET).

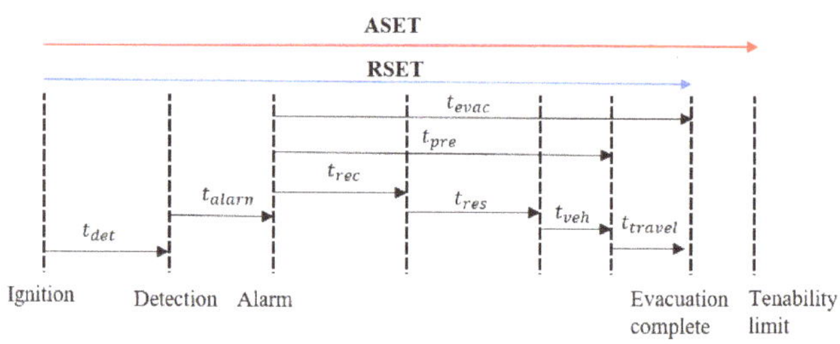

Figure 4. Four-stage evacuation process.

According to this description, RSET can be calculated through Equation (6) as below:

$$RSET = t_{det} + t_{alarm} + t_{pre} + t_{travel}, \quad (6)$$

where:

t_{det}: Detection time, time between when the fire starts and when the tunnel safety equipment detects the fire

t_{alarm}: Alarm time, the time between the first detection and the general alarm

t_{pre}: Pre-evacuation time, the time consists of three behavioural phases of each user, namely recognition time (t_{rec}), response time (t_{res}) and time to leave the vehicle (t_{veh}).

Detection time is different for each tunnel, and it depends on the type of detection system installed in the tunnel. The way of calculating alarm time includes a basic alarm time, 90 sec in this model, and safety measures influence it via their connection between them and their reliabilities.

Recognition time consists of a period between the activation of the alarm system and when occupants recognize the danger. The process of calculating the recognition time starts with calculating the basic recognition time, which is influenced by the distance users have from the accident and the severity of the fire. As a result, the basic recognition time will decrease for more severe accidents and consequently, people become aware of the threatening situation within a shorter time. Moreover, the basic recognition time depends on the distance occupants have from the accident as the closer you are to the accident, the sooner you will notice the accident. Considering the distance to the fire location, the area in the vicinity of the accident was divided into three zones first. Zone 1 is from where the accident happened to 150 m down the queue. Zone 2 spans from 150 m to 300 m down the queue and Zone 3 extends from 300 m to the tunnel entrance. Table 2 represents the recognition time allocated to each zone [28].

Table 2. Basic recognition time for various zones.

Zone No.	Basic Recognition Time (Min)
1	2
2	3
3	4

Regarding the severity of the fire, Table 3 shows the assumed influence of the fire severity on the recognition time as a function of fire heat release rate. The final recognition time is calculated by considering the influence of the fire safety systems.

Table 3. Basic recognition time reduction for various HRR range.

HRR (MW)	Reduction %
0–30	5
30–50	10
50–100	15

Once evacuees recognize the fire, they tend to collect their belongings and gather their family members and they may also be reluctant to leave the perceived safety of their vehicles. The variation of basic recognition time is opposite to that of the basic response time as those who are actively evacuating influence those still in their car. Table 4 presents the basic response time allocated to each zone [27].

Table 4. Basic response time for various zones.

Zone No.	Response Time (Min)
1	4
2	3
3	2

The final response time is also influenced by the available fire safety measures.

We assume that the flow capacity of a normal vehicle door is 1 person per 4 s [29]. Based on the average occupancy of each type of vehicle, the number of each type of vehicle in the queue, and the number of cells the average time of leaving the vehicle is calculated.

This model uses numerical calculation via Fire Dynamic Simulator (FDS) and the extinction coefficient is measured at each cell and at each time step. FDS + Evac [30], based on experiments conducted by Frantzich and Nilsson [31], uses Equation (7) to calculate the walking speed in smoke.

The travel time was calculated cell by cell considering the walking speed in the smoke and the evacuation distance. In other words, every time one evacuee travels one cell, his/her walking speed is calculated and then the travelling time is calculated (Equation (8)).

$$v_{s,i} = \mathrm{Max}\left\{v_{i,min}, v_{i,ave}\left(1 + \frac{\beta}{\alpha}K\right)\right\} \tag{7}$$

$$t_{travel,i} = \frac{L_C}{v_{s,i}}, \tag{8}$$

where $v_{s,i}$ is the moving speed in the smoke at a certain distance, the minimum walking speed is $v_{i,min} = 0.1 \times v_{i,ave}$, and $v_{i,ave}$ is average walking speed. The experimental parameters α and β are 0.069 and 0.015, respectively.

When the total time of evacuation, which is the sum of detection time, alarm time, pre-movement time, and movement time, is calculated, the survival of tunnel users should be evaluated to find out if evacuees can start and then continue the evacuation process. The time when untenable conditions rise is compared with the evacuation time (comparison between ASET and RSET). Users can survive if RSET < ASET; in the opposite case, the users are assumed dead.

The first evaluation is the comparison between the summation of detection time, alarm time, and pre-movement time in each cell and the time when the tenability thresholds exceed their limits for each cell placed in the i-th lane. If users start to evacuate very late (the summation of detection time, alarm time, and pre-movement time are more than ASET at each cell), they are therefore dead before they start the traveling phase.

The next step is the verification of users' egress process if evacuees can start the evacuation process. For users in each cell, the time taken to reach the next cell was contrasted with the development time of the impacts of each accidental fire. If they can reach the next cell before the time when untenable conditions arise, they will continue their traveling to another cell and the comparison is repeated. This process continues until evacuees reach the closest emergency exit.

Three tenability thresholds were measured at human height at the center of each cell to assess whether occupants can travel the evacuation path and reach a place of safety.

The tenability criteria used for this model were:

Fractional irritant concentration (FIC) < 1,

Fractional effective dose (FED) < 1,

Gas temperature < 60 °C

The extent of the damage of fire, the above criteria, was estimated with the support of computational fluid dynamics (CFD) simulation at each cell and at each time step. FDS, which solves the Navier–Stokes equations for low Mach numbers and performs the heat transfer and smoke propagation, was used to simulate tunnel fire scenarios. The tenability thresholds were measured at human height by FDS.

While at one stage of this process occupants reach a certain barycenter when conditions are untenable, users die, and they cannot continue the evacuation.

The total number of casualties is defined as the sum of fatalities of each cell where the tenability thresholds are exceeded (RSET > ASET). The total number of causalities of the whole tunnel is determined by the sum of fatalities in each lane.

However, the effect of fire incidents on emergency services (who have been identified as a sub-population of road users) will be greater compared to other road users due to their role in the rescue and recovery of persons from close to the scene of a fire. This is managed by their own safe systems of work and training. Figure 5 illustrates the process of the quantitative consequence analysis model.

2.2. *Quantitative Frequency Analysis*

The frequency of defined accident scenarios was calculated via an event tree. The first column of the frequency event tree is initial fire frequency which has been obtained from the historical statics of fire incidents in UK road tunnels. Then, the influence of tunnel length, location, traffic volume, and gradient on the basic fire rate was considered to derive the fire rate for a specific tunnel. The second column of the frequency event tree is "The Time of the Incidents". Since the fire rate is different in day and night, the influence of time on accidents should be considered in the frequency analysis. The data from UK roads have been used for the effect of time as the data from UK road tunnels were not ready at the time of drafting this paper. The third column of the frequency event tree is "Traffic Condition". As the fire rate is varied by traffic condition, the effect of traffic condition on fire rate was considered by considering the congested hours of the tunnel under study. Two types of fire incidents were considered in this model. Incidents that include 1 vehicle are Type 1 and collisions that include more than 1 vehicle are Type 2. The probability of type 1 and 2 incidents are derived from UK road data.

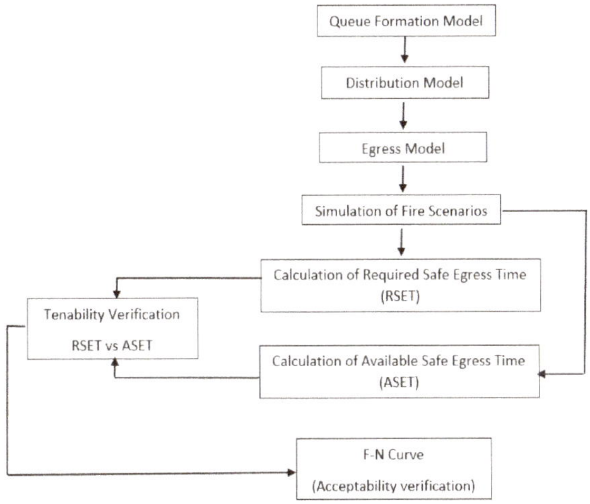

Figure 5. The process of quantitative consequence analysis model adopted from [24].

The fifth column of the frequency event tree is "Vehicle Type". According to tunnel fire data in PIARC 1999, the share of passenger cars, buses, vans, HGV, and trucks involved in the Type 1/Type 2 fire incidents listed in Table 5 were used to consider different vehicle types.

Table 5. Distribution of fire rates based on vehicle type.

Vehicle Type	Distribution
PC	13%
HGV & Truck	67%
Bus	17%
Van	3%

The last column of the frequency event tree is the Fire Source Location. To investigate the influence of various tunnel zones on the fire rate, the tunnel is divided into three zones. A typical sketch of the tunnel zones is presented in Figure 6.

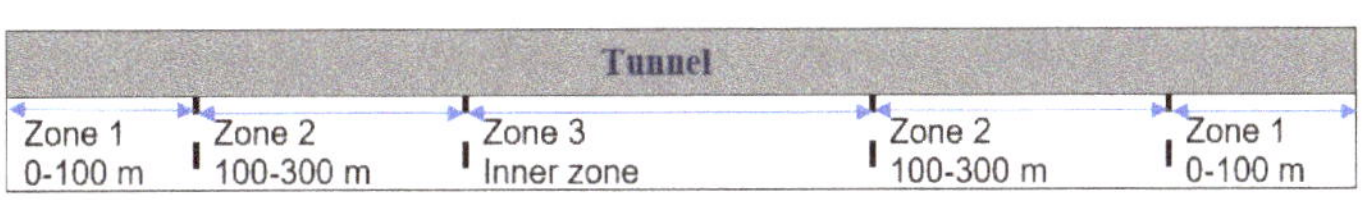

Figure 6. Tunnel zones for fire rate distribution.

The crash rates in Zone 1 are higher than in Zone 2 and Zone 3 because of the sudden change in the visual environment, and when drivers pass Zone 1 and Zone 2, they are more experienced and careful in their driving [32,33].

Based on a comparison of crash rate distribution along the tunnel zones (Zones 1–3), an average crash rate value was considered to quantitatively understand the safety level of each tunnel zone (Table 6).

Table 6. Comparison of tunnel crash rate in different zones.

Zone 1	Zone 2	Zone 3
0.23	0.2	0.15

The final fire frequency was calculated by multiplying the initial fire frequency by the below columns:

- Time of fire incidents
- Traffic condition
- Accident type
- Vehicle type
- Fire source location

The level of detail for an event tree is defined in such a way that the available data material can be used appropriately. The illustration is shown in Figure 7.

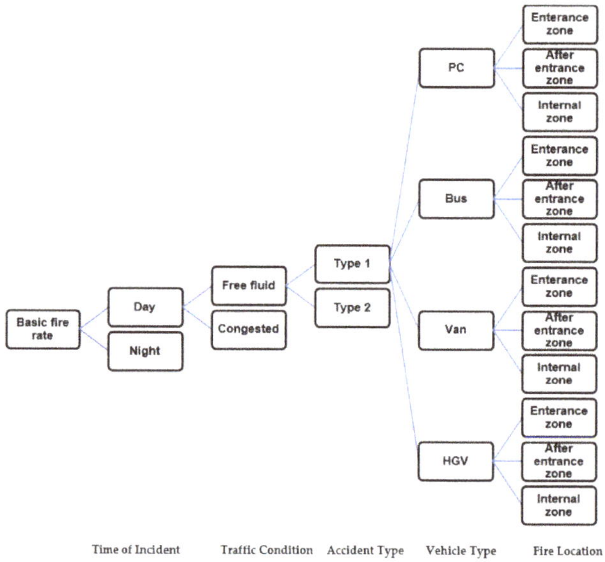

Figure 7. Illustration of a section of the frequency analysis event tree.

3. Case Study

UK road tunnels possess some characteristics which make them different from others in European countries. Therefore, in this paper, a QRA software named LBAQRAMo, which complies with the UK regulation and captures UK road tunnels' differences, is elaborated. The application of this model has been evaluated by applying the proposed model and assessing the risk of fire to tunnel users in the under-construction Gibraltar Airport Tunnel. This north to south tunnel is located along an urban road serving a seaside city. Therefore, it is affected by high traffic flow, especially in summer. The transportation of dangerous goods is not forbidden.

The Gibraltar Airport Tunnel is a rectangular uni-directional road tunnel with two bores. It is a 360 m length, 7.3 m width, and 5.3 m height tunnel. The annual average daily traffic density of this tunnel is 20,700 vehicles/day with an average percentage of 8% HGVs. There are three emergency exits with about 100 m distance between them.

The major safety systems include a smoke/fire detection system, traffic monitoring, and vehicle detection system, ventilation system, emergency panels, and a communication system.

The longitudinal ventilation system is constituted of four pairs of jet fans fixed on the tunnel ceiling. In the event of a fire incident, all fans are activated to remove and control smoke and toxic gases generated by fire. The west-east wind with an average hourly speed of 18.8 kilometers per hour is considered.

The quantitative fire risk analysis was performed by the methodology explained in Section 2.

The proposed QRA model requires determination of the fire scenarios to be examined and of the tunnel characteristics. In this case study, we estimate the societal risk for tunnel users. Different scenarios have been studied by changing the traffic flow, HRR, fire source location, and velocity and activation time of the ventilation system. Table 7 illustrates the list of scenarios used in this case study.

Table 7. List of scenarios in this study.

No.	HRR	Ventilation Velocity	Ventilation Activation Time	Traffic	Fire Location (%L)	Equipment Combination
1	30 MW	0	4	Free	0.5	1
2	30 MW	0	4	Free	0.85	1
3	158 MW	0	4	Free	0.5	1
4	158 MW	0	4	Free	0.85	1
5	30 MW	0	4	Stopped	0.5	1
6	30 MW	0	4	Stopped	0.85	1
7	158 MW	0	4	Stopped	0.5	1
8	158 MW	0	4	Stopped	0.85	1
9	30 MW	0	4	Congested	0.5	1
10	30 MW	0	4	Congested	0.85	1
11	158 MW	0	4	Congested	0.5	1
12	158 MW	0	4	Congested	0.85	1
13	30 MW	3	4	Free	0.5	1
14	30 MW	3	4	Free	0.85	1
15	158 MW	3	4	Free	0.5	1
16	158 MW	3	4	Free	0.85	1
17	30 MW	3	4	Stopped	0.5	1
18	30 MW	3	4	Stopped	0.85	1
19	158 MW	3	4	Stopped	0.5	1
20	158 MW	3	4	Stopped	0.85	1
21	30 MW	3	4	Congested	0.5	1
22	30 MW	3	4	Congested	0.85	1
23	158 MW	3	4	Congested	0.5	1
24	158 MW	3	4	Congested	0.85	1
25	158 MW	3.5	2	Stopped	0.5	1
26	158 MW	3.5	2	Stopped	0.85	1
27	158 MW	3.5	2	Congested	0.5	1
28	158 MW	3.5	2	Congested	0.85	1
29	158 MW	3.5	2	Free	0.5	1
30	158 MW	3.5	2	Free	0.85	1
31	30 MW	2	4	Congested	0.5	2
32	158 MW	2	4	Congested	0.5	2
33	30 MW	2	4	Congested	0.85	2
34	158 MW	2	4	Congested	0.85	2

After applying the characteristics of the tunnel under study and its traffic conditions, numerical simulation of fire scenarios by FDS was used to measure tenability conditions, temperature, FED, and FID, and visibility through the evacuation route. Based on RSET and

ASET calculations, the number of fatalities for each scenario was estimated. The frequency of each scenario was derived by the frequency event tree of this QRA tool.

Figure 8 shows the 3D simulation of fire spread for a fire scenario with 158 MW fire size, 3.5 m/s ventilation velocity, and stoppage traffic. Figure 9 depicts the F/N curve of scenarios with 0 m/s ventilation velocity. The F/N curve was evaluated by comparing it with the safety target set by the UK ALARP region. Figure 9 shows that the F/N curve can meet the safety target. Although the F/N curve locates in the acceptable region, a further analysis was carried out to study the effects of certain safety systems on the F/N curve.

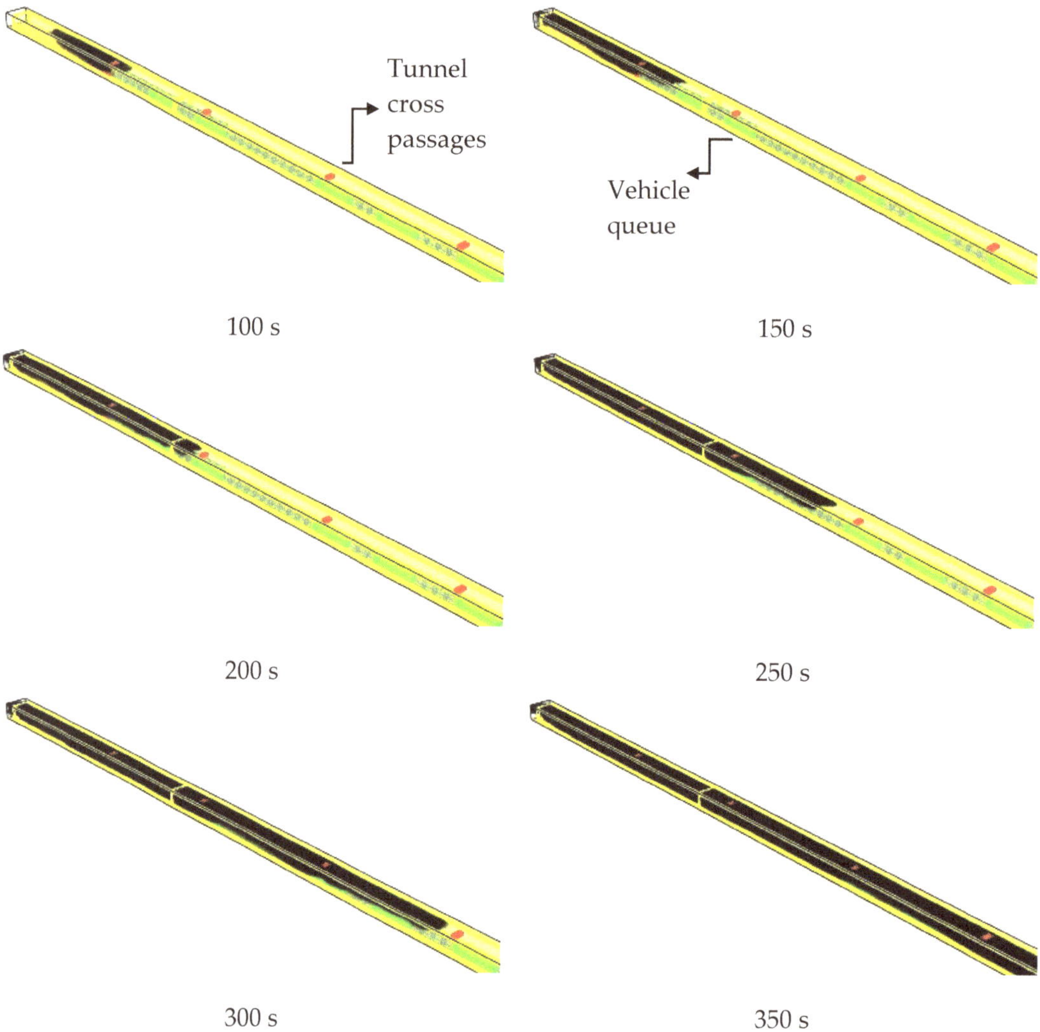

Figure 8. Three-dimensional simulation of fire spread for a fire scenario with 158 MW fire size, 3.5 m/s ventilation velocity, and stoppage traffic.

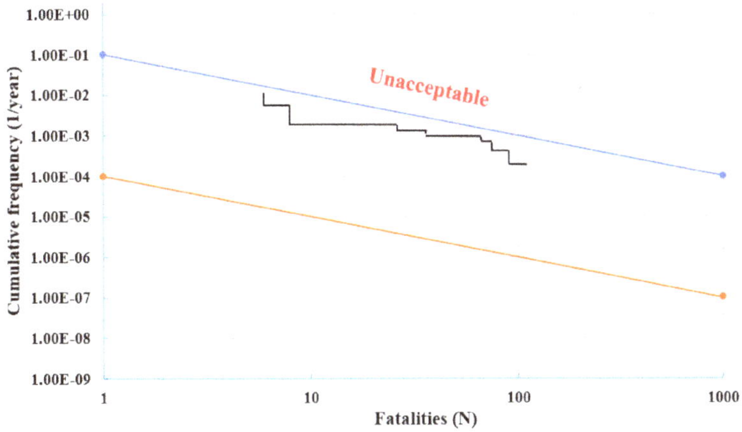

Figure 9. F/N curve for scenarios with 0 m/s ventilation velocity.

Figure 10 shows the F/N curve for ventilation velocities 0 m/s and 3 m/s for bus fire scenarios (a) and 0 m/s, 3 m/s, and 3.5 m/s for HGV fire scenarios (b). In these 34 scenarios, the tunnel can be considered safe. Although the F/N curve of 3 m/s scenarios is lower than 0 m/s scenarios for bus fire scenarios, which indicates the positive effect of the ventilation system, there is no significant difference between 0 m/s, 3 m/s, and 3.5 m/s HGV fire scenarios.

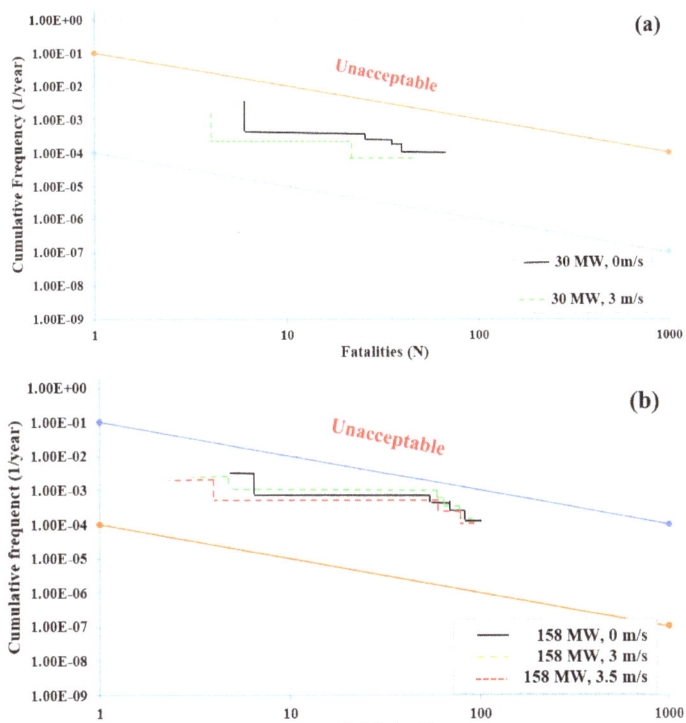

Figure 10. F/N curve for ventilation velocities 0 m/s and 3 m/s for bus fire scenarios (**a**) and 0 m/s, 3 m/s, and 3.5 m/s for HGV fire scenarios (**b**).

3.1. Sensitivity Analysis

3.1.1. Reference Recognition and Response Times, and Time Taken to Exit the Vehicle

Sensitivity analysis was carried out for 30 MW and 158 MW fire scenarios where there is no activate jet fans. Recognition time, response time, and time to leave the vehicle were determined pre-movement time of tunnel users 'evacuation and time before the evacuation process toward the emergency exits begins. In this model, these pre-movement times have basic values, $t_{rec,base}$, $t_{res,base}$, and t_{veh} (each type of vehicle). The higher these basic values, the more likely that the harmful effects of the fire incident reach the users before they start evacuation. Three modes, standard values of basic pre-movement parameters, halving of basic pre-movement parameters, and doubling basic pre-movement parameters, were considered to study the sensitivity of this model to pre-movement times. The expected damage value, EDV, which is calculated via Equation (9), is calculated for each mode.

$$EDV = \sum N_{si} \times F_{Si}, \tag{9}$$

where N_{si} is the number of fatalities for each fire scenario and F_{Si} is the cumulated frequency of each fire scenario. Table 8 illustrates the variation of EDV for halving and doubling pre-movement parameters in comparison with its value in the standard configuration.

Table 8. The variation of EDV for halving and doubling pre-movement parameters.

	Halved Parameters Zone 1, Zone 2, Zone 3	**Reference Parameters Zone 1, Zone 2, Zone 3**	**Doubled Parameters Zone 1, Zone 2, Zone 3**
$t_{rec,base}$ (min)	1, 1.5, 2	2, 3, 4	4, 6, 8
$t_{res,base}$ (min)	2, 1.5, 1	4, 3, 2	8, 6, 4
t_{veh} (s)	4 for PC, 8 for van, 30 for bus, and 2 for truck and HGV	8 for PC, 16 for van, 60 for bus, and 4 for truck and HGV	16 for PC, 32 for van, 120 for bus, and 8 for truck and HGV
EDV	2.6×10^{-1}	5.98×10^{-1}	6.57×10^{-1}
Variation %	−56.58%	-	9.85%

By halving these parameters, the EDV decreases by about 57%. On the other hand, by doubling these parameters, the EDV increases by 9.85% which indicates greater pre-movement time and consequently, a higher number of tunnel users implementing the evacuation process later and being in the intolerable condition for too long.

3.1.2. The Accidents Frequency Involving HGVs

To study the influence of fire incident frequency involving HGV, the frequency of related scenarios, which are 158 MW fire incidents, is decreased and increased by a factor of 10. These frequencies, 1/10× HGV and 10× HGV, respectively lead to lower and higher cumulative frequencies and lower and higher FN curves, compared to initial FN curves of the same scenarios (without ventilation) (Figure 11). These results explain that the frequency of occurrence of each scenario depends directly on the value of the incidental frequencies.

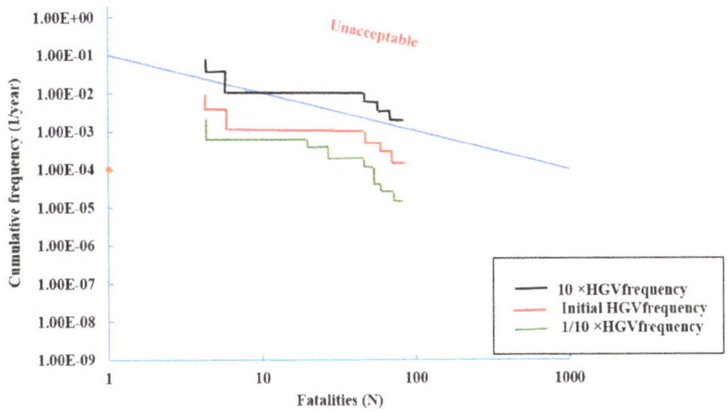

Figure 11. Sensitivity of the model to the parameter "HGV fire incident frequency".

4. Conclusions

Tunnel fires cause fatalities and property damage. This paper provides a new methodology to perform quantitative risk analysis for road tunnels. The integrated method includes quantitative consequence analysis and quantitative frequency analysis. The quantitative consequence analysis aspect is a general method to calculate the number of fatalities and can be used for a wide range of fire scenarios and road tunnels with various characteristics. The quantitative frequency analysis was based on UK road/road tunnels statistical data. It was also applied to Gibraltar Airport Tunnel to show how the model illustrates the social risk of various fire scenarios.

Generally, the model applies to all tunnels either with natural ventilation or mechanical ventilation. The benefits of the LBAQRAMo are:

- It has elements with high flexibility, and can be applied to almost every tunnel, ventilation, or traffic configuration. It can be used and supports decisionmakers to select safety measures for new tunnels or to upgrade measures for existing tunnels.
- New information can be implemented very easily.
- The impacts of important safety systems, their reliabilities, and the relationship between them can be included.
- The interaction between smoke and fire propagation inside the tunnel and the self-rescue operation can be modelled.
- The total distance of the evacuation path is calculated by considering both the longitudinal and lateral shares in the evacuation route
- Traffic distribution: considering lane utilization factor for each lane based on total traffic flow of vehicles (veh/hr) in two-lane and three-lane tunnels.
- The results of the model include information about the distribution of different accident consequence classes such as F/N curves.

Although LBAQRAMo has its limitations. Identification of relevant limitations is crucial for a correct interpretation of the results and the methods employed. This also makes it possible to extrapolate the findings to new applications. The main limitations of this project are:

- Quantitative measures must depend on the scope and accuracy of the defined measurement scale.
- Any errors in your setup or mistakes in execution can invalidate all your results. Even coming up with a hypothesis can be subjective, especially if you have a specific question that you already know you want to prove or disprove.
- There are no standards and universally accepted information for implementing this method.
- Only the lethal effects on humans have been of interest when analyzing the different scenarios, no consideration has been taken to material damage that might be caused because of the different scenarios that have been analyzed.

The Excel spreadsheet of the model has now been completed and has been successfully tested on three UK road tunnels.

Author Contributions: Conceptualization, R.K.H. and Z.H.; methodology, R.K.H.; validation, R.K.H. and Z.H.; formal analysis, R.K.H.; investigation, R.K.H.; resources, R.K.H.; data curation, Z.H.; writing—original draft preparation, R.K.H.; writing—review and editing, Z.H.; visualization, R.K.H.; supervision, Z.H.; funding acquisition, Z.H. All authors have read and agreed to the published version of the manuscript.

Funding: Transdisciplinary Research Grant Scheme TRGS/1/2020/UKM/02/1/1, UKM internal grant GUP-2018-012.

Institutional Review Board Statement: Not applicable.

Informed Consent Statement: Not applicable.

Data Availability Statement: Not applicable.

Acknowledgments: This research was supported by the Transdisciplinary Research Grant Scheme TRGS/1/2020/UKM/02/1/1, UKM internal grant GUP-2018-012 and London Bridge Associates Ltd.

Conflicts of Interest: The authors declare no conflict of interest.

References

1. Caliendo, C.; Guglielmo, M.L.D. Accident Rates in Road Tunnel and Social Costs Evaluation. In Proceedings of the SIIV—5th International Congress—Sustainability of Road Tunnels Infrastructures, Rome, Italy, 29–31 October 2012; pp. 166–177.
2. Meng, Q.; Qu, X.; Wang, X.; Vivi, Y.; Wong, S.C. Quantitative risk assessment modeling for non-homogeneous urban road tunnels. *Risk Anal.* **2011**, *31*, 382–403. [CrossRef] [PubMed]
3. Ntzeremes, P.; Kirytopoulos, K.; Benekos, I. Exploring the effect of national policies on the safety level of tunnels that belong to the trans-European road network: A comparative analysis. *Int. J. Crit. Infrastruct.* **2018**, *14*, 40–58. [CrossRef]
4. PIARC. Risk Analysis for Road Tunnels, Paris: Ref: 2008R02EN, ISBN: 2-84060. 2008, 202–204. Available online: https://www.piarc.org/en/order-library/5871-en-Risk%20analysis%20for%20road%20tunnels (accessed on 18 September 2022).
5. Yeung, J.S.; Wong, Y.D.; Xu, H. Driver perspectives of open and tunnel expressways. *J. Environ. Psychol.* **2013**, *36*, 248–256. [CrossRef]
6. Naevestad, T.-O.; Meyer, S.F. A survey of vehicle fires in Norwegian road tunnels 2008–2011. *Tunn. Undergr. Sp. Technol.* **2014**, *41*, 104–112. [CrossRef]
7. Kirytopoulos, K.; Kazaras, K.; Papapavlou, P.; Ntzeremes, P.; Tatsiopoulos, I. Exploring driving habits and safety critical behavioural intentions among road tunnel users: A questionnaire survey in Greece. *Tunn. Undergr. Sp. Technol.* **2017**, *63*, 244–251. [CrossRef]
8. Zhang, D.L.; Xiao, L.Y.; Wang, Y.; Huang, G.Z. Study on vehicle fire safety: Statistic, investigation methods and experimental analysis. *Saf. Sci.* **2019**, *117*, 194–204. [CrossRef]
9. Perard, M. Statistics on breakdowns, accidents and fires in french road tunnels. In Proceedings of the 1st International Conference on Tunnel Incident Management, Korsør, Denmark, 13–15 May 1996; pp. 347–365.
10. Beard, A.; Carvel, R. *The Handbook of Tunnel Fire Safety*; Thomas Telford: London, UK, 2005.
11. Haddad, R.K.; Maluk, C.; Reda, E.; Harun, Z. Critical velocity and backlayering conditions in rail tunnel fires: State-of-the-art review. *J. Combust.* **2019**, 3510245. [CrossRef]
12. Li, Y.Z.; Lei, B.; Ingason, H. The maximum temperature of buoyancy-driven smoke flow beneath the ceiling in tunnel fires. *Fire Saf. J.* **2011**, *46*, 204–210. [CrossRef]
13. Haddad, R.K.; Rasani, M.R.; Harun, Z. Fire-Induced Flow Temperature Distribution Beneath a Ceiling. *J. Kejuruter.* **2020**, *32*, 247–257.
14. Tang, W.; Hu, L.; Chen, L. Effect of blockage-fire distance on buoyancy driven back-layering length and critical velocity in a tunnel: An experimental investigation and global correlations. *Appl. Therm. Eng.* **2013**, *60*, 7–14. [CrossRef]
15. Haddad, R.K.; Harun, Z.; Maluk, C.; Rasani, M.R. Experimental study of the influence of blockage on critical velocity and backlayering length in a longitudinally ventilated tunnel. *JP J. Heat. Mass. Transf.* **2019**, *17*, 451–476. [CrossRef]
16. Harun, Z. An Experimental Investigation on Vehicular Blockage Effect on the Maximum Smoke Temperature in Tunnel. *J. Inst. Eng.* **2020**, *81*, 31–37.
17. Yamamoto, K.; Sawaguchi, Y.; Nishiki, S. Simulation of tunnel fire for evacuation safety assessment. *Safety* **2018**, *4*, 12. [CrossRef]
18. Khaksari, R.; Harun, Z.; Fielding, L.; Aldridge, J. Numerical Simulation of the Evacuation Process in a Tunnel during Contraflow Traffic Operations. *Symmetry* **2021**, *13*, 2392. [CrossRef]
19. PIARC Technical Committee on Road Tunnel Operation (C3.3). Risk Analysis for Road Tunnels. *Natl. Acad. Sci. Eng. Med.* **2008**, *245*.

20. Qu, X.; Meng, Q.; Vivi, Y.; Wong, Y.H. Design and Implementation of a quantitative risk assessment software tool for Singapore road tunnels. *Expert Syst. Appl.* **2011**, *38*, 13827–13834. [CrossRef]
21. Meng, Q.; Qu, X.; Yong, K.T.; Wong, Y.H. QRA model based risk impact analysis of traffic flow in urban road tunnels. *Risk Anal.* **2011**, *31*, 1872–1882. [CrossRef]
22. Organization for Economic Cooperation and Development (OECD). Safety in tunnels transport of dangerous goods through road tunnels. *Int. Transp. Forum* **2001**.
23. Cassini, P. Road transportation of dangerous goods: Quantitative risk assessment and route comparison. *J. Hazard. Mater.* **1998**, *61*, 133–138. [CrossRef]
24. Borghetti, F.; Cerean, P.; Derudi, M.; Frassoldati, A. *Road Tunnels: An Analytical Model for Risk Analysis*; Springer: Cham, Switzerland, 2019. Available online: https://link.springer.com/book/10.1007/978-3-030-00569-6 (accessed on 19 August 2022).
25. CD 352 Design of road tunnels. *Natl. Highw.* **2020**, 170. Available online: https://www.standardsforhighways.co.uk/prod/attachments/987a669b-13a1-40b9-94da-1ea4e4604fdd?inline=true (accessed on 18 September 2022).
26. Yousif, S.; Al-Obaedi, J.; Henson, R. Drivers' Lane utilization for United Kingdom motorways. *J. Transp. Eng.* **2013**, *139*, 441–447. [CrossRef]
27. PIARC. *Vehicle Emissions Air Demand Environment Longitudinal Ventilation*; World Road Association (PIARC): Paris, France, 2003; ISBN 978-84060-500-3. Available online: https://www.piarc.org/en/order-library/4476-en-Road%20Tunnels:%20Vehicle%20Emissions%20and%20Air%20Demand%20for%20Ventilation (accessed on 18 September 2022).
28. Persson, M. *Quantitative Risk Analysis Procedure for the Fire Evacuation of a Road Tunnel*; Report 5096; Department of Fire Safety Engineering, Lund University: Lund, Sweden, 2002.
29. Bryan, J.L. Behavioral response to fire and smoke. *SFPE Handb. Fire Prot. Eng.* **2002**, *2*, 42.
30. Korhonen, T.; Hostikka, S.; Heliövaara, S.; Ehtamo, H. FDS+ Evac: An agent based fire evacuation model. In *Pedestrian and Evacuation Dynamics*; Klingsch, W.W.F., Rogsch, C., Schadschneider, A., Schreckenberg, M., Eds.; Spinger: Berlin/Heidelberg, Germany, 2008; pp. 109–120.
31. Frantzich, H.; Nilsson, D. *Utrymning Genom Tät Rök: Beteende Och Förflyttning*; Lund University: Lund, Sweden, 2003; p. 3126.
32. Lemke, K. Road safety in tunnels. *Transp. Res. Rec.* **2000**, *1740*, 170–174. [CrossRef]
33. Yeung, J.S.; Wong, Y.D. Road traffic accidents in Singapore expressway tunnels. *Tunn. Undergr. Sp. Technol.* **2013**, *38*, 534–541. [CrossRef]

fire

MDPI

Article

The Ignition Frequency of Structural Fires in Australia from 2012 to 2019

Samson Tan *, Khalid Moinuddin and Paul Joseph

Institute of Sustainable Industries and Liveable Cities, Victoria University, Melbourne, VIC 8001, Australia
* Correspondence: samson.tan@live.vu.edu.au

Abstract: Appropriate estimates of ignition frequency derived from fire statistics are crucial for quantifying fire risks, given that ignition frequency underpins all probabilistic fire risk assessments for buildings. Rahikainen et al. (Fire Technol 2004; 40:335–53) utilized the generalized Barrois model to evaluate ignition frequencies for different buildings in Finland. The Barrois model provides a good prediction of the trend of the ignition frequency; however, it can underestimate the ignition frequency depending on the building type. In this study, an analysis of the Australian fire statistical data from 2012 to 2019 was performed and compared with studies from Finland. A new coefficient is proposed to improve the Barrois model for a better fit for buildings in Australia. Several categories, such as hotels and hospitals, which were absent in previous studies, have been included as separate categories in this study. Office and retail spaces in Finland have an ignition frequency one order of magnitude lower than in Australia. On the other hand, other buildings (retail and apartments in particular) are much more prone to fire ignition in Australia than in Finland. The improved generalized Barrois model based on the Australian fire statistical data will be useful for determining ignition frequency for risk quantification in the Australian context.

Keywords: ignition frequency; probabilistic risk analysis; fire statistics; risk quantification; building categories; Barrois model

check for updates

Citation: Tan, S.; Moinuddin, K.; Joseph, P. The Ignition Frequency of Structural Fires in Australia from 2012 to 2019. *Fire* **2023**, *6*, 35. https://doi.org/10.3390/fire6010035

Academic Editors: Tiago Miguel Ferreira and Grant Williamson

Received: 21 November 2022
Revised: 3 January 2023
Accepted: 9 January 2023
Published: 16 January 2023

1. Introduction

Ignition frequency is a critical variable in probabilistic fire risk quantification in buildings. The expected value of the fire loss is the product of the ignition frequency and the consequences added to the distribution of burned buildings from the studied building stock. Ignition frequency is always a linear multiplying factor for different fire losses, such as life, economic, or societal [1]. Hence, adequate knowledge of the ignition frequency derived from fire statistics must be made available to quantify fire risks. The fire-starting probability is the most crucial factor in probabilistic fire risk analyses for performance-based fire engineering designs.

The Australian Building Codes Board (ABCB) has been moving towards the adoption of risk quantification in their National Construction Codes (NCC) [2]. It is envisaged that a Quantitative Probabilistic Risk Assessment (QPRA) will be adopted, which typically uses the Fault Tree (FT) [3] and Event Tree (ET) to quantify risk, and a critical input is fire ignition frequency which forms the first node of the ET. While statistical data is generally scarce, it will be useful to have a correlation to aid the quick determination of fire ignition frequency for QPRA. The generalized Barrois model developed by Rahikainen et al. [1,4–6] based on fire statistics from Finland is commonly adopted by fire researchers [7–12]. However, it has been shown in some studies that the Barrois model tends to underestimate fire ignition when applied to various contexts [13,14].

Rahikainen et al. [1,4–6] determined ignition frequencies and ignition frequencies per floor area for various building categories in Finland as combined groups and as a function of the building floor area. Their results show that differences between building categories or locations within the country are so minor that a universal curve for the whole country could be established. For small buildings, a strong dependence on size is observed for ignition frequency per floor area; however, it remained more or less constant for large buildings. In addition, periodic variations of building ignition frequency by month and week of the year, day of the week, and time of the day were calculated, and limited tests of the generality of the results were made on the theoretical models.

Ramachandran [15] studied the statistical methods typically used in the fire risk assessment of an industry. Ramachandran defined two primary components of the fire risk that include (i) the probability of a fire starting and (ii) the probability of damage reaching various levels in a fire event. The values of the distribution parameters and the fire-starting probability vary depending on factors that include ignition sources, property size, and the presence/absence of sprinklers. The results show that a set of similar properties could be altered for the risk evaluation in an individual property. He further proposed that a stochastic model be developed to predict the fire spread in a building and, in turn, outlined a model, specified the data required for estimation, and validated the parameters.

Sandberg [16] performed a comparative case study analysis to review and identify the existing models and data relevant to determining ignition frequencies. Sandberg underlined the factors that affect the frequency of ignition, which depends on the number of ignition sources and increases with the building size. The results showed that the ignition frequency varied with the total building floor area. Several other factors seemed to influence the ignition frequency, including building occupancy, equipment faults, human activity, and other natural causes. Furthermore, the author applied a different approach by analyzing the social, demographic, and economic variables. The first method is based on the maximum likelihood estimator and is used when data is available both for buildings exposed to fire and structures at risk. Although this method is more accurate, getting hold of the necessary data is challenging. Such detailed information is typically not collected at the national level. However, insurance companies may have collected this data but often keep it confidential.

The second method applied by Sandberg [16] was also based on the maximum likelihood estimator and is used when data is available only for buildings exposed to fire. Data for facilities at risk, which have not been involved in a fire incident, may not exist. The other methods the author studied were based on Bayesian data analysis. The Bayesian approach is suitable when there is little data or information available, such as in the case of the safety of nuclear power plants. The author suggested that these methods must be reserved for events with low probability and high consequences. An average ignition frequency was estimated in different occupancies in this study. The work collected data on various buildings in different categories and estimated their parameters. This data was then used to examine the ignition frequency and apply the developed models.

D'Este et al. [17] analyzed and compared climatic, topographic, anthropic, and landscape drivers to investigate the patterns of fire ignition in terms of frequency and fire occurrence in European regions. In order to achieve this, they mapped the probability of fire ignition occurrence and frequency using negative binomial hurdle models, while the performance of models was assessed using metrics such as AUC, prediction accuracy, RMSE, and the Pearson correlation coefficient. Their results revealed an inverse correlation between distance from infrastructures (e.g., urban roads and areas) and fire occurrence in all the study regions. Furthermore, a positive correlation was found between fire occurrence and landscape drivers relevant to regions. They concluded that anthropic activity, compared to the climatic, topographic, and landscape drivers, influences fire ignition and frequency more significantly in all the regions. The probability of fire ignition occurrence and frequency were found to increase when the distance from urban roads and areas decreased. One of the conclusions is that it is essential to implement long- to medium-term intervention plans to suppress the proximity between potential ignition points and fuels.

Traditionally, building fire probability analysis is performed based on either statistics or fire science. Hu et al. [18] combined the two approaches and improved the statistical method for determining building ignition probability. The factors that affect the probability of fire ignition are divided into humans, ignition sources, combustibles and environments. Given the factor classification, they developed a Bayesian network of the ignition probability of buildings and introduced the nodes and detailed conditional probability table in the Bayesian network, according to which the ignition probability of the building was quantified. Finally, they chose some typical buildings as examples to test the model's applicability, estimated the posterior probability value by obtaining the relevant building data, and incorporated it into the Bayesian network. They demonstrated that ignition probability is dynamic, and the comparison with the statistical data of building fire is rational.

It is evident from the above literature review that the ignition frequency of a building is primarily affected by its floor area with a weak correlation to other factors. However, due to the limited data available, many other potential factors were not fully examined.

This study aims to further investigate ignition frequency using Australian historical fire data from 2012 to 2019. A modified Barrois model assisted in accurate calculations of the ignition frequency for the buildings, which will be described later. The results obtained are then compared with those from the study conducted by Tillander et al. [5,6] to determine any similarities and differences between the two jurisdictions. The current study covers building fires from 2012 to 2019. The data are drawn from the Australian Incident Reporting System (AIRS) Database managed by the Australasian Fire and Emergency Service Authorities Council (AFAC). The calculation of fire frequencies for Australia, based on historical data, is expected to fill an important gap in probabilistic fire risk analysis that is currently unavailable.

Finally, some internal data validations were conducted to estimate the influence of data deficiencies on the obtained ignition frequencies. The following sections present the materials and methods used in the analysis, followed by results, discussion, conclusion and recommendations in the final section.

2. Materials and Methods

2.1. Statistical Building Data

Quantitative Probabilistic Risk Assessment applications require high-quality statistical data sets related to fire ignitions in buildings [19]. ARUP and the University of Queensland [20] obtained raw data from the Australian Bureau of Statistics (ABS) for fires and other sources from 2012 to 2019 to determine the rate of fire starts for various building categories. The Australian Bureau of Statistics provided the total floor areas for residential buildings, while the Department of Climate Change and Energy Efficiency provided data for hotels, offices, retail spaces, hospitals, and schools. There is no data available for Class 4, 7b, and 8 buildings. It should be noted that the total number of parking spaces in Australia, as reported by Colliers, was also not included in the comparison as it was not included in the Finnish studies [4–6]. Results are presented in two forms: rate of ignition per square meter per year and rate per unit per year. The Australian National Construction Code (NCC) defines the following classes of buildings:

Class 1: Houses, standalone domestic, or residential dwellings.

Class 1a: Detached houses or attached dwellings such as townhomes or row houses.

Class 1b: Small boarding houses, guesthouses, or hostels with less than 12 residents or short-term holiday accommodation with four or more single dwellings on one allotment.

Class 2: Multi-unit residential buildings where people live above and below each other or single-storey attached dwellings with a common space below.

Class 3: Long-term or transient living for a number of unrelated people, including larger boarding houses, guest houses, hostels, or accommodation for backpackers; residential care buildings; and residential parts of hotels, motels, schools, or jails.

Class 4: Sole dwellings or residences within a non-residential building.

Class 5: Office buildings for professional or commercial purposes.

Class 6: Retail establishments such as shops, restaurants, and cafes.

Class 7: Storage-type buildings, divided into Class 7a (carparks) and Class 7b (warehouses, storage buildings, or wholesale display buildings).

Class 8: Buildings for production, assembly, alteration, repair, finishing, packing, or cleaning of goods or produce, including mechanic's workshops and food processing buildings.

Class 9: Public buildings, divided into Class 9a (hospitals and clinics), Class 9b (assembly buildings such as schools, universities, sporting facilities, and public transport buildings), and Class 9c (residential care buildings with 10% or more residents needing physical assistance in daily activities and evacuation during an emergency, such as aged care facilities).

Data for Class 1 and 2 occupancies are available from the Australian Bureau of Statistics (ABS), and aggregate floor areas for Class 1 and Class 2 occupancies within Australia were obtained from their census reports (ABS—2019 Census, from https://quickstats.censusdata.abs.gov.au/census_services/getproduct/census/2019/quickstat/036?op, accessed on 28 December 2022).

The statistical floor areas of various building stocks in Australia between 2012 and 2019 are shown in Table 1. Structural fire incidents for various occupancies in Australia from 2012 to 2019 are presented in Table 2. The ignition frequency for each class between 2012 and 2019 derived from Tables 1 and 2 are shown in Table 3, and a summary of the ignition frequency of structural fires in Australia from 2012 to 2019 is presented in Table 4. The percentage floor areas and fire ignitions are shown in Figure 1 as pie charts.

Table 1. Statistical floor areas of various building stocks in Australia from 2012 to 2019 extracted from the Australian Bureau of Statistics and [20].

Class	Occupancy	2012 [m^2]	2013 [m^2]	2014 [m^2]	2015 [m^2]	2016 [m^2]	2017 [m^2]	2018 [m^2]	2019 [m^2]	Average [m^2]
1a	Houses	1.81×10^9	1.84×10^9	1.86×10^9	1.83×10^9	1.85×10^9	1.82×10^9	1.88×10^9	1.90×10^9	1.85×10^9
2	Apartments	2.30×10^8	2.23×10^8	2.22×10^8	2.32×10^8	2.33×10^8	2.31×10^8	2.33×10^8	2.33×10^8	2.30×10^8
3	Hotel	1.10×10^7	1.09×10^7	1.13×10^7	1.14×10^7	1.15×10^7	1.17×10^7	1.19×10^7	1.21×10^7	1.15×10^7
5	Offices	3.83×10^7	4.02×10^7	4.06×10^7	3.95×10^7	4.25×10^7	4.17×10^7	4.06×10^7	4.54×10^7	4.11×10^7
6	Retail spaces	4.22×10^7	4.46×10^7	4.54×10^7	4.68×10^7	4.69×10^7	4.88×10^7	5.00×10^7	5.15×10^7	4.70×10^7
9a	Hospitals	1.30×10^7	1.30×10^7	1.35×10^7	1.34×10^7	1.36×10^7	1.41×10^7	1.44×10^7	1.42×10^7	1.36×10^7
9b	Schools	4.13×10^7	4.15×10^7	4.20×10^7	4.28×10^7	4.32×10^7	4.39×10^7	4.46×10^7	4.54×10^7	4.31×10^7

Table 2. Structural fire incidents for NCC Classes in Australia from 2012 to 2019 [fires/year] extracted from [20].

Class	Occupancy	2012	2013	2014	2015	2016	2017	2018	2019	Average
1a	Houses	8977	9689	8332	7494	8335	8668	8613	8627	8592
2	Apartments	2531	2678	2445	2109	2026	1986	2023	1979	2222
3	Hotels	438	458	476	412	459	445	441	434	445
5	Offices	536	603	487	434	468	500	402	454	486
6	Retail spaces	1393	1427	1271	1077	1078	1074	1100	1029	1181
9a	Hospitals	298	273	283	215	190	198	187	184	229
9b	Schools	380	456	349	325	298	312	254	295	334

Table 3. Ignition frequency of structural fires for NCC classes in Australia from 2012 to 2019 [fire/m^2·year] taken from the Australian Bureau of Statistics, the Department of Climate Change and Energy Efficiency, and [20].

Class	Occupancy	2012	2013	2014	2015	2016	2017	2018	2019	Average
1a	Houses	4.96×10^{-6}	5.27×10^{-6}	4.48×10^{-6}	4.11×10^{-6}	4.49×10^{-6}	4.76×10^{-6}	4.58×10^{-6}	4.53×10^{-6}	4.65×10^{-6}
2	Apartments	1.10×10^{-5}	1.20×10^{-5}	1.10×10^{-5}	9.10×10^{-6}	8.70×10^{-6}	8.60×10^{-6}	8.70×10^{-6}	8.50×10^{-6}	9.60×10^{-6}
3	Hotels	4.00×10^{-5}	4.20×10^{-5}	4.20×10^{-5}	3.60×10^{-5}	4.00×10^{-5}	3.80×10^{-5}	3.70×10^{-5}	3.60×10^{-5}	3.90×10^{-5}
5	Offices	1.40×10^{-5}	1.50×10^{-5}	1.20×10^{-5}	1.10×10^{-5}	1.10×10^{-5}	1.20×10^{-5}	9.90×10^{-6}	1.00×10^{-5}	1.20×10^{-5}
6	Retail spaces	3.30×10^{-5}	3.20×10^{-5}	2.80×10^{-5}	2.30×10^{-5}	2.30×10^{-5}	2.20×10^{-5}	2.20×10^{-5}	2.00×10^{-5}	2.50×10^{-5}
9a	Hospitals	2.30×10^{-5}	2.10×10^{-5}	2.10×10^{-5}	1.60×10^{-5}	1.40×10^{-5}	1.40×10^{-5}	1.30×10^{-5}	1.30×10^{-5}	1.70×10^{-5}
9b	Schools	9.20×10^{-6}	1.10×10^{-5}	8.30×10^{-6}	7.60×10^{-6}	6.90×10^{-6}	7.10×10^{-6}	5.70×10^{-6}	6.50×10^{-6}	7.79×10^{-6}

Table 4. Statistical data for ignition frequency of structural fires in Australia from 2012 to 2019 from the Australian Bureau of Statistics and the Department of Climate Change and Energy Efficiency [20].

Class	Occupancy	Floor Area (m^2)	Number of Fires	Frequency (fires/m^2·year)
1a	Houses	1,849,263,351	8592	4.65×10^{-6}
2	Apartment	229,555,576	2222	9.60×10^{-6}
3	Hotels	11,474,068	445	3.90×10^{-5}
5	Offices	41,092,722	486	1.20×10^{-5}
6	Retail spaces	47,020,320	1181	2.50×10^{-5}
9a	Hospital	47,020,320	229	1.70×10^{-5}
9b	Schools	43,081,041	334	7.80×10^{-6}

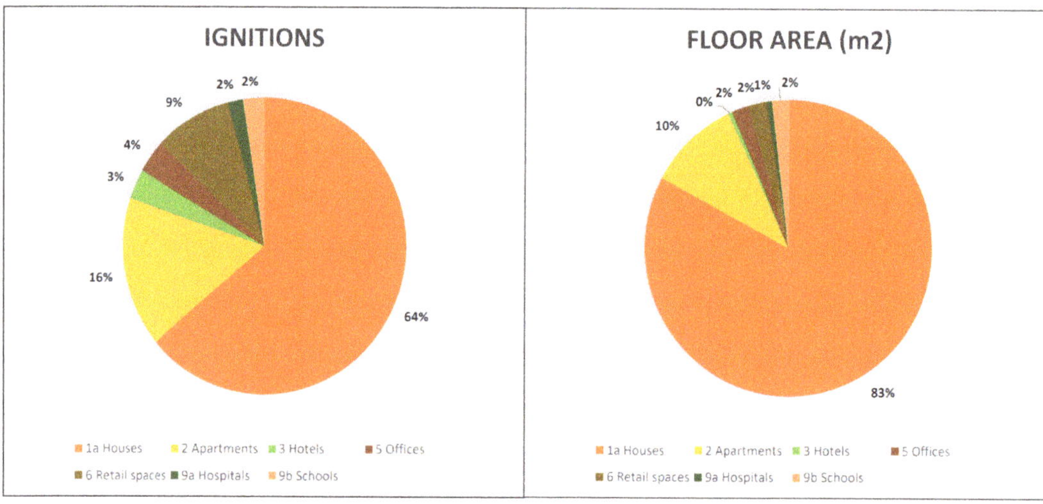

Figure 1. Percentage of ignitions from 2012 to 2019 and floor areas in different building categories in Australia.

2.2. Barrois Model

For any reliable probabilistic fire risk assessment to be conducted, it is necessary to have reliable ignition frequency data which is largely based on the type of building (referred to as "class" in this study), while the ignition frequency within each building class is dependent on the floor area of the building. A practical method to model the dependence of the average annual probability of fire ignition in a building of a particular class on the floor area of the building was originally proposed by Barrois in 1835 [21]. The generalized Barrois model can be described as the sum of two power law functions. The equation for the Barrois curve is given below [5,6]:

$$f'' = c_1\, A^r + c_2\, A^s \tag{1}$$

where f'' is the ignition frequency of a building with a floor area A within 12 months, c_1, c_2, r, and s are coefficients. Table 5 presents the parameters of the generalized Barrois model for different building categories proposed by Tillander [6]. The parameter R_2 represents the coefficient of determination—it indicates the proportion of variance of the dependent variable that is related to the independent variable. This statistical measure shows how well the regression model represents the data. Higher values of this indicator imply a better fit for the model. These parameters are useful for determining the ignition frequency of buildings with floor areas between 100 m^2 and 20,000 m^2. In the generalized Barrois model equation (Equation (1)), the coefficients r and s are both less than zero. This means that as the floor area A approaches 0, the limit of the equation produces an unreality. However, in the context of our study, the floor area A is always at least 100 square meters and typically falls within the range of 100 square meters to 20,000 square meters. Therefore, the values of r and s being less than zero do not significantly impact the accuracy or reliability of the statistical analysis, as the model is not being applied to very small values of A.

Table 5. Parameters of the generalized Barrois model extracted from Tillander [6].

Building Category	c_1	c_2	r	s	R_2 [%]
Residential buildings	1.00×10^{-02}	5.00×10^{-06}	−1.83	−0.05	84
Commercial buildings	7.00×10^{-05}	6.00×10^{-06}	−0.65	−0.05	26
Office buildings	5.60×10^{-02}	3.00×10^{-06}	−2.00	−0.05	74
Transport and firefighting and rescue-service buildings	7.00×10^{-05}	1.00×10^{-06}	−0.65	−0.05	75
Buildings for institutional care	2.00×10^{-04}	5.00×10^{-06}	−0.61	−0.05	68
Assembly buildings	3.00×10^{-03}	2.00×10^{-06}	−1.14	−0.05	85
Educational buildings	3.00×10^{-03}	3.00×10^{-06}	−1.26	−0.05	46
Industrial buildings	3.00×10^{-04}	5.00×10^{-06}	−0.61	−0.05	90
Warehouses	3.82	2.00×10^{-06}	−2.08	−0.05	98
Other buildings	1.18	1.00×10^{-04}	−1.87	−0.20	95

To further demonstrate the robustness of the statistical analysis, we performed additional analyses and simulations using a range of different values for r and s. We found that the fitted models remained accurate and reliable for building floor areas within the range of 100 square meters to 20,000 square meters, regardless of the specific values of r and s. This suggests that the values of r and s being less than zero do not significantly impact the results of the statistical evaluation within this range of building floor areas.

In order to improve the flexibility and applicability of the generalized Barrois model equation (Equation (1)) across all building categories, we have introduced a new coefficient, c_3, that depends on the specific building being considered (see Equation (2)). Table 6 presents the values of coefficient c_3 for different building categories. These values were computed by assigning different values to c_3 and comparing the ignition frequencies predicted by the modified model with those obtained from the AIRS Database. When the data points fit the modified curve well, the value of c_3 for the coefficient is considered a valid assumption.

$$f'' = c_1 A^r + c_2 A^s + c_3 \tag{2}$$

Table 6. Values assumed for the new coefficient c_3 in the improved Barrois model.

Building Category	c_3
Others (Hotels)	4.05×10^{-05}
Others (Offices)	1.05×10^{-05}
Residential (Houses)	4.16×10^{-06}
Residential (Apartments)	1.05×10^{-05}
Others (Schools)	4.05×10^{-05}
Others (Hospitals)	4.05×10^{-05}

It is important to note that these values of c_3 are used to shift the original curve upwards so that the modified model curve fits the aggregated statistical data from the AIRS Database more accurately. This allows the model to more accurately represent the ignition frequencies of different building categories and to be more flexible and applicable across a wider range of building sizes and types in Australia. The coefficients c_3 significantly impact the ignition frequencies of buildings with larger floor areas, while their impact is reduced considerably for buildings with small floor areas. One issue with this curve-fitting approach is that there are relatively few data points available, which can limit the accuracy and reliability of the model. To improve the outcomes of the analysis, it would be beneficial to have access to non-aggregated data from the statistics, such as records that include the floor area of each individual building [22]. This would allow for a more detailed and accurate evaluation of the model's performance across a range of building sizes and types.

2.3. Limitations of the Study

The limitations of the study are as follows:

- The correlations are limited by the statistical data available;
- There is limited knowledge about the acceptable range of change for the parameters introduced into the model. Such knowledge would provide greater certainty in the Barrois model predictions;
- There is an uncertainty in the model correlations due to inconsistencies in the AIRS database;
- Some data are missing from the reported years and the database is cumbersome, which affects the accuracy of the model parameters.

3. Results and Discussion

3.1. Ignition Frequencies Based on Improved Barrois Model and Comparison between Australia and Finland

The average ignition frequency ($1/m^2/year$), defined as the probability per floor area per unit time of a building exposed to fire, can be determined as the ratio of the number of fires in a specific building category during a year and its combined floor area. The results obtained by Tillander and Keski-Rahkonen [5] for Finland are presented in Figure 2 with the following categories: residential (A), commercial (C), office (D), transport and communication buildings, buildings for institutional care (F), assembly (G), educational (H), industrial (J), warehouses (K), firefighting and rescue service buildings (L), and other buildings (N). Following their methodology, the resulting data for Australia are represented in Figure 3 and compared to those from Finland. Data in both Figures 2 and 3 are presented in logarithmic scale. The thick blue horizontal line represents the average of all categories for Finland. The comparison of the two graphs indicates that the ignition frequency for most categories is lower in Finland than in Australia. The average value is higher for Finland because its most impacting class (other buildings, N) with an ignition frequency of 2.7×10^{-4} shifts the average value upwards.

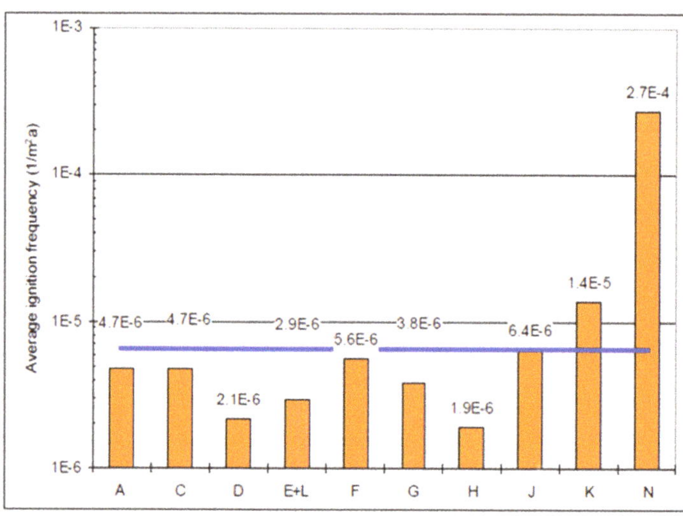

Figure 2. Average ignition frequencies of different building categories in Finland from 1996 to 1999 (extracted from Tillander and Keski-Rahkonen [5]).

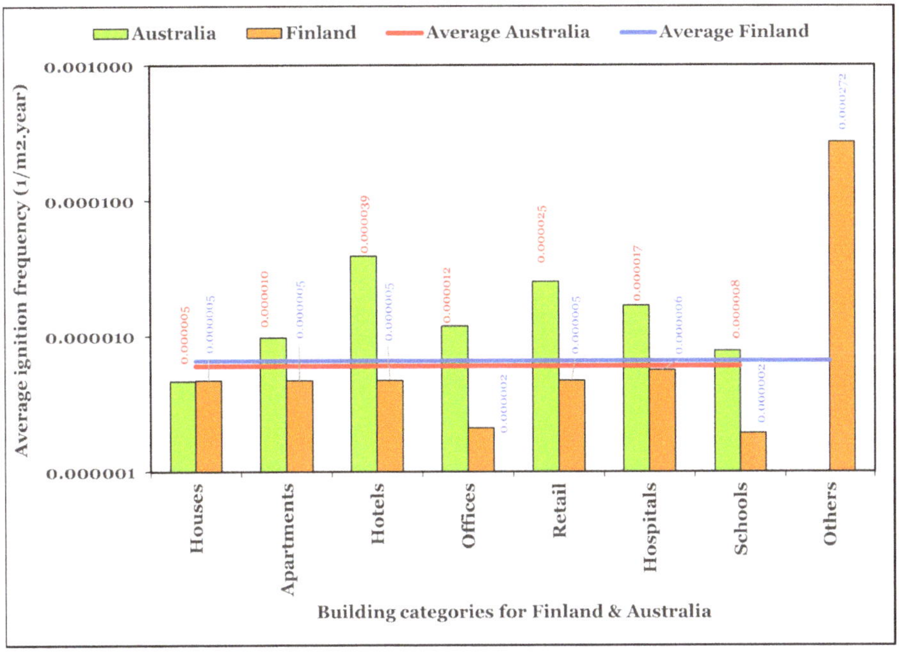

Figure 3. Ignition frequencies of different building categories in Australia compared to Finland.

It can also be noted that there are some categories (hotels and hospitals) that have no specific fire frequency values for Finland. These data categories are included in the 'Others' group. Furthermore, it is not possible to distinguish between houses and apartments in the Finnish dataset, so the same ignition frequency is used in the graph, assuming there are no significant differences in risk between apartments and houses. Nevertheless, it is evident that apartment fires in Australia occur with a higher frequency (9.60×10^{-6}) when compared with those in single houses (4.65×10^{-6}), an increase of about 100%. This suggests that separating the two categories of buildings would make sense.

As a general observation, it can be said that the expected ignition frequency for all buildings is lower in Australia than in Finland. The opposite occurs when specific categories are considered; for example, office spaces have a fire frequency of 1.19×10^{-5} in Australia, while it is only 2.14×10^{-6} in Finland, almost five times lower. The same can be argued for retail spaces, with a 2.50×10^{-5} value for Australia against a fire frequency of 4.70×10^{-6} in Finland. The frequency of fires in school is higher in Australia, with a value of around 7.79×10^{-6} fire/m^2·year compared to 1.93×10^{-6} in Finland.

The reasons for these can be many. Firstly, one can observe that in Finland, timber is primarily adopted as a building material, while in Australia, houses are constructed in either timber or concrete frames with internal plasterboard walls and external facing bricks. The widespread use of timber in Finland would lead to a greater probability of ignition, as timber is combustible, whereas concrete is not. On the other hand, specific categories of buildings (office, retail, and apartments in particular) are way more prone to fires in Australia; jurisdiction-specific rules about electrical installation, fire loads, fire alarm systems, and other factors can also influence the spreading of fires, as well as the differences in climatic conditions (relatively higher temperatures and frequent occurrence of droughts in Australia can act in favor of fire ignition in built environments).

For the sake of a more reliable degree of comparison, Figure 4 presents the normalized floor area values for the two countries; it is evident that the most relevant category is residential in both jurisdictions.

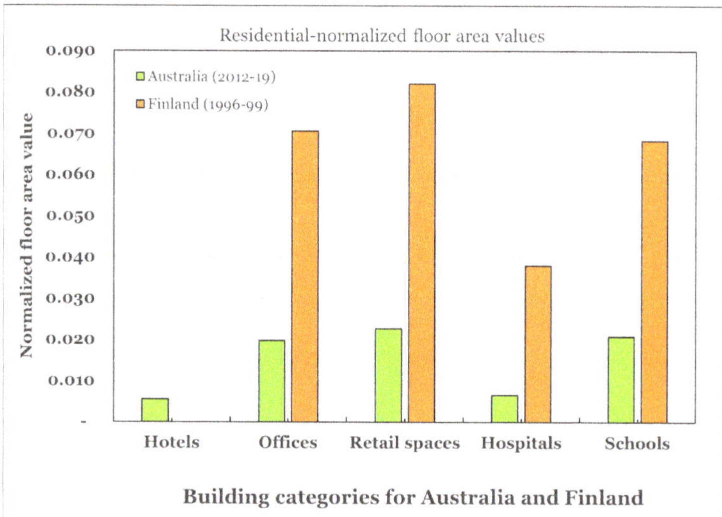

Figure 4. Normalized values of floor area in Australia compared to Finland (reference value: residential area).

The ignition frequency curve in Figure 5 for Class 3 (Hotels) is derived using the generalized Barrois model (Equation (1)). The data follows the behavior of the Barrois model (fire frequency descending with the area), but the Barrois line (red) underestimates the frequencies from statistics (blue triangles).

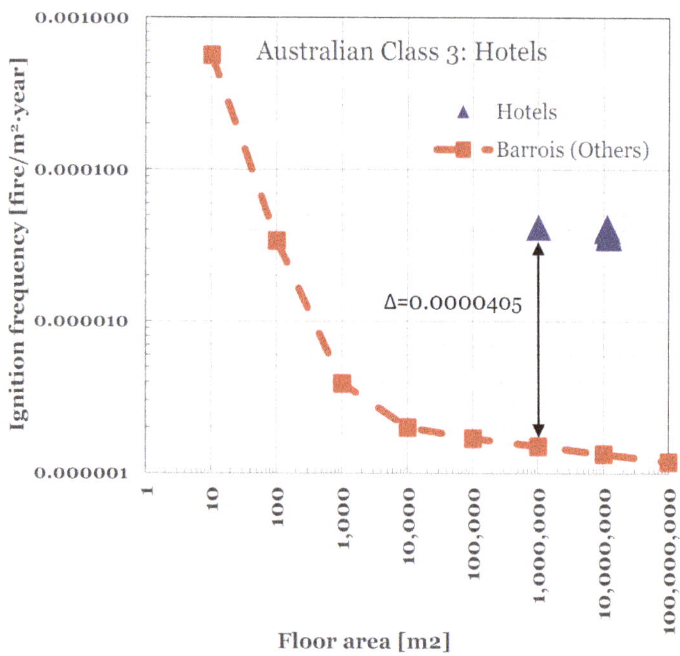

Figure 5. The ignition frequency curve for Class 3 (Hotels) in Australia from 2012 to 2019 was derived using the generalized Barrois model and the ignition frequency observations (blue triangles).

The correlation between ignition frequency and building floor area was modified by adding a fixed term, or coefficient c_3, to the original line (as described in Section 3.2). The value of c_3 is 4.05×10^{-5}, and the red line in Figure 6 represents the modified curve fitting that deviates from previous studies based on single building floor area statistics ($<1.20 \times 10^5$ m^2), rather than the nationwide aggregate floor area of above 1.00×10^7 m^2. Unfortunately, we do not have specific data for hotels to verify the shape of the red curve for floor areas below 1.00×10^6. However, we do have data for other building categories that shows that the ignition frequency follows the 'inverted hockey stick' fire trend phenomenon for building floor areas between 100 m^2 and 20,000 m^2, as demonstrated by Tillander [6]. This trend is statistically reliable and consistent with the data from both Finland and Australia.

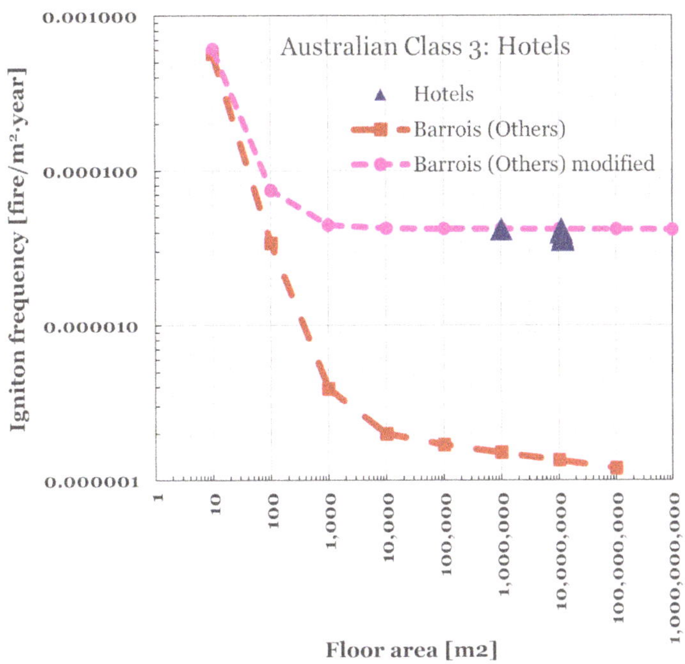

Figure 6. Ignition frequency curves for Class 3 (Hotels) in Australia from 2012 to 2019 derived using the generalized Barrois model and the modified Barrois version.

Given this information, we believe that the constant shift by c_3 is consistent with the 'inverted hockey stick' curve phenomenon for both jurisdictions. While we recognize that this may not be applicable to all building categories, particularly for smaller floor areas, it is a reasonable assumption based on the data we have available. The inclusion of the coefficient serves as a correction factor to account for the increased fire risk in areas with certain environmental and weather conditions, such as hot and dry climates, that may increase the probability of ignition.

It should be noted that the curve behavior has already been observed in Tillander and Keski-Rahkonen's study [5], where an underestimation of the ignition frequency for floor areas above 1.00×10^{-4} is given (see Figure 7). The error bars in the figure are an indication of statistical noise. The point value furthest from the blue curve in Figure 7 is not a result of statistical inaccuracy, as similar deviations were also observed for other building groups. However, due to the need for more sufficient observations in buildings with the largest floor area, it is impossible to establish the ignition frequency of buildings with a floor area exceeding 20,000 m^2 based on this data.

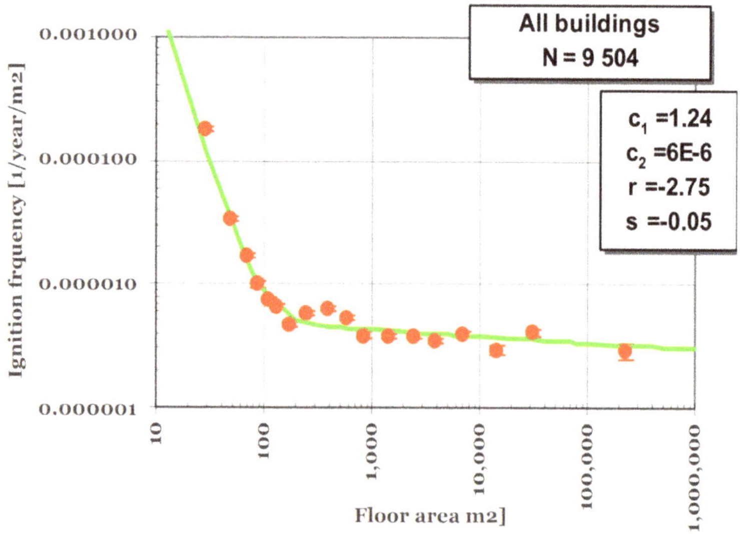

Figure 7. Ignition frequency observations (orange dots) in all building categories in Finland from 1996 to 1999 and a generalized Barrois model fitted to the data (solid green line) (extracted from [6]).

Likewise, an underestimation of the ignition frequency emerges from the comparison between the Australian data and the Barrois model for office buildings, as depicted in Figure 8. Here again, a correction factor of 1.05×10^{-5} is used for the Barrois curve to fit the statistical data. The gap is now four times lower than in the previous case.

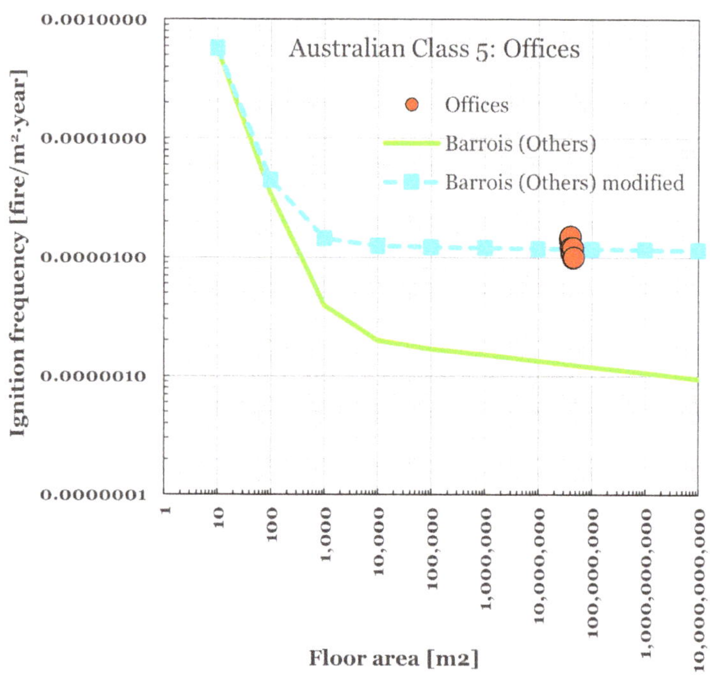

Figure 8. Ignition frequency curves for Australian Class 5—Offices derived from the Barrois model.

The revised curve is also presented in Figure 9. It can be noted that the curve for the 'Others' class of building has been updated for a better fit with the Australian data. The original line (in orange) underestimates the ignition frequency for that type of building, while the revised red line more accurately represents the actual data.

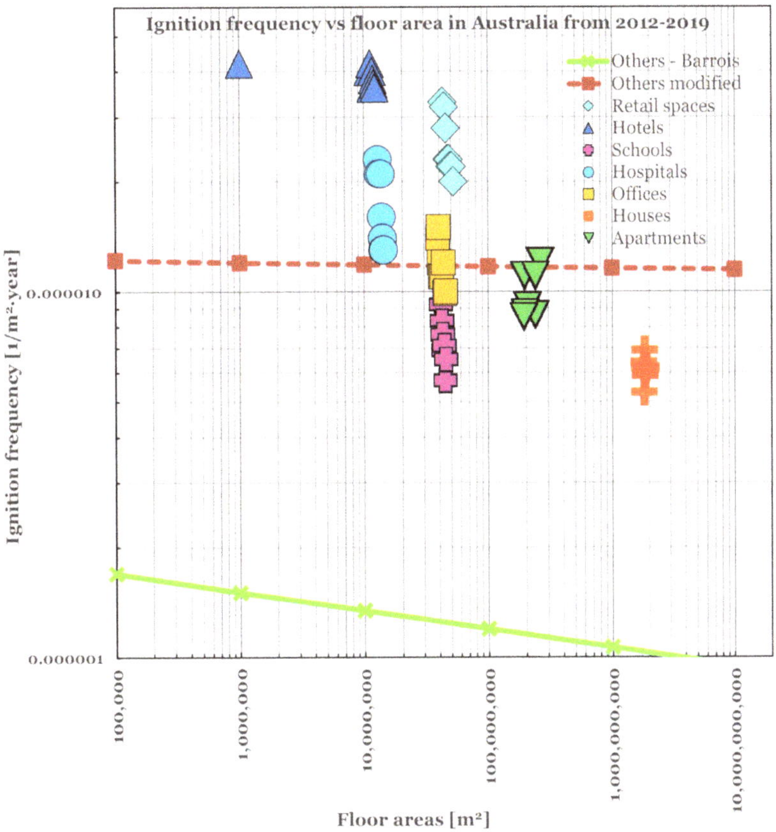

Figure 9. The average ignition frequency of different building classes as a function of floor areas in Australia between 2012 and 2019. (In Figure 9, only two Barrois curves are presented as the focus is on the 'Others' group of buildings).

For the Class 2 (apartments) group, the comparison with the theoretical output from the Barrois model shows similar results to the previous case, with a slightly higher gap. The same correction factor is adopted here.

3.2. Comparison of the Generalized BARROIS Model with the Australian Historical Data

To compare the statistical data for the different building categories, it is necessary to collect the historical data from 2012 to 2019. This is performed for all categories, as shown in Figure 9. It can be seen that the highest ignition frequency is for 'Hotels', with an average value of 3.89×10^{-5} fire/m^2·year, and the lowest fire frequency is for 'Houses', with a value of 4.65×10^{-6}, followed by 'Schools' at 7.79×10^{-6} fire/m^2·year. In general, the yearly variations have minimal impact on the average value for each category.

It is important to highlight that the average ignition frequency [fire/m^2·year], the probability per floor area and time unit in the year of fire incident in a building were obtained by dividing the annual number of fires in the specific building class by its combined floor area, which is similar to the methodology used by Rahikainen et al. [4–6]. Tillander [6] has shown that all building classes generally have high ignition frequency values for small buildings but level off to a much lower ignition frequency value for large buildings. This 'inverse hockey stick' phenomenon, where the trend line starts with a steep decrease, followed by a relatively flat trend line, is statistically reliable for all building classes with floor areas between 100 m^2 and 20,000 m^2 [6]. In our case, the average ignition frequency for hotels is 3.89×10^{-5} fire/m^2·year. To compare the data for the single occupancy with the calculated values, the mean value of the calculated data is compared with the historical series, as shown in Figure 10. The ignition frequency is highest for hotels and lowest for houses, while the length of the error bars indicates that the frequencies for retail buildings vary the most, followed by hospitals, hotels, schools, offices and apartments.

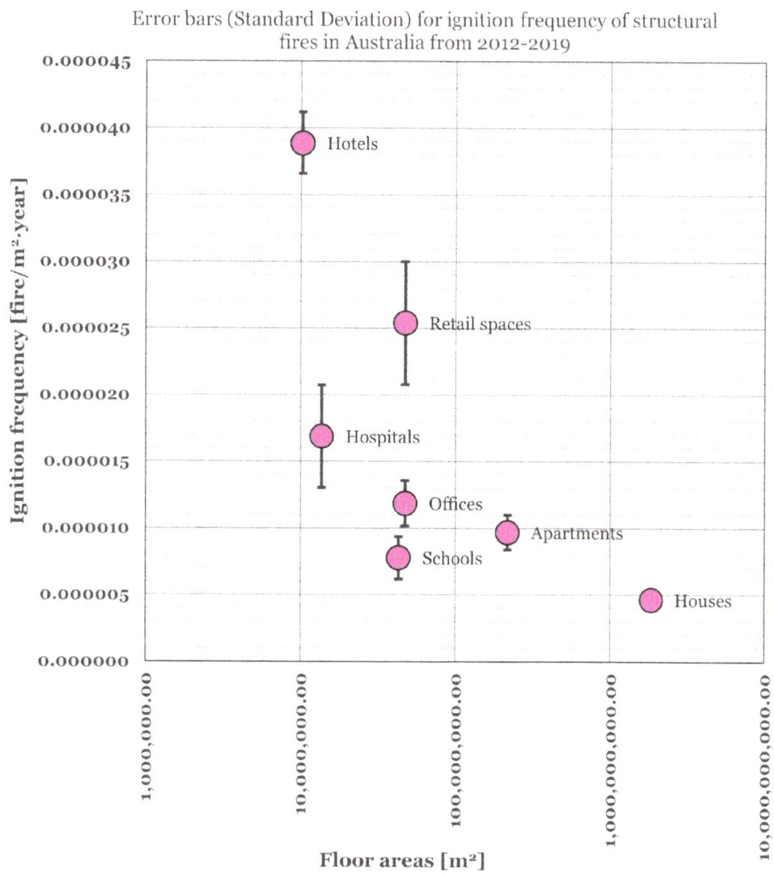

Figure 10. Error bar graphs from the historical data series for ignition frequency of structural fires in Australia from 2012 to 2019.

A detailed comparison of the ignition frequency in each building category is presented below.

Hotels: The Barrois model underestimates the ignition frequency in the Australian context, with an average calculated value of 5.29×10^{-6} fire/m^2·year against an average of 3.89×10^{-5} fire/m^2·year resulting from statistics. For the curve to fit the Barrois data, it should be shifted upwards to about 4.05×10^{-5}, as shown in Figure 11. From this perspective, it can be noted that fires in Australia are more frequent than those predicted by the Barrois model.

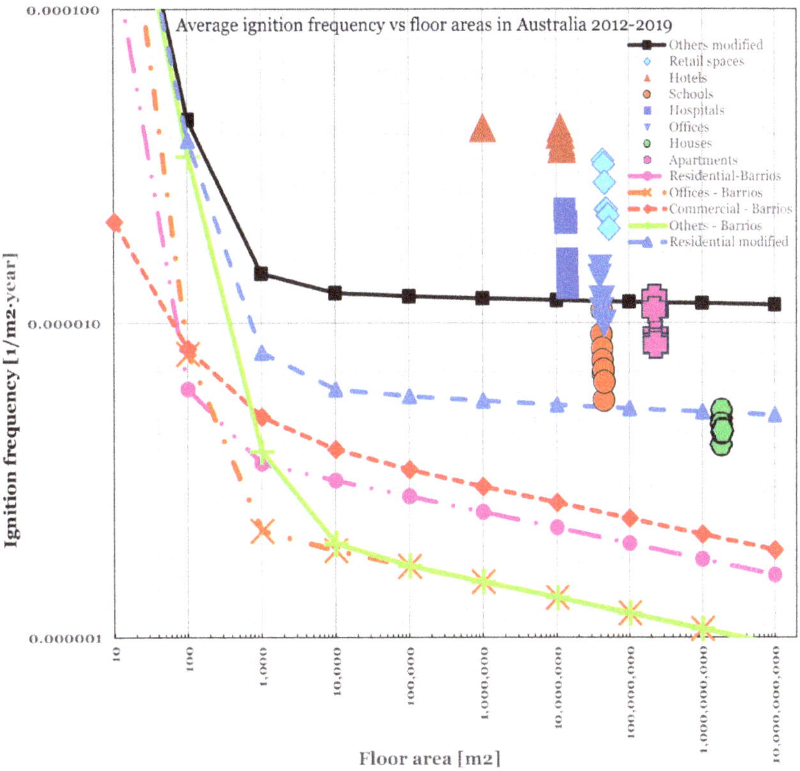

Figure 11. Comparison of the statistical ignition frequency in hotels in Australia with generalized and improved Barrois models.

3.3. Comparison between Ignition Frequency of Structural Fires in Australia with Other Models/Statistics

Table 7 provides a comparison of the ignition frequency for hotels according to different models or statistics. It is clear that the average ignition frequency for hotels varies significantly among the different models and statistics. The Italian statistics show the lowest average ignition frequency at 1.84×10^{-6} fire/m^2·year, while the Ramachandran model shows the highest average ignition frequency at 8.00×10^{-5} fire/m^2·year. The generalized Barrois model and the Finnish statistics fall on the lower end of the spectrum, with average ignition frequencies of 4.20×10^{-6} and 4.70×10^{-6} fire/m^2·year, respectively. The Australian statistics also show a relatively high average ignition frequency of 3.89×10^{-5} fire/m^2·year. It should be noted that these values may be influenced by various factors such as the age and type of the building, fire protection measures in place, and jurisdictional fire codes and regulations.

Table 7. Hotel average ignition frequency according to different models or statistics.

Model/Statistics	Ignition Frequency 1×10^{-06} [fire/m^2·year]	Source
Italian statistics (hotels)	1.84	[13] Malagnino
Generalized Barrois model	4.2	[6] Tillander
Finnish statistics (commercial buildings)	4.7	[5] Tillander & Keshi-Rahkonen.
Australian statistics (hotel)	38.9	[20] Arup & UQ
Ramachandran model	80	[15] Ramachandran

The actual fire probability for different building categories can also be compared with the values provided in the British Standards BSI PD7974-7-2019 [23]. The BSI provides the value in fires/year per building; it is, therefore, necessary to have the total number of hotels in Australia. From Australian statistical data, the average number of hotels between 2011 and 2016 is 4337. In the same period, the average number of fires in hotels is 445. Therefore, the frequency of fire is 0.103 fires/year. This value can be compared with the value provided in BSI PD7974:2019, which is 0.046 fires/year. This indicates that the PD7974:2019 standard underestimates fire risk for this particular category. Or put simply, ignition frequency for hotels is lower in the UK than in Australia.

Houses: There are sixteen statistical observations for the residential occupancy (2012–2019 values for 'Houses' and 'Apartments'). The average fire ignition frequency for Australia is 4.65×10^{-6} for 'Houses' and 9.60×10^{-6} for 'Apartments'. For the residential category the average value of the Barrois model is 2.86×10^{-6}, so again, the model underestimates the statistical values considerably. The comparison with the British Standard is based on the number of buildings in that category, given that the probability of fire is expressed in those terms and not floor area units. These values are reported in Table 8.

Table 8. Comparison of ignition frequencies with the British standard.

Year	Number of Residential Buildings	Number of Fires	Frequency	PD 7974-7
2011	1,798,878	11,654	6.47×10^{-3}	0.13×10^{-2}
2016	2,206,875	10,442	4.73×10^{-3}	0.13×10^{-2}

Compared to both 2011 and 2016 Australian data, BSI PD 7974:2019 largely overestimates the ignition frequency and hence, the fire risk for residential occupancy.

Offices: The average value for Barrois model calculations is far below the average value extracted from the Australian data, as shown in Figure 11; in fact, the average calculated value is 2.20×10^{-6} fire/m^2·year against an average value of 1.19×10^{-5} fire/m^2·year for the years between 2012 and 2019.

Retail spaces: Figure 11 shows that the generalized Barrois curve for 'Commercial' buildings is below the Australian data. This indicates that the methodology underestimates the ignition frequency in this particular case, and hence, the curve must be shifted upwards to match the statistical data. The average statistical value is 5.29×10^{-6}, and the calculated value is around 2.54×10^{-5}.

Hospitals and Schools: For 'hospital' buildings, the average value from the generalized Barrois (5.29×10^{-6}) is lower than the Australian statistical data (1.69×10^{-5}), indicating an underestimation of the ignition frequency value for the Australian context (1.53×10^{-5}).

Also, in schools, the Barrois model underestimates the fire frequency, with an average value of 5.29×10^{-6} against the statistical value of 7.79×10^{-6}. The difference is about 2.50×10^{-6}.

The analysis above is summarized in Table 9.

Table 9. Summary of the ignition frequency comparison for different building stocks.

Type of Occupancy	Average Frequency from Barrois Model (fB)	Average Frequency from Statistical Data (fs)	Variations (fB − fs)
Residential	2.86×10^{-6}	5.20×10^{-6}	-2.33×10^{-5}
Hotels (Others)	5.29×10^{-6}	3.89×10^{-5}	-3.36×10^{-5}
Offices	2.20×10^{-6}	1.19×10^{-5}	-9.70×10^{-5}
Retail spaces (Others)	5.29×10^{-6}	2.54×10^{-5}	-2.01×10^{-5}
Hospitals (Others)	5.29×10^{-6}	1.69×10^{-5}	-1.16×10^{-5}
Schools (Others)	5.29×10^{-6}	7.79×10^{-6}	-2.50×10^{-6}

Figure 11 shows the comparison between the generalized and improved Barrois curves for all building categories in Australia, while Figure 12 shows the generalized Barrois curves for building categories in Finland.

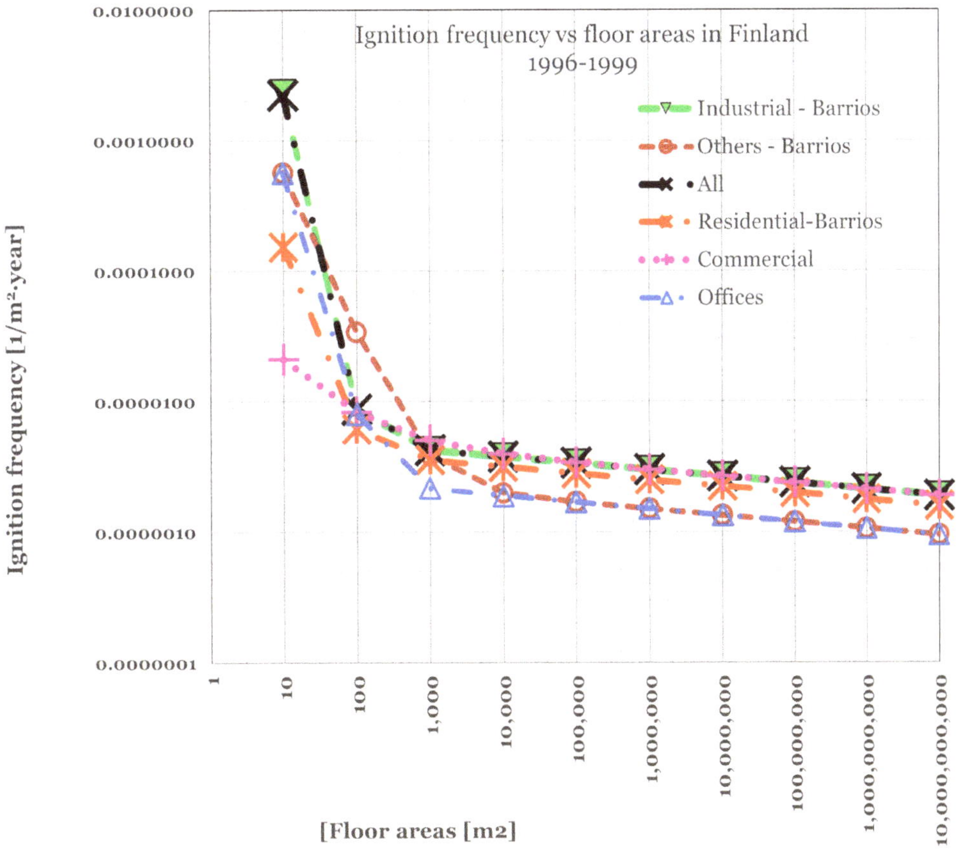

Figure 12. Generalized Barrois curves for building categories in Finland 1996–1999.

4. Main Conclusions

Historically, it has been shown that ignition frequency is dependent on the floor area of the building, with a weak dependency on other factors. This study examined the ignition frequency of structural fires in Australia between 2012 and 2019, using considerable statistical data drawn from the AIRS Database. The main conclusions are as follows:

1 The Barrois model, as found in the literature and used in other previous studies, cannot be fully applied to the Australian context; typically, the statistical data would have a reasonably good fit with the Barrois model, but in this case, the Barrois curve underestimates the ignition frequencies when compared with the Australian fire statistics;
2 Some categories, such as hotels and hospitals, were not dealt with as separate categories, resulting in deviations. In this study, both are treated as different categories;
3 When the fire ignition frequency for structural fires in Australia is compared with Finland, several conclusions can be made:
    - As a general observation, it can be said that the expected fire ignition frequency for all buildings is lower in Australia than in Finland;

- The opposite occurs when specific categories are considered: for example, office spaces in Finland have an ignition frequency which is five times lower than in Australia. The same can be argued for retail spaces. The ignition frequency for schools is four times lower in Finland (1.93×10^{-6}) than in Australia (7.79×10^{-6});
- There are several reasons for this. First, one can observe that in Finland, timber is primarily adopted as a building material, while in Australia, houses are constructed with timber and/or concrete frames, internal plasterboard walls and external facing bricks. This would lead to a greater probability of fire ignition for homes in Finland, as timber is combustible, whereas concrete is not. On the other hand, specific categories of buildings (office, retail, and apartments, in particular) are way more prone to fire ignition in Australia; jurisdiction-specific rules about electrical installations, fire loads, and fire alarm systems, among others, can influence the spreading of fires, as well as climatic differences (higher temperature and droughts in Australia can act in favor of fire ignition). The most relevant category is residential in both regions.

4. In this study, we analyzed the aggregate national value for each building class in different years. For example, we found that the average fire frequency for hotels is 3.89×10^{-5} fire/m^2·year, and it is assumed to remain constant for both small and large floor area values. We then compared the mean value of the calculated data with the historical series to determine the accuracy of the model. Our analysis showed that the ignition frequency is the highest for hotels and the lowest for houses. The error bars indicate that the frequencies vary the most for retail buildings, followed by hospitals, hotels, schools, offices, and apartments;
5. We found that the Barrois curve tends to underestimate the ignition frequency when compared with statistical data from Australia for a variety of building categories, including houses, hotels, offices, hospitals, and schools. By introducing an additional coefficient c_3 to the generalized Barrois equation, we were able to obtain a better fit for these different categories. Our proposed improvements to the generalized Barrois model for calculating fire ignition frequency in Australia are an important contribution to the field. The improved model would be valuable for risk quantification in ABCB's NCC and for quick determination of fire ignition frequency for QPRA. Overall, our results suggest that the generalized Barrois model with the additional coefficient c_3 provides a more accurate and reliable tool for predicting fire ignition frequencies for various building categories in Australia;
6. This study updates and improves on the generalized Barrois model proposed by Rahikainen et al. [4–6] for a better fit to the Australian context with the inclusion of a new coefficient c_3; for future research, this can be tested and validated in other jurisdictions and building categories.

More real AIRS data needs to be collected so that we can calibrate the model to a finer level of granularity.

Author Contributions: Conceptualization, S.T. and K.M.; methodology, S.T. and K.M.; software, S.T.; validation, S.T., K.M. and P.J.; formal analysis, S.T. and K.M.; resources, S.T. and K.M.; data curation, S.T. and K.M.; writing—original draft preparation, S.T.; writing—review and editing, S.T., K.M. and P.J.; visualization, S.T. and K.M.; project administration, K.M. and P.J.; funding acquisition, S.T. and K.M. All authors have read and agreed to the published version of the manuscript.

Funding: This research received no external funding.

Institutional Review Board Statement: Not applicable.

Informed Consent Statement: Not applicable.

Data Availability Statement: Not applicable.

Conflicts of Interest: The authors declare no conflict of interest.

References

1. Rahikainen, J.; Keski-Rahkonen, O. Statistical determination of ignition frequency of structural fires in different premises in Finland. *Fire Technol.* **2004**, *40*, 335–353. [CrossRef]
2. ABCB. *Building Code of Australia Volume 2, Class 1 and Class 10 Buildings, Australia Building Codes Board*; ABCB: Canberra, Australia, 2016.
3. MacLeod, J.; Tan, S.; Moinuddin, K. Reliability of fire (point) detection system in office buildings in Australia—A fault tree analysis. *Fire Saf. J.* **2020**, *115*, 103150. [CrossRef]
4. Rahikainen, J.; Keski-Rahkonen, O. Determination of ignition frequency of fires in different premises in Finland. *Fire Eng. J.* **1998**, *58*, 33–37.
5. Tillander, K.; Keski-Rahkonen, O. The ignition frequency of structural fires in Finland 1996–1999. *Fire Saf. Sci.* **2003**, *7*, 1051–1062. [CrossRef]
6. Tillander, K. Utilisation of Statistics to Assess Fire Risks in Buildings. Ph.D. Thesis, VTT Technical Research Centre of Finland, Espoo, Finland, 2004.
7. Krasuski, A.; Hostikka, S. AAMKS—Integrated cloud-based application for probabilistic fire risk assessment. *Fire Mater.* **2021**, *45*, 744–756. [CrossRef]
8. Xin, J.; Huang, C. Fire risk analysis of residential buildings based on scenario clusters and its application in fire risk management. *Fire Saf. J.* **2013**, *62*, 72–78. [CrossRef]
9. Tan, S.; Weinert, D.; Joseph, P.; Moinuddin, K.A.M. Incorporation of technical, human and organizational risks in a dynamic probabilistic fire risk model for high-rise residential buildings. *Fire Mater.* **2021**, *45*, 779–810. [CrossRef]
10. Tan, S.; Weinert, D.; Joseph, P.; Moinuddin, K. Impact of technical, human, and organizational risks on reliability of fire safety systems in high-rise residential buildings—Applications of an integrated probabilistic risk assessment model. *Appl. Sci.* **2020**, *10*, 8918. [CrossRef]
11. Tan, S.; Weinert, D.; Joseph, P.; Moinuddin, K. Sensitivity and uncertainty analyses of human and organizational risks in fire safety systems for high-rise residential buildings with probabilistic T-H-O-risk methodology. *Appl. Sci.* **2021**, *11*, 2590. [CrossRef]
12. Tan, S.; Weinert, D.; Joseph, P.; Moinuddin, K.A.M. A Dynamic Probabilistic Fire Risk Model Incorporating Technical, Human and Organizational Risks for High-Rise Residential Buildings. In Proceedings of the 2019 Interflam Fire Science and Engineering Conference, London, UK, 1–3 July 2019; Volume 1, pp. 937–950.
13. Malagnino, A. Integrating Statistics based Fire Risk Assessment with Building Life-Cycle Management. Ph.D. Thesis, University of Salento, Lecce, Italy, 2020; pp. 1–251.
14. Johansson, U.; Hui, M.C. *Fire Safety of Early Childhood Centres in High Rise Buildings in Australia*; ABCB: Canberra, Australia, 2019.
15. Ramachandran, G. Statistical methods in risk evaluation. *Fire Saf. J.* **1980**, *2*, 125–415. [CrossRef]
16. Sandberg, M. Statistical Determination of Ignition Frequency. Master's Thesis, Lund University, Lund, Sweden, 2004.
17. D'Este, M.; Ganga, A.; Elia, M.; Lovreglio, R.; Giannico, V.; Spano, G. Modeling fire ignition probability and frequency using Hurdle models: A cross-regional study in Southern Europe. *Ecol. Process.* **2020**, *9*, 54. [CrossRef]
18. Hu, J.; Shu, X.; Shen, S.; Yan, J.; Tian, F.; He, S. A method to improve the determination of ignition probability in buildings based on Bayesian network. *Fire Mater.* **2021**, *46*, 666–676. [CrossRef]
19. Tan, S.; Moinuddin, K. Systematic review of human and organizational risks for probabilistic risk analysis in high-rise buildings. *Reliab. Eng. Syst. Saf.* **2019**, *188*, 233–250. [CrossRef]
20. Arup; The University of Queensland. *Risk Metrics Data Study—Australian Building Codes Board*; ABCB: Canberra, Australia, 2021.
21. Barrois, T.J. Essai sur L'application du Calcul des Probabilités aux Assurances Contre les Incendies [A Proposition on the Application of Probability Theory on Fire Insurance]. In *Mémoires de la Société Royale des Sciences, de L'agriculture et des Arts de Lille*; 1834; Volume 11, pp. 85–282. (In French)
22. Moinuddin, K.; Tan, S. *Future Data Collection Strategy for the Quantification of Fire Safety Performance Requirement*; Australian Building Codes Board: Canberra, Australia, 2020.
23. *BSI, PD 7974-7:2019*; Application of Fire Safety Engineering Principles to the Design of Buildings—Part 7: Probabilistic Risk Assessment, British Standards Published Document. British Standards Institution: London, UK, 2019.

fire

MDPI

Article

Understanding Building Resistance to Wildfires: A Multi-Factor Approach

André Samora-Arvela [1,*], José Aranha [2], Fernando Correia [1], Diogo M. Pinto [1], Cláudia Magalhães [3] and Fantina Tedim [1]

1 Research Centre in Geography and Spatial Planning, CEGOT, Geography Department, Faculty of Arts and Humanities, University of Porto, Via Panorâmica, 4150-564 Porto, Portugal
2 Research Centre for the Research and Technology of Agroenvironmental and Biological Sciences (CITAB), University of Trás-os-Montes and Alto Douro, 5001-801 Vila Real, Portugal
3 Faculty of Arts and Humanities, University of Porto, Via Panorâmica, 4150-564 Porto, Portugal
* Correspondence: anesamora@letras.up.pt

Abstract: In terms of researching fire-related structure loss, various factors can affect structure survival during a wildfire. This paper aims to assess which factors were determinants in house resistance in the specific context of a case study of an extreme wildfire in the Central Region of Portugal and therefore which factors should be taken into account in the definition of a municipal mitigation strategy to defend buildings against wildfires. In this context, it is possible to conclude that various factors presented a predominant influence, some in building destruction and others in building survival. The existence of overhanging vegetation and lack of defensible space constitute major factors for structure destruction. the inherent wildfire severity, the location in the forest area, and the structure's isolation from major roads were equally important factors that induced house destruction. Building survival was determined by its increasing distance from the forest and by its location in a dense urban agglomeration. Thus, a strategy to enhance resilience should include the prohibition of roof overhanging vegetation and the restriction of building permits in forest areas, in isolated locations, and/or very far from major roads. These orientations can be extrapolated to municipalities with similar susceptibility and vulnerability to wildfires.

Keywords: wildfire; building survival; building destruction; factors

check for updates

Citation: Samora-Arvela, A.; Aranha, J.; Correia, F.; Pinto, D.M.; Magalhães, C.; Tedim, F. Understanding Building Resistance to Wildfires: A Multi-Factor Approach. *Fire* **2023**, *6*, 32. https://doi.org/10.3390/fire6010032

Academic Editor: Tiago Miguel Ferreira

Received: 21 December 2022
Revised: 10 January 2023
Accepted: 11 January 2023
Published: 13 January 2023

1. Introduction

In a context where their frequency and intensity are expected to increase, wildfires pose major challenges to the existence of people and buildings in hazardous wildlands. The implantation of such buildings in highly flammable areas is tendentially allowed by the spatial planning decision-making process and is seen as a factor in increasing wildfire exposure.

As such, an urgent need to understand why and how structures are being destroyed during wildfires is becoming paramount. Therefore, in terms of understanding why buildings are destroyed, there is an emerging literature regarding the various factors that can determine structure destruction during a wildfire [1–10], namely those related to ornamental vegetation in the building's surroundings, defensible space, landscape scale, building construction materials, and building use, among others.

1.1. Ornamental Vegetation in Building Surroundings

Ornamental vegetation and edges around the properties and along roads can provide fuel continuity to fire propagation [2]. The lack of clear-cutting around the house and the existence of tall trees overhanging the roof area are major causes of wildfire building destruction [2,5].

Structures are more likely to be destroyed if they are surrounded by wildland vegetation than by urban or impervious space [7]. Nonetheless, in some cases, property loss is more or as likely within herbaceous areas when compared to wooded areas with higher fuel volume [7]. This is because low-fuel-volume grasslands can facilitate high wind speed and fire spread, and can promote longer fire seasons and high fire frequency given its low fuel moisture and low heat requirements to ignition [7].

Therefore, Syphard et al. [8] concluded for the case of San Diego County, California that the most effective actions were the reduction of wooden cover by up to 40% immediately adjacent to the building and the assurance that vegetation does not overhang or touch the house.

1.2. Defensible Space Dimension

When researching the role of defensible space in building destruction, the number of empirical case studies is growing. Several studies have empirically tested the relative benefits of defensible space and concluded that vegetation reduction up to 30 m around the houses increases the probability of building survival [8,11–13]. Regarding the 2013–2018 California wildfires, Syphard and Keeley [10] found that most of the buildings that had more than 30 m of defensible space were destroyed, meaning that the optimum defensible space is considered to be much shorter since there is no protection gained beyond the 30 m of defensible space [8,10,14].

In this context, the state of California requires a minimum of 30 m of defensible space around structures. However, some Californian localities are requiring a lot more: namely a minimum of 60 m in certain circumstances [10].

Regarding this factor, the Portuguese wildfire management framework regulates a defensible space of 50 m around structures [6], while the effectiveness of a 30 m defensible space has never been tested.

1.3. Landscape-Scale Factors

In this line of thought, other factors stand out, namely those at the landscape-scale, such as: housing location, building density, slope, and firefighter access given by road proximity. These determine the spatial arrangement of buildings and its significant influence on the likelihood of building damage and destruction [7,14]. Since many of the relationships between these landscape-scale variables are nonlinear [7], it is recommended to test their performance in each case study.

As for building density, various studies have empirically found that the probability of building loss due to wildfires is highest in small and isolated building clusters, considering their low-medium house density and the existence of sparse roads [1,14].

The strong correlation between low–medium structure density and fire destruction can derive from interspersed houses, which have a larger perimeter of contact with wildland vegetation and, because of that, are more exposed to wildfire [1].

Another justification can be grounded on the more difficult and expensive access of these low-density rural buildings that may not be prioritized in comparison to high-density and high-value developments. This is the reason why distance to major roads is of prime importance for low-density building survival [9,15]. Isolated structures can therefore be more difficult for firefighters to access and defend [7,16–19]. Another explanation could be that scattered buildings are, in some cases, located in high-slope areas with high fire risk and low accessibility.

Apart from low- to medium-density structure destruction, high-density communities are also not completely resistant during a wildfire. Small clusters with a large number of buildings can have a high probability of structure destruction due to the closeness of houses and the inherent susceptibility of home-to-home fire spread [20].

In the Santa Monica Mountains, structures located on the edge of housing developments and housing clusters on steep slopes were found to be highly susceptible [7].

Since the main construction type in Portugal is masonry, 'structure to structure' ignition is not common when compared to other countries with predominantly wooden or metal construction materials [6,21,22].

Given this, Ribeiro et al. [6] conclude that structures in compact urban areas are protected, especially those situated at the cluster's core.

1.4. Construction Materials

In terms of exterior construction materials, stucco was associated with the highest proportion of building survival over masonry, wood, or metal. Regarding roofing material, tile offers more building resistance, being responsible for the highest house survival proportion when compared to composite, shake, metal, or shingle [9].

As a factor in wildfire house survival, roofing is considered more important than a home's siding [20]. However, beyond the roof covering, various sub-factors should be taken into account when studying roofing and its wildfire resistance capacity, namely: the edge, the roof complexity, whether or not the roof was treated with fire retardant [20], the accumulation of combustible leaf litter or debris on the roof, and— since if the roofs are not sealed embers could have a point of entry through the roof breaches [9]—the degree of roof maintenance [20].

Regarding building loss by ember flow, Syphard and Keeley [10] stated that factors such as exterior siding or roofing material were much less important than exposed eaves, vents, or windows. Bowditch et al. [22] assessed that windows, especially framing materials and panes, were more important than roofs or siding concerning structure survival since they can be an easy point of entry for firebrands.

During Portugal's Pedrógão Grande extreme fire [6], the majority of ignition points were on the structure's roofs (61.8%), and the windows were the second most representative point of ignition (16%), where vents without particle retention systems, mainly in older houses, were the point of entry for firebrands.

In this context, multi-pane windows are more resistant to thermal exposure than single-pane windows [23]. Also important is the type of glass in the window, as different types offer different levels of resistance to radiant heat [22,23].

Vinyl is characterized by a lower melting point and lower ignition temperature than aluminum [9]. The highest proportion of building survival was reported in structures using vinyl as a window framing material when compared to metal and wood, and especially when comparing dual-pane with single-pane [9]. It should be acknowledged that vinyl and dual-paned windows are more often present in newer constructions; this is one reason why structure age can be related to building survival during a wildfire [9], as sometimes these construction innovations are a consequence of implementing newer building codes.

1.5. Building Use

For Ribeiro et al. [6], in the Pedrógão Grande wildfire, the low to medium degree of building preservation was related to the high level of structure damage. Building use was also a factor that determined building resistance. The main groups of the destroyed and damaged structures were shed/storage (38.6%), which cannot be prioritized by firefighters dealing with an extreme wildfire. Uninhabited houses (12.7% of the total damaged buildings), and secondary housing (11.9%) can be less well defended when compared to primary and permanent housing, which only represented 13.3% of the damaged buildings.

Active defense driven by residents can also decrease the likelihood of structure destruction [5,24]. For example, the Australian government's wildfire safety policy advocated for a 'Prepare, stay and defend, or leave early' strategy in the period 2005–2009 [24,25]. However, after the February 2009 Victorian bushfires this policy was repudiated and replaced with the 'Prepare, Act, Survive' policy, which emphasizes the evacuation to the detriment of householder response to bushfire threat [26] (pp. 491–492).

1.6. Combining Factors

Focusing on the study area of San Diego County, CA, USA, Syphard et al. [9] found that the most important factors for explaining structure loss to wildfires were mainly: structure density, building age, slope, presence of vegetation overhanging the roof, distance to major roads, window framing and windowpanes, roofing and siding material, and defensible space dimension.

Price et al. [5] examined the wildfire determinants of damage for Eastern Australia, listing, in this context, six main themes: preparedness actions (management of fuels up to 30 m around the house, namely the defensible space); responsive actions (undertaken to defend from wildfires); house construction (materials, shape, gaps); landscape fuels (i.e., beyond the defensible space of 30 m from houses); and topography and weather. They concluded that the major factors of building survival were preparedness by the provision of defensible space and defensive actions, since the houses that were not defended had almost three times the probability of being damaged or destroyed [5].

1.7. Research Questions

This paper aims to assess which factors were determinants in house resistance in the specific context of Portugal's Central Region through the case study of the extreme wildfire of 15 October 2017 in Santa Comba Dão municipality, reflecting about which factors should be taken into account in a mitigation strategy for house losses. We defined three research questions (RQs), namely:

RQ1: What factors most determine the destruction of buildings by wildfire in the example of the municipality of Santa Comba Dão?

RQ2: Knowing from the scientific literature that there is no gain of protection beyond the 30 m of defensible space, what is the importance of the 30 m defensible space for building destruction in Santa Comba Dão municipality?

RQ3: Is the 50 m dimension of defensible space regulated for Portuguese territory justified?

2. Materials and Methods

2.1. Study Area

Fifteen people died and a total of 54,407 hectares (ha) were burned as a result of the Lousã extreme wildfire in the Central Region of Portugal [27] that started at 08:41 (a.m.) on 15 October 2017 in the village of Prilhão, before quickly spreading to Santa Comba Dão municipality (Figure 1).

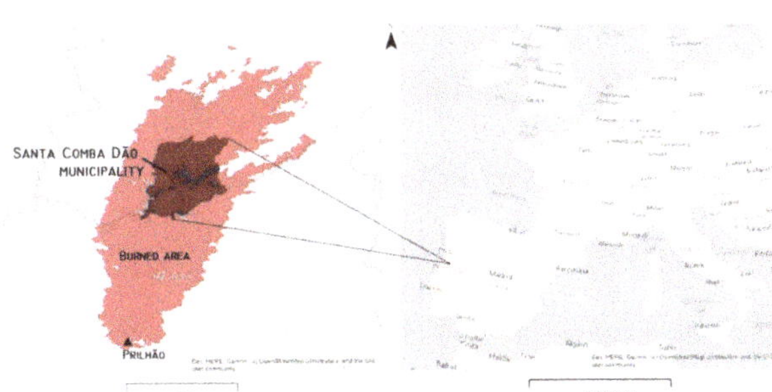

Figure 1. Santa Comba Dão municipality and 15 October 2017 extreme wildfire.

This municipality was chosen as a case study because of its variety of land uses, its large burned area, and the high number of buildings impacted by the 2017 wildfire. Santa Comba Dão is an 11,195 ha town and municipality partially bordered by the Mondego River in the Viseu District, Central Region. It had a population of 11,597 inhabitants in

2011 [28] and 10,641 inhabitants in 2021 [29], corresponding to a population decrease of 8.4% in ten years.

The municipality is divided into six civil parishes (Ovoa and Vimieiro; Pinheiro de Ázere; Santa Comba Dão and Couto do Mosteiro; São Joaninho; São João de Areias;and Treixedo and Nagozela). The main settlements are interconnected by a web of roads of different importance (Figure 2).

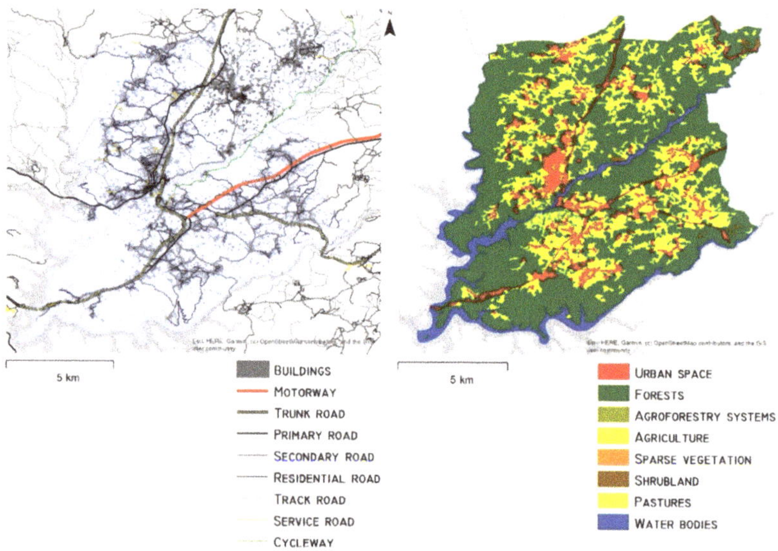

Figure 2. Municipality road and land use structure [30].

Forests represent 60.03% of the municipal territory, with about 43% % of the forest area occupied by *Eucalyptus* globulus [30]. According to the Land Use/Cover Map (COS 2015), the remaining landscape structure is constituted by agriculture (23.03%), urban space (8.08%), water bodies (7.37%), pastures (0.41%), agroforestry systems (0.18%), and sparse vegetation (0.02%) [31] (Figure 2).

Eucalyptus globulus is present in recent plantations for pulp and paper; frequently on terraces in fragmented areas of poor management and land abandonment. In such forests, *Eucalyptus* is mixed with other tree species; mostly pines (*Pinus pinaster*), oaks (*Quercus spp.*), and the exotic *Acacia dealbata*.

The proliferation of eucalyptus in Portugal is in part explained by the occurrence of wildfires, which encourages forest owners to replace pine with species with shorter life cycles which are therefore compatible with recurrent fires [32].

The areas burned on 15 October 2017 now exhibit massive *Eucalyptus* wildlings, reaching hundreds of plants per square meter in some areas with limited presence of native shrubs (*Cistus psilosepalus*, *Rubus sp.*, *Ulex sp.*) and herbaceous species (mostly *Pteridium aquilinum* and *Rubia peregrina*) [30]. The presence of wildlings adds to the current dangerous naturalization of *Eucalyptus* that is gradually increasing the likelihood of fire in many areas of Portugal [30].

2.2. Data Assembly and Analysis

To study which factors determined building resistance in Santa Comba Dão, we ran exploratory regression models in which various data were assembled to assign a numerical value to each variable influencing structure survival.

Firstly, the building damage and loss during the 2017 wildfire was assessed using digital image interpretation in Google Earth [33]. Vectorial data on primary housing reconstructed by the Portuguese Centre's Coordination and Development Commission (CCDR-Centro) was given by the Santa Comba Dão City Council [34]. To finalize the burned buildings database, some field visits were made to clarify doubtful identifications made during digital image interpretation.

Afterward, we defined three categories of house damage and loss from the resulting building damage and loss vectorial database: 0—Unaffected buildings; 1—Buildings little affected; 2—Buildings partially destroyed; and 3—Buildings completely destroyed) using a Geographic Information System (GIS).

Secondly, data referring to the definition of explanatory variables were assembled, before being extracted as multi-values to building damage and loss shapefile polygons in GIS. The variables were organized according to the following themes:

The variable regarding defensible space was:

- *Presence of trees and shrubs in a 30 m radius around the buildings affected*: using digital image interpretation, we identified the affected buildings that had surrounding trees and shrubs before the 2017 wildfire. This variable was measured using digital image interpretation with the aim of studying the importance of non-compliance with the 30-meter defensible space in explaining the destruction of buildings in the municipality studied.

The landscape-scale variables were:

- *Distance to the forest*: We used the Portuguese Land Use/Cover Map for 2015 (before the 2017 wildfire: COS 2015), provided by the Portuguese General Directory of Territory (DGT), to identify forested areas. The distance to forest areas was calculated in GIS, being the distance values integrated into each vectorial building polygon. This variable was chosen since it was assumed that buildings farther from forest areas would be more resilient to a rural fire.
- *Distance to major roads*: the distance to major roads was calculated in GIS using a road shapefile, and values were extracted to each polygon of the building shapefile. The source of the road shapefile was Open Street Map [35]. This variable was chosen because a greater distance of buildings from roads can result in less accessibility for firefighters, leading to more destruction of buildings.
- *Distance to minor roads*: calculated in GIS as with major roads.
- *Building density*: I was possible to calculate building density using the GIS Kernel density module and the building damage and loss shapefile. This variable was chosen because the scientific literature indicated that denser urban areas are more resistant to fires.
- *Euclidean nearest neighbor building*: this landscape metric of building isolation was calculated from the housing database using the *Fragstats 4.2.* software [36], with results automatically aggregated to building polygons. This is an indicator of dispersion of buildings in the territory. As such, this variable was chosen in order to study whether more dispersed buildings were more damaged by the fire.
- *Slope*: the slope map was determined in GIS from a digital elevation model (DEM) derived from satellite data (STRM) with a spatial resolution of 25 m [37]. The slope values were then extracted through GIS for each building polygon. This variable was chosen in order to reveal its influence on the destruction of buildings by fire.

The type of construction materials used and the presence of concrete slabs can determine the destruction of a building. As such, related independent variables were included in the multiple linear regression. The variables related to building construction materials were:

- *Concrete structure*: the number of buildings with reinforced a concrete structure was obtained from the shapefile of the Geographical Base for Referencing Information (BGRI) subsections of the 2011 Portuguese Census [28]. The values were integrated into the study area building database using GIS.
- *Structures with slabs*: the number of buildings constructed from masonry walls with slabs was calculated as with concrete structure.

- *Adobe structure*: the number of buildings with constructed from adobe walls or loose stone masonry was calculated as above.
- *Other structure*: the number of buildings with other types of structure was calculated as above;
- *Structures without slabs*: the number of buildings with masonry walls without slabs was calculated as above.

The implantation area and the number of rooms in a building can affect the likelihood of destruction of the building. To test this influence in a multiple linear regression, we included the following variables based on building area:

- *Up to 50 m^2*: the number of buildings with an area up to 50 m^2 within each BGRI subsection shapefile was calculated using the previous method.
- *50–100 m^2*: the number of buildings with an area between 50 and100 m^2 within each BGRI subsection shapefile was calculated using the previous method.
- *100–200 m^2*: the number of buildings with an area between 100 and200 m^2 within each BGRI subsection shapefile was calculated using the previous method.
- *More than 200 m^2*: the number of buildings with an area above 200 m^2 within each BGRI subsection shapefile was calculated using the previous method.
- *1–2 rooms*: the number of buildings with 1 or 2 rooms within each BGRI subsection shapefile was calculated using the previous method.
- *3–4 rooms*: the number of buildings with 3 or 4 rooms within each BGRI subsection shapefile was calculated using the previous method.

In order to test the influence of the predominant land-use type in the place where each building is located on its destruction by fire, the following variables were selected for inclusion in the multiple linear regression:

- *Urban space*: we distinguished between spaces listed as urban areas in COS 2015 and those that were not using binary differentiation in GIS (0–No urban space; 1–Urban space). These binary attribute values were then indexed to each building polygon in the study area.
- *Agriculture*: the previous method was repeated for buildings located in areas described by COS 2015 as agriculture areas.
- *Forests*: the previous method was repeated about the buildings located in areas described by COS 2015 as forest areas.
- *Agroforestry systems*: the previous method was repeated for the buildings located in areas described by COS 2015 as agroforestry systems.
- *Pastures*: the previous method was repeated for the buildings located in areas described by COS 2015 as pasture areas.
- *Shrubland*: the previous method was repeated for the buildings located in areas described by COS 2015 as shrubland areas.
- *Sparse vegetation*: the previous method was repeated about the buildings located in areas described by COS 2015 as containing sparse vegetation.

2.3. Multiple Regression Analysis

The entire previous data assembly was intended to identify the influence of each factor on the destruction of buildings by wildfire. In this context, the most robust and adequate method for studying this is multiple linear regression [38]. For that reason, it was chosen as a main method.

The previous definition of three categories of house damage and loss expresses the degree of building damage. This served as the dependent variable in the multiple regression.

Five thematic multiple linear regression models were conducted using GIS and Statistical Package for the Social Sciences (SPSS) for Windows (Version 26) [39] to identify the explanatory factors of Santa Comba Dão's structural resistance to wildfire, based on explanatory variables on the following themes:

- Defensible space.

- Landscape-scale factors.
- Building construction materials.
- Building area.
- Land use type at building location.

The multiple regression results for the five theme factors are presented and discussed in the next sections.

3. Results

3.1. Number of Destroyed and Affected Buildings

From a total of 26854 buildings in Santa Comba Dão municipality, 2.9% (782) were completely destroyed, 0.12% (32) were partially destroyed (up to 50% of the implantation area affected), 0.03% (8) were slightly affected, and 96.90% (25,762) were not affected (Figure 3).

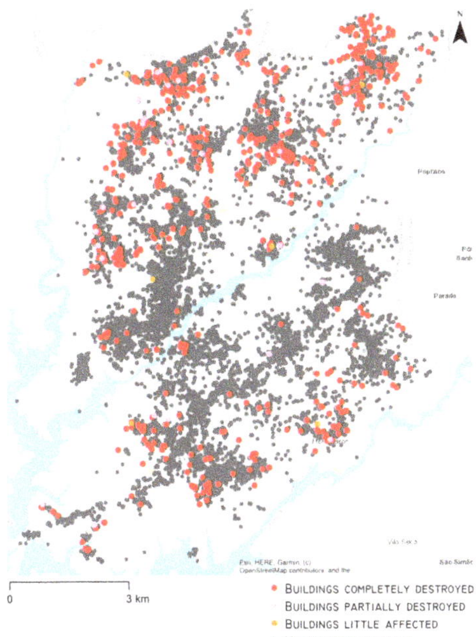

Figure 3. Degree of damage/destruction of buildings after the 2017 wildfire in Santa Comba Dão municipality.

3.2. Multiple Regression Regarding the Defensible Space

The presence of trees and bushes within a radius of 30 m from the buildings was the most important determining factor in the degree of damage/destruction of the buildings ($\beta = 0.881$; p-value = 0). This variable accounted for 77.60% of the damage/destruction-degree variance (Table 1).

Table 1. Summary of the multiple regression significance of defensible space explanatory variables.

Independent Variables	Non-standardized Coefficients		Standardized Coefficient	t	*p*-Value
	B	Error	Beta (β)		
Constant	0.019	0.002		12.795	0.000
Buildings with trees and shrubs in a 30 m buffer	2.902	0.010	0.881	303.263	0.000

No. of data points = 26854; F = 91968.188; R = 0.881; Adjusted R^2 = 0.776; *p*-value = 0.000.

3.3. Multiple Regression Regarding Landscape-Scale Factors

The influence of distance to forest areas (variable: distance to forest) had a negative influence (β = −0.042; *p*-value = 0). Therefore, it is possible to infer that the further away the buildings are from the forest areas, the less serious their destruction and thus the greater their resistance. The increasing distance to major (variable: distance to major roads) (β = 0.122; *p*-value = 0) and minor (variable: distance to minor roads) (β = 0.045; *p*-value = 0) roads had a statistically significant positive influence on damage and destruction. The greater the density of buildings in Santa Comba Dão (variable: building density), the less serious their destruction; the variable of building density displayed a negative influence (β = −0.023; *p*-value = 0.003). Neither the Euclidean nearest neighbor building, as a measure of isolation/proximity between buildings (variable: ENN building) nor the slope (variable: slope) presented any statistical significance for analysis purposes.

These independent variables only represented 2.4% (Adjusted R^2 = 0.024) of the variance of the dependent variable (degree of damage/destruction of buildings) (Table 2).

Table 2. Summary of the significance of building scale explanatory variables in the multiple regression.

Independent Variables	Non-standardized Coefficients		Standardized Coefficient	t	*p*-Value
	B	Error	Beta (β)		
Constant	0.045	0.009		5.318	0
Distance to forest	0	0	−0.042	−5.612	0
Distance to major roads	8.440×10^{-5}	0	0.122	19.626	0
Distance to minor roads	0.000	0	0.045	6.528	0
Building density	-1.819×10^{-5}	0	−0.023	−2.973	0.003
ENN building	0	0	0.008	1.232	0.218
Slope	-3.101×10^{-5}	0	−0.008	−1.374	0.169

No. of data points = 26584; F = 110.48; R = 0.156; Adjusted R^2 = 0.024; *p*-value = 0.

3.4. Multiple Regression Regarding Building Construction Materials

A regression model was carried out to research the relationship between the degree of damage and the type of construction material of the buildings in the municipality under study at the scale of the BGRI subsection. It can be seen that the number of buildings with a reinforced concrete structure (variable: concrete structure) (β = −0.041; *p*-value = 0), buildings with a structure of masonry walls with slabs (variable: structure with slab) (β = −0.064; *p*-value = 0), buildings with a structure of adobe walls or loose stone masonry (variable: adobe structure) (β = −0.016; *p*-value = 0.011), and buildings with other types of structure (variable: other structure) (β = −0.020; *p*-value = 0.001), all have a clearly significant inverse relationship with the degree of damage. On the other hand, the number of buildings with masonry walls without slabs (variable: structure without slab) varied positively with the degree of damage to buildings affected by the 2017 wildfire (β = 0.020; p-value = 0.003).

These independent variables only represented 0.5% (Adjusted R^2 = 0.005) of the variance of the dependent variable (degree of damage/destruction of buildings) (Table 3).

Table 3. Summary of multiple regression significance of building materials explanatory variables.

Independent Variables	Non-standardized Coefficients		Standardized Coefficient	t	p-Value
	B	Error	Beta (β)		
Constant	0.128	0.005		26.876	0.000
Concrete structure	−0.002	0.000	−0.041	−6.093	0.000
Structure with slab	−0.005	0.000	−0.064	−9.874	0.000
Structure without slab	0.002	0.001	0.020	2.923	0.003
Adobe structure	−0.004	0.001	−0.016	−2.533	0.011
Other structure	−0.023	0.007	−0.020	−3.239	0.001

No. of data points = 26854; F = 28.983; R = 0.074; Adjusted R^2 = 0.005; p-value = 0.000.

3.5. Multiple Regression Regarding Building Area

The BGRI number of houses with an area of 50 to 100 m^2 (variable: 50–100 m^2) (β = −0.080; p-value = 0), with an area of 100 to 200 m^2 (variable: 100–200 m^2) (β = −0.036; p-value = 0), with an area greater than 200 m^2 (variable: more than 200 m^2) (β = −0.024; p-value = 0.001), and those that were a primary residence with one or two rooms (variable: 1–2 rooms) (β = −0.016; p-value = 0.012), had an inverse influence concerning the degree of damage to the building (Table 4). Smaller dwellings (with an area up to 50 m^2; variable: up to 50 m^2) (β = 0.019; p-value = 0.026) and those with more rooms (variable: 3–4 rooms) (β = 0.051; p-value = 0) seem to have less resistance due to their positive impact on the degree of damage.

Table 4. Summary of the significance of building area explanatory variables in the multiple regression.

Independent Variables	Non-standardized Coefficients		Standardized Coefficient	t	p-Value
	B	Error	Beta (β)		
Constant	0.136	0.005		28.272	0.000
Up to 50 m^2	0.006	0.003	0.019	2.225	0.026
50–100 m^2	−0.008	0.001	−0.080	−6.129	0.000
100–200 m^2	−0.003	0.001	−0.036	−3.797	0.000
More than 200 m^2	−0.008	0.002	−0.024	−3.381	0.001
1–2 rooms	−0.026	0.010	−0.016	−2.523	0.012
3–4 rooms	0.006	0.002	0.051	3.870	0.000

No. of data points = 26854; F = 24.511; R = 0.080; Adjusted R^2 = 0.006; p-value = 0.000.

These independent variables only represented 0.6% (Adjusted R^2 = 0.006) of the variance of the dependent variable (degree of damage/destruction of buildings) (Table 4).

3.6. Multiple Regression Regarding Land Use Type at Building Location

According to the results of the multiple regression, the buildings located in impervious areas of urban space as defined in the COS 2015 map, show resistance in the event of a wildfire due to the negative influence of this variable on the degree of damage (variable: urban space) (β = −0.050; p-value = 0) (Table 5). Dispersed buildings, located in patches of forest (variable: forests) are more susceptible to damage; this is confirmed by the positive variation between this variable and the degree of damage/destruction (β = 0.079; *p*-value = 0).

Table 5. Summary of significance of land use explanatory variables (COS, 2015) in the multiple regression.

Independent Variables	Non-standardized Coefficients		Standardized Coefficient	t	*p*-Value
	B	Error	Beta (β)		
Constant	0.115	0.013		8.525	0.000
Urban space	−0.054	0.014	−0.050	−3.990	0.000
Agriculture	0.004	0.014	0.003	0.294	0.769
Forests	0.152	0.016	0.079	9.607	0.000
Agroforestry systems	−0.115	0.114	−0.006	−1.005	0.315
Pastures	−0.080	0.060	−0.008	−1.331	0.183
Shrublands	0.052	0.124	0.003	0.423	0.672
Sparse vegetation	0.313	0.193	0.010	1.627	0.104

No. of data points = 26854; F = 45.904; R = 0.109; Adjusted R^2 = 0.012; *p*-value = 0.000.

The other variables did not present statistical significance in this regression.

These independent variables only represented 1.2% (Adjusted R^2 = 0.012) of the variance of the dependent variable (degree of damage/destruction of buildings) (Table 5).

4. Discussion

From the results of the Santa Comba Dão case study, it is possible to state that various factors present a predominant influence in building destruction, with some corroborating the influence of factors identified in the scientific literature.

Compared with the Australian and Californian studies [5,7–10,24,25], it appears that non-compliance with a defensible space of 30 m is the most important factor determining building destruction. This factor was responsible for 77.60% of the variance in the degree of damage/destruction of buildings due to the 2017 wildfire. As such, the defensible space of 50 m regulated for Portuguese territory is excessive, leading to a herculean effort in human and financial resources that often do not exist.

An increasing distance to major and minor roads is also an influential variable for the degree of damage/destruction, reducing building resistance. It is noted that buildings located further away from main and secondary roads are less resistant due to the clear isolation from access to rescue means during the wildfire, as stated in other case studies [7,9,15–19].

The distance from forest areas variable was elucidative, allowing us to say that the further away buildings are from forests, the more protected they will be.

An increase in building density was related to a decrease in the degree of building damage/destruction, reiterating the observation made in other case studies that high density of buildings is a factor in building resistance to wildfires [1,6,14].

However, the distance to the forest, distance to major and minor roads, and building density only represented 2.4% of the factors responsible for the destruction of the buildings.

Innovative in this study is the proof that building dispersion is not a factor in resistance to wildfires; this was proven by the fact that the Euclidean nearest neighbor building explanatory variable did not show statistical significance. As such, this study does not corroborate the building's vulnerability due to its isolation, as was observed in other case studies [7,16–19].

Similarly, the slope was not a variable that showed statistical significance in explaining the buildings' destruction, contrary to what has been verified in other empirical studies [7,9].

Regarding the building materials and building area variables, it appears that their contribution is not relevant to determine the destruction of the building.

Nonetheless, despite the independent variables relating to building materials only representing 0.5% of building-destruction variance, it should be noted that buildings with a concrete structure, with slabs, adobe structures, and other structures, all have more resistance; this conforms with the influence of some of these factors calculated by Syphard, Brennan and Keeley [8]. On the other hand, buildings without slabs, which commonly have wood in their structure, have less resistance.

The building area variables also represented a tiny part (0.6%) of the variance of the building damage/destruction variable. In this context, buildings with a footprint of up to 50 m^2 (sheds and garages) were less resilient to fire when compared to buildings with a larger footprint.

In turn, the land-use type where each building is located explained 1.2% of the variance of the degree of building damage/destruction. Thus, buildings located in consolidated urban spaces seem to be more protected from a wildfire impact. Considering that the minimum mappable unit of the COS is 1 hectare (ha), it is also possible to affirm that the buildings that form clusters of urban space with an area greater than 1 ha are characterized by greater resistance to damage or destruction to rural wildfire. In contrast, the existing buildings in forest areas reveal themselves to be deeply vulnerable.

Given what has been exposed so far, it appears that the explanatory variables together are responsible for 82.30% of the variance of the building damage/destruction degree by wildfires.

The non-compliance with the defensible space of a dimension of 30 m was responsible in this case study for 77.60% of the building destruction. From the point of view of public policies, these results show that the defensible space for existing buildings in Portuguese territory could perhaps be rethought to 30 m following a pilot project in the municipality studied or in other municipalities exposed to wildfires.

Given the results, spatial planning policies and integrated management of rural fires are correct to restrict new building permits in forest areas, but the building dispersion and the slope were not verified as factors involved in increasing the vulnerability of buildings to wildfires. Despite this, buildings located in consolidated urban areas are more resistant to wildfires. That said, the criterion to be taken into account for the granting of new building permits should be that they are not located in a forest area or near to a forest edge, regardless of whether it is in a consolidated or dispersed urban area.

Finally, although its results were not very significant, it appears that the Portuguese building codes regarding construction materials are supporting factors in building resistance.

5. Limitations

Since the multiple linear regressions only explained 82.30% of building destruction, it still remains to find the factors responsible for the other 17.70%.

Regarding construction materials and building area, there are no differences regarding the strength of the buildings. The same was found with the construction ages of the buildings, regression data relating to which are not presented since they did not show statistical significance. This may derive from the fact that the analysis of these variables was carried out at the scale of the statistical BGRI subsection and not at the building scale.

Over the years, masonry and tile roofs have been the main construction materials in Portuguese housing, including in Santa Comba Dão. These materials have a medium to high resistance to wildfires [6], which was not studied in this case study. It was also not possible to confirm the influence of single- and multi-pane windows on Santa Comba Dão's housing resistance to the 2017 wildfire.

Likewise, it was not possible to assess which buildings were occupied in order to define variables related to building use that could contribute to explaining the remaining variance in the degree of building damage/destruction.

Therefore, a research opportunity for the future will be to collect more detailed data about the construction materials, age, state of conservation, and use of each building before the wildfire, which will perhaps explain the missing factors that represent 17.70% of the variance of the building damage/destruction variable.

6. Conclusions

This research identified the main factors that should be taken into account in the definition of a municipal mitigation strategy to defend buildings against wildfires that can be extrapolated to municipalities with similar susceptibility and vulnerability to wildfires.

The main factors involved in the destruction of buildings by fire were identified, emphasizing that the non-compliance with a defensible space of 30 m around buildings is the factor that most determines the destruction of buildings. This finding corroborates what has been proven by the scientific literature regarding this subject, and such knowledge should have implications for public policies on fuel management around buildings.

Given that the human and financial resources of communities and institutions to comply with all fuel management strips are scarce and their use must therefore be well allocated, the Portuguese legal requirement of complying with 50 m of defensible space around buildings should be reduced to 30 m in favor of greater effectiveness and efficiency.

Therefore, the strategy should also include the prohibition of roof overhanging vegetation and the restriction of building permits in forest areas, and/or areas very far from major roads. Although there is no data regarding the differentiated presence of single- or multi-pane windows (with or without vinyl framing) on the study area buildings, it will always be recommendable, given the literature, to create incentives for their introduction into structures. All this should be combined with an efficient program to oversee and monitor its implementation.

Author Contributions: Conceptualization, A.S.-A. and F.T.; methodology, A.S.-A. and F.T.; formal analysis, A.S.-A. and F.T.; investigation, A.S.-A., J.A. and F.T.; resources, A.S.-A. and F.T.; data curation, A.S.-A. and F.T., writing—original draft preparation, A.S.-A. and F.T.; writing—review and editing, A.S.-A., F.T., J.A., F.C., D.M.P. and C.M.; supervision, F.T.; project coordination, F.T. All authors have read and agreed to the published version of the manuscript.

Funding: This research was supported by the project 'AVODIS—Understanding and building on the social context of rural Portugal to prevent wildfire disasters' (FCT Ref: PCIF/AGT/0054/2017), financed by national funds through Foundation for Science and Technology (FCT), Portugal. This research received dissemination support from the Centre of Studies in Geography and Spatial Planning (CEGOT), funded by national funds through the Foundation for Science and Technology (FCT) under the reference UIDB/04084/2020.

Data Availability Statement: Data sharing is not applicable to this article. The data are not publicly available due to the privacy of affected people and respective houses.

Conflicts of Interest: The authors declare no conflict of interest.

References

1. Alexandre, P.M.; Stewart, S.I.; Mockrin, M.H.; Keuler, N.S.; Syphard, A.D.; Bar-Massada, A.; Clayton, M.K.; Radeloff, V.C. The relative impacts of vegetation, topography and spatial arrangement on building loss to wildfires in case studies of California and Colorado. *Landsc. Ecol.* **2016**, *31*, 415–430. [CrossRef]
2. Ganteaume, A.; Barbero, R.; Jappiot, M.; Maillé, E. Understanding future changes to fires in southern Europe and their impacts on the wildland-urban interface. *J. Saf. Sci. Resil.* **2021**, *2*, 20–29. [CrossRef]
3. Mockrin, M.H.; Stewart, S.I.; Radeloff, V.C.; Hammer, R.B.; Alexandre, P.M. Adapting to Wildfire: Rebuilding After Home Loss. *Soc. Nat. Resour.* **2015**, *28*, 839–856. [CrossRef]
4. Penman, T.D.; Collins, L.; Syphard, A.D.; Keeley, J.E.; Bradstock, R.A. Influence of fuels, weather and the built environment on the exposure of property to wildfire. *PLoS ONE* **2014**, *9*, e111414. [CrossRef]
5. Price, O.F.; Whittaker, J.; Gibbons, P.; Bradstock, R. Comprehensive examination of the determinants of damage to houses in two wildfires in eastern australia in 2013. *Fire* **2021**, *4*, 44. [CrossRef]
6. Ribeiro, L.M.; Rodrigues, A.; Lucas, D.; Viegas, D.X. The impact on structures of the Pedrógão Grande fire complex in June 2017 (Portugal). *Fire* **2020**, *3*, 57. [CrossRef]

7. Syphard, A.D.; Keeley, J.E.; Massada, A.B.; Brennan, T.J.; Radeloff, V.C. Housing arrangement and location determine the likelihood of housing loss due to wildfire. *PLoS ONE* **2012**, *7*, e33954. [CrossRef] [PubMed]
8. Syphard, A.D.; Brennan, T.J.; Keeley, J.E. The role of defensible space for residential structure protection during wildfires. *Int. J. Wildland Fire* **2014**, *23*, 1165. [CrossRef]
9. Syphard, A.D.; Brennan, T.J.; Keeley, J.E. The importance of building construction materials relative to other factors affecting structure survival during wildfire. *Int. J. Disaster Risk Reduct.* **2017**, *21*, 140–147. [CrossRef]
10. Syphard, A.D.; Keeley, J.E. Factors associated with structure loss in the 2013–2018 California wildfires. *Fire* **2019**, *2*, 49. [CrossRef]
11. Cohen, J.D. Preventing Disaster: Home Ignitability in the Wildland-Urban Interface. *J. For.* **2000**, *98*, 3–15.
12. Cohen, J.D. Relating flame radiation to home ignition using modeling and experimental crown fires. *Can. J. For. Res.* **2004**, *34*, 1616–1626. [CrossRef]
13. Penman, S.H.; Price, O.F.; Penman, T.D.; Bradstock, R.A. The role of defensible space on the likelihood of house impact from wildfires in forested landscapes of southeastern Australia. *Int. J. Wildland Fire* **2019**, *28*, 4. [CrossRef]
14. Miner, A. *Defensible Space Optimization for Preventing Wildfire Structure Loss in the Santa Monica Mountains*; Johns Hopkins University: Baltimore, MD, USA, 2014.
15. Keeley, J.E.; Safford, H.; Fotheringham, C.J.; Franklin, J.; Moritz, M. The 2007 Southern California Wildfires: Lessons in Complexity. *J. For.* **2009**, *107*, 287–296.
16. Keeley, J.E.; Fotheringham, C.J.; Morais, M. Reexamining Fire Suppression Impacts on Brushland Fire Regimes. *Science* **1999**, *284*, 1829–1832. [CrossRef]
17. Lampin-Maillet, C.; Jappiot, M.; Long, M.; Bouillon, C.; Morge, D.; Ferrier, J.P. Mapping wildland-urban interfaces at large scales integrating housing density and vegetation aggregation for fire prevention in the South of France. *J. Environ. Manag.* **2010**, *91*, 732–741. [CrossRef]
18. Syphard, A.D.; Radeloff, V.C.; Hawbaker, T.J.; Stewart, S.I. Conservation Threats Due to Human-Caused Increases in Fire Frequency in Mediterranean-Climate Ecosystems. *Conserv. Biol.* **2009**, *23*, 758–769. [CrossRef]
19. Syphard, A.D.; Radeloff, V.C.; Keeley, J.E.; Hawbaker, T.J.; Clayton, M.K.; Stewart, S.I.; Hammer, R.B. Human influence on california fire regimes. *Ecol. Appl.* **2007**, *17*, 1388–1402. [CrossRef] [PubMed]
20. Quarles, S.L.; Valachovic, Y.; Nakamura, G.M.; Nader, G.A.; de Lasaux, M.J. *Home Survival in Wildfire-Prone Areas: Building Materials and Design Considerations*; University of California, Agriculture and Natural Resources: Richmond, CA, USA, 2010. [CrossRef]
21. Caton, S.E.; Hakes, R.S.P.; Gorham, D.J.; Zhou, A.; Gollner, M.J. Review of Pathways for Building Fire Spread in the Wildland Urban Interface Part I: Exposure Conditions. *Fire Technol.* **2017**, *53*, 429–473. [CrossRef]
22. Bowditch, P.A.; Sargeant, A.J.; Leonard, J.E.; Macindoe, L. *Window and Glazing Exposure to Laboratory-Simulated Bushfires*; Report to the Bushfire CRC, A Bushfire CRC initiative; Bushfire CRC: Melbourne, Victoria, Australia, 2006.
23. Cuzzillo, B.R.; Pagni, P.J. Thermal Breakage of Double-Pane Glazing By Fire. *J. Fire Prot. Eng.* **1998**, *9*, 1–11. [CrossRef]
24. Whittaker, J.; Haynes, K.; Handmer, J.; McLennan, J. Community safety during the 2009 Australian "Black Saturday" bushfires: An analysis of household preparedness and response. *Int. J. Wildland Fire* **2013**, *22*, 841. [CrossRef]
25. McLennan, J.; Ryan, B.; Bearman, C.; Toh, K. "Should we leave now? *": Behavioral factors in wildfire evacuation. Fire Technology* **2018**, *55*, 487–516. [CrossRef]
26. Mccaffrey, S.M.; Rhodes, A. Public Response to Wildfire: Is the Australian "Stay and Defend or Leave Early" Approach an Option for Wildfire Management in the United States? *J. For.* **2009**, *107*, 9–15.
27. Viegas, D.X. *Análise dos Incêndios Florestais ocorridos a 15 de Outubro de 2017*; Centro de Estudos sobre Incêndios Florestais, Departamento de Engenharia Mecânica, Faculdade de Ciências e Tecnologia, Universidade de Coimbra: Coimbra, Portugal, 2019.
28. INE (2013). Population by Parish. Census 2011 (Definitive Results). Portuguese National Institute of Statistics (INE): Portugal, Lisbon. Available online: https://www.ine.pt/xportal/xmain?xpid=INE&xpgid=ine_publicacoes&PUBLICACOESpub_boui=122103956&PUBLICACOESmodo=2 (accessed on 7 November 2021).
29. INE (2021). Population by Parish. Census 2021 (Preliminary Results). Portuguese National Institute of Statistics (INE): Portugal, Lisbon. Available online: https://www.ine.pt/scripts/db_censos_2021.html (accessed on 9 November 2021).
30. Silva, J.S.; Nereu, M.; Pinho, S.; Queirós, L.; Jesús, C.; Deus, E. Post-Fire Demography, Growth, and Control of Eucalyptus globulus Wildlings. *Forests* **2021**, *12*, 156. [CrossRef]
31. DGT. Direção Geral do Território. DGT: Portugal, Lisboa. Available online: https://www.dgterritorio.gov.pt/dados-abertos (accessed on 24 October 2021).
32. Turco, M.; Jerez, S.; Augusto, S.; Tarín-Carrasco, P.; Ratola, N.; Jiménez-Guerrero, P.; Trigo, R.M. Climate drivers of the 2017 devastating fires in Portugal. *Sci. Rep.* **2019**, *9*, 13886. [CrossRef]
33. Google Earth Pro. Available online: http://www.google.com/earth/ (accessed on 25 October 2021).
34. CMSCD. Câmara Municipal de Santa Comba Dão. CMSCD: Santa Comba Dão, Portugal. Available online: https://cm-santacombadao.pt/ (accessed on 25 October 2021).
35. OpenStreetMap. Available online: https://www.openstreetmap.org/#map=13/40.4581/-8.0662 (accessed on 25 October 2021).
36. McGarigal, K.; Marks, B.J. *FRAGSTATS: Spatial Pattern Analysis Program for Quantifying Landscape Structure*; General Technical Report; Department of Agriculture, Forest Service, Pacific Northwest Research Station: Portland, OR, USA, 1995. [CrossRef]

37. CIIMAR. STRM Digital Elevation Model with Spatial Resolution of 25 m. Coastal Monitoring and Management Group. Available online: https://www.fc.up.pt/pessoas/jagoncal/dems/ (accessed on 9 November 2021).
38. Tabachnick, B.G.; Fidell, L.S.; Ullman, J.B. *Using Multivariate Statistics*; Pearson: Boston, MA, USA, 2007.
39. *Statistical Package for the Social Sciences (SPSS) for Windows*, Version 26; IBM: Armonk, NY, USA, 2019.

fire

Article

Reduced Scale Experiments on Fire Spread Involving Multiple Informal Settlement Dwellings

Vigneshwaran Narayanan, Anene Oguaka and Richard Shaun Walls *

Department of Civil Engineering, Stellenbosch University, Stellenbosch 7599, South Africa
* Correspondence: rwalls@sun.ac.za

Abstract: Fire disasters in informal settlements (also referred to as slums, shantytowns, favelas, etc.) are a major challenge worldwide, with a single incident being able to displace thousands of people. Numerous factors including dwelling spacing, material type, topography, weather, fuel loads, roads, and more influence fire spread. Conducting large-scale experiments to quantify and understand these phenomena is difficult and costly. Hence, it would be beneficial if Reduced Scale Experiments (RSE) could be developed to study the influence of these phenomena. Previous research has demonstrated that a 1/4th scale informal settlement dwelling (ISD) RSE can sufficiently capture the fire behaviour and fire dynamics within dwellings. The objective of this work is to develop a methodology for multi-dwelling ISD scaling such that large-scale spread phenomena can be captured. This paper carries out a series of RSEs to study the influence of (a) the number of dwellings, (b) orientation of dwellings, windows, and door openings, (c) cladding material, (d) wind effects, (e) the distance between dwellings and (f) fuel load on spread. Results are compared to previous large-scale experiments. It is shown that the geometric scaling of distance between dwellings is suitable for capturing spread. It was found that wind and the fuel load contribute significantly to the fire spread, but the type of cladding, distance between dwellings, dwelling orientation, and type of structural members used also affects fire spread rates. The comparative results with full-scale experiments (FSEs) shows that the peak temperatures were comparable and had similar profiles. A good correlation exists between FSEs and RSEs in terms of fire dynamics and spread characteristics, but the spread time (scaled or unscaled) does not correlate well with FSEs. Further work is needed before the work can be reliably used for predicting multi-dwelling spread, especially when wind is involved, due to the complex interaction of parameters and difficulty in scaling flame impingement.

Keywords: informal settlements; enclosure fire dynamics; reduced scale fire experiment; fire spread

Citation: Narayanan, V.; Oguaka, A.; Walls, R.S. Reduced Scale Experiments on Fire Spread Involving Multiple Informal Settlement Dwellings. *Fire* **2022**, *5*, 199. https://doi.org/10.3390/fire5060199

Academic Editor: Tiago Miguel Ferreira

Received: 11 October 2022
Accepted: 21 November 2022
Published: 23 November 2022

1. Introduction

Informal settlements (ISs) can be found all over the world. Nearly one billion people, most of them belonging to lower income bracket, live in these areas [1]. These settlements are usually unplanned, are built over unoccupied lands [2] and the informal settlement dwellings (ISDs) are predominantly make-shift enclosures constructed from thermally thin steel (corrugated) sheets, timber materials, masonry, plastic sheets, or any readily available materials [3,4].

Fires in ISDs are largely underestimated despite being a frequently occurring event. ISs have a wide range of potential ignition sources such as open flames, faulty electrical installations, candles, arson, "other" and "undetermined" causes [4] (the latter two are typically for when the fire services are unable to identify the cause). Fires primarily spread through three main processes: radiation, direct flame impingement, and fire branding [5]. The mitigation of the consequences of fire spread in these settlements demands good understanding of fire development and spread between the dwellings. Fire spread through informal settlement dwellings (ISDs) is a complex process that involves various parameters such as ISD density, fuel load, topography, wind, materials between dwellings, firefighting operations, and construction materials. Thus, it becomes expensive and time consuming to perform a large-scale experimental study on fire spread between ISDs. As a result, reduced-scale experiments (RSEs) need to be developed to simulate fire spread on large-scale ISDs.

In recent years, multiple full-scale fire experiments (FSEs) have been conducted on ISDs. The experimental study on single IS fires has focussed on fire development, and the influence of combustibles on the fire spread in the dwellings [4,6]. Studies on multiple ISs have focussed on the influence of various cladding materials, type of fuel, effect of heat fluxes and building layouts on fire spread between the dwellings [7–12]. In addition, various numerical simulations have been developed especially with steel-clad models. A preliminary FDS model was developed to understand the fire dynamics of ISD (with temperature, flame behaviour, and air velocity near the door) and found good agreement with experimental results [13]. Numerical modelling for ISDs was further developed to predict the fire parameter (such as temperatures, heat fluxes, etc) using FDS and OZone [6]. The paper discusses challenges in addressing compartment fires and its impact in estimating heat fluxes. The OZone model showed good agreement with gas temperatures but failed to predict heat fluxes. A FDS model developed by Cicione et al. [14] to predict fire spread between multiple full-scale ISDs showed good correlation with experimental works, but fire spread rates were seen to be extremely sensitive to the input parameters (such as ignition temperature, specific heat, conductivity, emissivity of the lining material, emissivity of the compartment boundaries, soot yield and radiative fraction). Numerical modelling for combustible cladding was a challenging task because of continuously changing ventilation conditions (i.e., as the side cladding burnt through). Predicting large scale informal settlement fires spread using numerical simulations needs significant research to develop a simple and practical approach. A fire spread analysis was simulated using B-Risk that resulted in good correlation in baseline scenario with the experimental results however, the software overestimated the fire spread rate for model for real informal settlement [15]. A study on fire spread rate has been briefly discussed by Flores Quiroz et al. [16] where lateral fire spread in large urban fires and ISs has been analysed and the average fire spread rate were between 1.2 to 3.6 m/min. Hence, based on the aforementioned discussions, it can be noted that studying fire spread using FSEs and numerical models still pose a challenge and significant variations can be obtained in results.

Reduced scaled dwelling (RSD) fire experiments have been developed to mitigate various challenges linked with full scale ISDs fire experiments and to gain wider understanding of various factors. Scaling of fire behaviour is a complex task as phenomena such as temperature, heat release rates, heat fluxes, convection, mass flows and geometry interact in a non-linear manner. An article from the primary author, investigated the application of scaling methods for thermally thin compartment fires, such as ISDs, through RSEs and Fire Dynamics Simulator (FDS) modelling [17]. Different sizes, namely: 1/4 scale, 1/5 scale, 1/7.5 scale, 1/10 scale, 1/15 scale of reduced scale dwellings were used to assess the behaviour of the RSEs. The results from this study indicated that reduced-scale modelling with RSE models of 1/4 scale and 1/5 scale can be used to replicate an ISD fire with a reasonable level of certainty, depending on the parameter being studied. The results from FDS model predicted the general trend in experimental results and larger RSD models had better agreement with RSE results than smaller RSD model. Another study on reduced scale informal settlement dwelling by Beshir et al. [18], discussed the effect of horizontal opening in the fire dynamics and fire spread of the reduced scale compartment. In addition, a FDS model was developed. The Beshir study concludes that the horizontal opening in the centre of the roof reduces the heat flux from the door opening by a substantial degree. The FDS results reproduced the main trends and fire dynamics of the compartments, but the combustion efficiency and gas temperatures were not effectively captured.

In this paper, 19 RSEs were performed on 1/4th scale dwellings involving multiple ISDs that ranged between two and six dwellings in each experiment. The RSE series aimed to replicate or model previous FSEs in the literature that studied the effect of separation distance on fire spread, fire spread between multiple ISDs, and large-scale fire spread. The experiment series was conducted to investigate the effect of the following parameters: (a) number of dwellings, (b) orientation of dwellings, (c) cladding material, (d) wind effects, and (e) the distance between dwellings. The data from RSEs are analysed and compared with previous full-scale experiments (FSE) on fire spread and based on the comparative results fire spread mechanisms in RSEs are identified and quantified. Initially the paper introduces previous FSEs upon which the work was based. Thereafter an overview of the experimental regime and methodology is given. Results from the different RSEs are presented by comparing time-temperature curves and spread behaviour. The results of the FSEs and RSEs are compared, followed finally with recommendations and conclusions. An extensive database of experimental results and associated analysis are presented below. These have not been split into multiple papers as it is important to compare results between the different experiments such that the influence of parameters and associated trends can be more clearly identified.

2. Overview of the Literature on Full-Scale Experiments Used as Benchmarks

The RSDs and related experiments in this study are based on various FSEs conducted in the past. A brief discussion on each full-scale ISD experiments, namely double and triple dwelling experiments by Cicione et al. [7,9] and a 20-dwelling large scale experiment by de Koker et al. [10] are described in the following sections.

2.1. Double Dwelling FSE Focussing on Separation Distance

Two double dwelling FSEs were conducted by Cicione et al. [9] to investigate the effect of separation distance on fire spread rates between the dwellings. In the experiments two dwellings were placed adjacent to each other and the separating distances between them were varied as shown in Figure 1, along with the varying of parameters such as window position and wall lining material. The full-scale steel-clad dwellings [9] had a floor area identical to the ISO 9705 room of 3.6 m $\times$ 2.4 m, but with a height of 2.3 m (rather than 2.4 m). The dwelling of fire origin, denoted as ISD1, had a single door opening of 2.05 m (height) $\times$ 0.85 m (width) on the longer side. The other dwelling denoted as ISD2 had two openings, a door of 2.05 m (height) $\times$ 0.85 m (width) on the 3.6 m long side and a window of 0.6 m (height) and 0.85 m (width) on the 2.4 m long side 1.25 m from the ground level (sill height). The ISDs were positioned in such a way that the door of ISD-1 was facing the backwall of ISD-2, and the dwellings during these experiments were separated by 1 m and 1.75 m, and the same can be seen in Figure 1.

During the experiments, the ignition wood crib in ISD-1 was ignited, and the impact of flame and heat flux emanating from ISD-1 on ISD-2 was analysed. Parameters such as heat release rate (HRR), gas temperatures inside the compartment at various levels using thermocouple (TC) trees, surface temperatures on the claddings, velocity and temperatures at the openings, and incident heat flux were measured. Each dwelling was loaded with approximately 130 kg of fuel consisting of wood cribs and cardboard lining contributing to a surface-controlled heat release rate of approximately 6.7 MW. The maximum recorded HRR for ISD-1 and ISD-2, based on mass loss data, were 6.5 MW and 8.4 MW, respectively. It was found, as expected, that an increase in separation distance from 1 m to 1.75 m decreased the incident heat flux from ISD-1 onto ISD-2 from 33 kW/m^2 to 27 kW/m^2. Consequently, the ignition time of ISD-2 also increased exponentially as the separating distance increased, consistent with fundamental radiation equations.

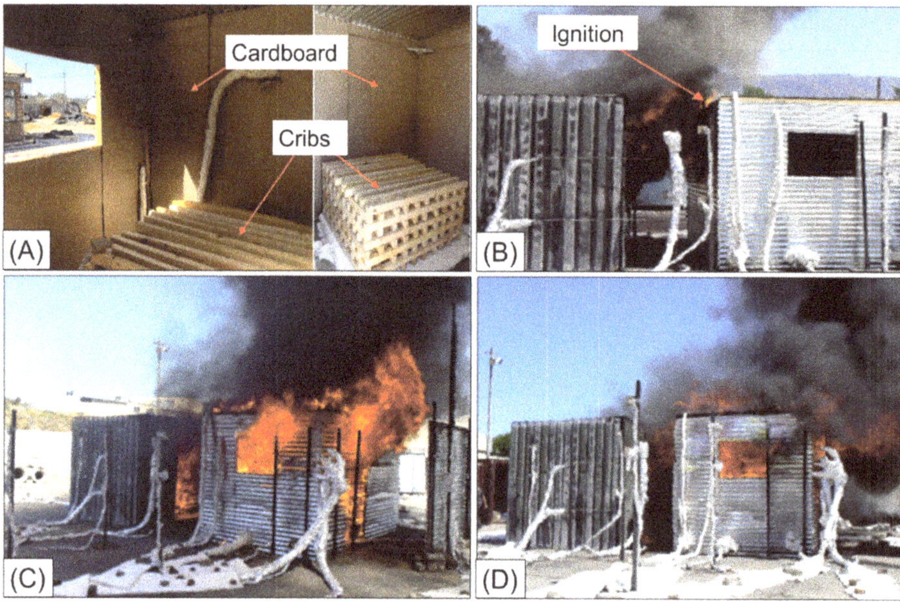

Figure 1. The effect of separation distance between informal dwellings on fire spread rates (**A**) Interior and wood crib arrangement; (**B**) ignition from ISD1 to ISD2; (**C**,**D**) fire spread between dwellings at 1 m and 1.75 m [9].

2.2. Triple Dwelling FSE Focussing on Separation Distance

Similar large-scale studies on ISDs were conducted to understand the fire spread between multiple steel and timber-clad informal settlement dwellings (ISDs) [7] as shown in Figure 2. Two full-scale ISD burn experiments were conducted with (a) three ISDs clad with steel sheeting and (b) three ISDs clad with timber. The heat fluxes and temperatures of both experiments were compared (steel-clad vs. timber-clad) to understand the effect of cladding materials on fire spread. The dwelling size used was 3 m × 3 m × 2.3 m with a door of 0.86 m × 2.03 m and a window of 0.6 m × 0.6 m. To limit the amount of setup variability and number of parameters studied in this paper the RSEs developed below are based on the rectangular 2.4 m × 3.6 m floor plan rather than the 3 m × 3 m floor plan. The paper addressed fire spread behaviour in between different ISDs and primarily focused providing a technical basis with regards to experienced temperatures, fire spread rate, heat fluxes, etc. steel-clad and timber-clad dwellings had similar internal temperature profile. The heat fluxes from timber-clad dwellings ranges from 140 kW/m^2 to 240 kW/m^2 which approximately 1.5 times higher than steel-clad dwellings. The cardboard linings for timber-clad dwellings had reduced effect on fire spread because of combustibility of cladding itself. During full developed phase, the HRR in the steel-clad dwellings was ventilation controlled, whereas for timber-clad dwelling, HRR was limited by the fuel availability. Fire spread times (from the start of flashover in the first dwelling to the end of flashover in the last dwelling) for both experiments (timber and steel) ranged between 4 and 9 min. The paper concludes that the timber-clad dwellings are clearly a higher risk to fire spread defines a critical separation of 3–5 m, based on the studied geometry and fuel specifications, to prevent fire spread between dwellings.

Figure 2. Fire spread between multiple full-scale informal dwellings. (**Top**) Steel-clad dwellings, (**bottom**) wooden-clad dwelling [7].

2.3. Twenty Dwelling Large-Scale Experiment

This experiment aimed at understanding the settlement-scale fire spread behaviour of informal settlement dwellings (ISDs) [10], using a mock twenty dwelling test settlement. The dwellings were built as simple timber cross member assembled from 48 × 48 mm square pine sections and cladding was attached to these cross members. There were fourteen steel-clad dwellings of 0.5 mm galvanised steel sheets, and the remaining six dwellings were clad with 14 mm thick timber planks. The roof panels of all dwellings were provided with 0.5 mm galvanised steel sheeting. A video of the experiment can be viewed at https://youtu.be/kkXr6ueakAU (accessed on 1 May 2021).

All dwellings were constructed with a floor area of 3.6 m × 2.4 m (length × width) and a height of 2.2 m. The mock settlement consists of four dwellings spaced 1.0 m apart in a row and there were five rows of dwellings spaced 1.2 m apart, except for four instances where the spacing was 2.2 m. Doors or windows were located on the left-hand or right-hand side of each longitudinal dwelling wall (i.e., not on the short edge), and alternated to cover door–door, window–window, window–door, and door–window facing wall configurations across transverse alleyways as in Figure 3. Interiors of the dwellings were covered with a cardboard lining to represent insulation of ISDs.

Cribs from South African pine were chosen as the primary fuel. Each dwelling was loaded with 450 MJ/m^2 of fuel consisting of 1.0 m lengths of the same 48 mm × 48 mm timber arranged into six cribs per dwelling, each stacked as seven alternating transverse layers of four timber lengths.

A "fire line" scenario was created by simultaneously igniting four dwellings in a row, and then allowing the fire to propagate through the settlement to replicate fire disasters involving large numbers of homes. Results highlight the critical hazard posed by the proximity of neighbouring dwellings (1–2 m), with the wind playing a primary role in directing and driving the spread process. Even with a mild wind speed of 15–25 km/h, the fire spread through the entire mock settlement within a short 5 min period. Following ignition of a given dwelling, flashover is reached very quickly, with the temperatures reaching more than 1000 °C within 1 min, and downwind neighbour structures igniting in less than a minute thereafter. The results suggest that multi-dwelling effects are not dominant in these types of fires but may become meaningful at a larger scale when branding and topography play a role.

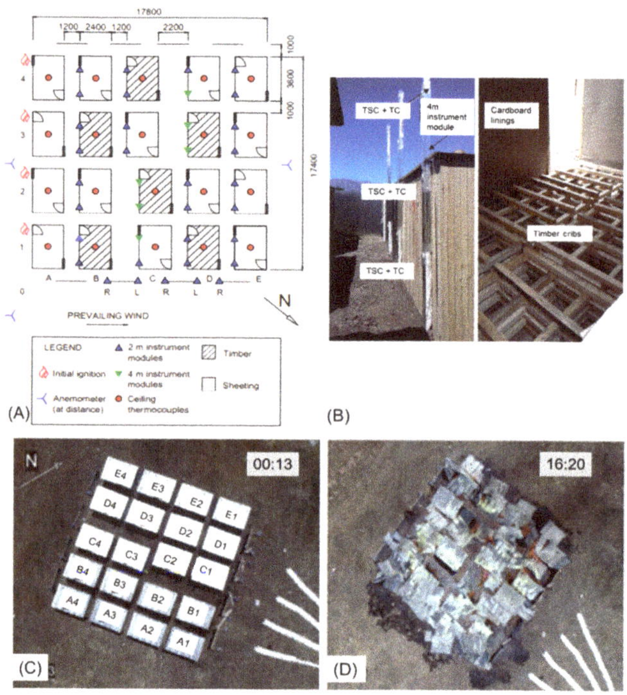

Figure 3. Full-scale fire test on 20 informal dwellings. (**A**) Experiment layout, (**B**) instrumentation and fuel load arrangement, (**C**) aerial image of fire experiment soon after ignition, (**D**) image of fire experiment at 16 min 20 s [10].

3. Reduced Scaled Modelling Methodology

3.1. Scaling Considerations for Ignition

When developing RSEs it must be considered what parameter and behaviour should be captured through an experiment, as discussed above. For informal settlements the occurrence of fire spread between dwellings is of primary concern such that interventions for reducing spread, improving home layouts or enhancing construction materials used can be identified, whilst also enhancing data available for the development of numerical models. Ignition of dwellings typically occurs due to (1) radiation onto target dwellings exceeding critical heat fluxes (or flux-time products for time-variable fluxes), or (2) flame impingement where flames are able to impart sufficient flux to cause ignition. The two mechanisms cannot easily be separated and always occur simultaneously. The first item ignited is typically a material such as cardboard, curtains, plastic linings, and timber cross members in its vicinity due the energy emitted by the flames above openings, cladding, and from fire within a dwelling (i.e., openings).

In terms of quantifying fluxes: (1) the total radiative flux received by a dwelling is based on the summation of the radiative flux emitted from a burning dwelling's flames (i.e., above openings), openings and sidewall cladding (especially for heated steel or burning timber): $\dot{q}''_{radiation} = \dot{q}''_{flames} + \dot{q}''_{openings} + \dot{q}''_{cladding}$. The radiation received from each is a function of the view/configuration factor, $\varnothing$, temperature difference of the emitter and receiver to the fourth power, $\left(T_e^4 - T_r^4\right)$, and emissivity of each component, ϵ. (2) For localised flame impingement there is convective and radiative transfer to the target surface

$\dot{q}''_{impingement} = \dot{q}''_{convective} + \dot{q}''_{radiative}$ responsible for piloted ignition of already pyrolysing combustibles, which is dependent upon flame length and flame height.

From the equations above, it can be observed that even if the only criterion that one seeks to match in RSEs is whether ignition occurs, it is a task involving multiple parameters that interact in a non-linear manner. In some ways scaling radiation is easier, provided that temperatures of emitting items are known, as equations can be more readily applied. The scaling of flame length and height, along with the flux that a smaller flame imparts, is a greater challenge. Hence, this work seeks to identify what can, and cannot, be accomplished through scaling methodologies to capture spread behaviour.

3.2. Wind Considerations

Following discussions that flame behaviour and impingement is important, in real informal settlement incidents wind plays a significant role in directing fire spread. Due to climatic conditions, local terrain, densely packed dwellings and even vegetation there is often significant turbulence and gusting of winds. Furthermore, wind movements around dwellings can lead to localised changes in wind direction and speed. This leads to wind causing flames to pulse from dwelling to dwelling and fluctuate from side to side. Ignition can occur at any time, even during a short-lived wind direction reversal. Such behaviour is extremely difficult to capture. In the tests conducted wind effects were included by testing outside and measuring the average wind speed. However, increased wind speeds may not always lead to increased spread as they can also lead to convective cooling of dwellings, thereby causing increased times to ignition, rather than more rapid ignition, even under favourable wind direction conditions.

3.3. Reduced Scale Dwelling Design and Methodology

The RSEs designed in this paper were based on Froude scaling technique. The scaling correlations of dimensional groups from the conservation equations are based on Quintiere [19]. In this work, the geometry of the Reduced-Scale Dwellings (RSDs) is based on the full-scale ISD experiment introduced above. The RSDs are geometrically scaled according to respective geometric ratios and the vents (doors and windows) are scaled based on ventilation factor scaling according to $A(h)^{0.5} \sim s^2$ as used by Bryner et al. [20] where 's' represents the scaling ratio (i.e., 1/4 as discussed below). For the RSEs, the HRRs were scaled according to $\dot{Q} \sim s^{5/2}$ and subsequently the average burning times were scaled according to $t \sim s^{1/2}$. Wood cribs were used as fuel to represent the anticipated fire load. The design details of the wood cribs for RSEs are discussed in the sections that follow. The steel sheets are considered a thermally thin material (Biot number in order of $10^{-2} < 0.1$), with a thickness of 0.6–0.8 mm, and the thermal properties can be assumed to be homogeneous. This property allows the thickness of steel sheets to be scaled geometrically and the influence of thermal inertia ($\kappa\rho c$) to be neglected.

In this research informal settlement dwellings (ISDs) at $\frac{1}{4}^{\text{th}}$ scale of the full-scale experiment (FSE) were constructed for the reduced scale experiments (RSEs). The scaling principles, methodology, and scaling correlations are described in [17]. In this experiment series, two types of reduced-scale dwellings (RSDs) were constructed: (a) Type "D1" with internal dimensions of 0.9 m (length) × 0.6 m (width) × 0.58 m (height) to replicate the double dwelling setup of Cicione et al., and (b) Type "D2" with internal dimensions of 0.9 m × 0.6 m × 0.55 m but also with different ventilation details to represent the 20-dwelling setup of de Koker et al. Each RSD had a door and a window opening which is shown in Figure 4. The dimensions of the openings of full-scale dwellings and RSD along with their associated ventilation condition are provided in Table 1.

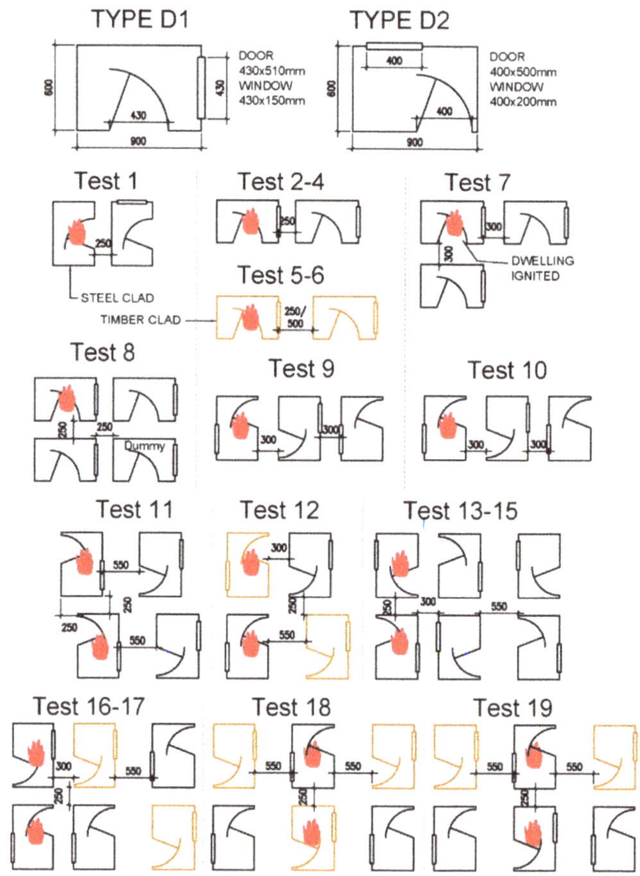

Figure 4. Pictorial representation of 19 test setups showing dwelling layouts (top) and test setups. Steel-clad dwellings are shown in black, timber-clad dwellings in brown and ignited dwellings with the fire symbol.

Table 1. Dimensions of $\frac{1}{4}$th scale reduced scale dwellings.

Scale	Parameter	Dimensions (m)			Total Opening Area (m^2)	Total Opening Height (m)	Total Ventilation Factor (m$^{3/2}$)	Ventilation-Controlled HRR (kW)
		Room	Door	Windows				
Full Scale-1	Length	3.60	-	-	2.253	1.722	2.956	6029
	Width	2.40	0.85	0.85				
	Height	2.30	2.05	0.60				
"D1"—RSD Type-1	Length	0.90	-	-	0.282	0.430	0.185	498
	Width	0.60	0.43	0.43				
	Height	0.58	0.51	0.15				
Full Scale-2	Length	3.60	-	-	2.240	1.657	2.884	7785
	Width	2.40	0.80	0.80				
	Height	2.20	2.00	0.80				
"D2"—RSD Type-2	Length	0.90	-	-	0.280	0.471	0.192	519
	Width	0.60	0.40	0.40				
	Height	0.55	0.50	0.20				

3.4. Reduced Scale Experiment Configurations

The RSDs in these experiments were varied in terms of the cladding materials (14 mm timber strips or 0.5 mm galvanised steel sheet), type of cross member (25 mm × 25 mm timber or steel angles), and positions of windows and doors. All the RSDs were provided with a 1.5 mm thick cardboard layer on the inner side to replicate combustible thermal insulation often used for ISDs. The presence of cardboard significantly influences fire dynamics within dwellings.

An RSD configuration in which the fire spread is observed from the side i.e., from left RSD to right RSD, it is defined as "lateral" and if fire spread is observed from the front door to backwall of a RSD, then it is considered as "longitudinal. The wind directions during experiments were: (1) "favourable" meaning wind that aids fire spread, (2) "opposing" refers to a wind direction that opposes fire spread, (3) "cross wind" means a wind direction that is approximately perpendicular to fire spread, and (4) "circular" refers to a fluctuating wind does not have a specific direction, or which changes its direction all the time. Experiments were conducted outside meaning that it was not possible to control the wind speed and direction. The flashover time during the test was determined by the ceiling temperature of RSDs exceeding 600 °C or the time when the flames were visible through the openings, whichever was observed first. The experimental fire time of an RSD was considered to be a duration between ignition until collapse or until fuel burn out. Ignition is defined based on either observations or a distinct increase in temperature above ambient evidenced by thermocouple data.

Table 2 lists the testing regime which includes all experiments, relevant variable factors that can affect the fire spread in-between the dwellings, and a unique test identification number consisting of all the data provided in the table. The experiment ID is unique for all test setups to identify dwellings arrangement to assist the reader in sections that follow. For example, "1-2D-S-250-F" represents the test number (e.g., 1); number of 'D'wellings involved (2D to 6D); type of cladding ('S'teel or 'T'imber); spacing of RSDs (e.g., 250 mm); and wind condition ('F'avourable, 'O'pposing, Crosswind 'K', 'C'ircular, not applicable or negligible '/'). Figure 4 provides the pictorial representation of the 19 setups with unique RSD ID that are provided on the top of each test setups. In Experiment 8, a dummy dwelling was included which had no fuel load or cladding, and simply served as a flame barrier.

Table 2. Summary of testing regime.

<table>
<tr><th>Experiment Id</th><th>Test No</th><th>No of RSD</th><th>Cladding</th><th>Fire Load (kW)</th><th>RSDs Spacing (mm)</th><th>Type of RSDs</th><th>Wind Condition</th></tr>
<tr><td>1-2D-S-250-/</td><td>1</td><td rowspan="6">2</td><td rowspan="3">S</td><td rowspan="3">520</td><td rowspan="3">250</td><td rowspan="6">D1</td><td>/</td></tr>
<tr><td>2-2D-S-250-K</td><td>2</td><td>Cross</td></tr>
<tr><td>3-2D-S-250-C</td><td>3</td><td>Circular</td></tr>
<tr><td>4-2D-S-250-F</td><td>4</td><td rowspan="3">T</td><td rowspan="3">402</td><td rowspan="3">500</td><td>Favourable</td></tr>
<tr><td>5-2D-T-250-C</td><td>5</td><td>Circular</td></tr>
<tr><td>6-2D-T-500-O</td><td>6</td><td>Opposing</td></tr>
<tr><td>7-3D-S-300-F</td><td>7</td><td rowspan="4">3</td><td rowspan="4">S</td><td rowspan="4">520</td><td>300</td><td rowspan="2">D1</td><td>Favourable</td></tr>
<tr><td>8-3D-S-250-O</td><td>8</td><td>250</td><td>Opposing</td></tr>
<tr><td>9-3D-S-300-O</td><td>9</td><td rowspan="2">300</td><td rowspan="2">D2</td><td>Opposing</td></tr>
<tr><td>10-3D-S-300-/</td><td>10</td><td>/</td></tr>
<tr><td>11-4D-S-550-/</td><td>11</td><td rowspan="2">4</td><td>S</td><td>520</td><td>550</td><td>D1/D2</td><td>/</td></tr>
<tr><td>12-4D-2S/2T-250/550-F</td><td>12</td><td>2S/2T</td><td>402</td><td>250/500</td><td>D2</td><td>Favourable</td></tr>
</table>

Table 2. *Cont.*

<table>
<tr><th>Experiment Id</th><th>Test No</th><th>No of RSD</th><th>Cladding</th><th>Fire Load (kW)</th><th>RSDs Spacing (mm)</th><th>Type of RSDs</th><th>Wind Condition</th></tr>
<tr><td>13-6D-S-300/550-O</td><td>13</td><td rowspan="7">6</td><td rowspan="3">S</td><td rowspan="3">402</td><td rowspan="3">300/550</td><td>D2</td><td>Opposing</td></tr>
<tr><td>14-6D-S-300/550-F</td><td>14</td><td>D2</td><td>Favourable</td></tr>
<tr><td>15-6D-S-300/550-K</td><td>15</td><td>D2</td><td>Cross</td></tr>
<tr><td>16-6D-4S/2T-300/550-O</td><td>16</td><td rowspan="2">4S/2T</td><td rowspan="2">402</td><td rowspan="2">300/550</td><td>D2</td><td>Opposing</td></tr>
<tr><td>17-6D-4S/2T-300/550-O</td><td>17</td><td>D2</td><td>Opposing</td></tr>
<tr><td>18-6D-3S/3T-550-/</td><td>18</td><td>3S/3T</td><td rowspan="2">402</td><td rowspan="2">500</td><td>D2</td><td>/</td></tr>
<tr><td>19-6D-4S/2T-550-/</td><td>19</td><td>4S/2T</td><td>D2</td><td>/</td></tr>
</table>

3.5. Fuel Source

From the previous work on ISDs, it was seen that the fuel load in an informal dwelling range between 400 MJ/m^2 to 2000 MJ/m^2 [4]. Although the fuel load range varies significantly, fires in these dwellings are typically ventilation controlled and they collapse before the fuel burns out, indicating the role of other factors that could play a substantial effect on fire dynamics inside an ISD. For all RSEs, the cross-sectional (b_w) dimension, number of stick levels (n_{level}) of wood cribs were obtained by scaling according to $b_w \sim s^{1/3}$ and $n_{level} \sim s^{1/3}$, respectively, whereas the length of each stick and number of sticks per level were changed to obtain the desired HRR. The HRR (Q) of the cribs were calculated as per [21].

In this work, untreated and kiln dried wood cribs from South African pine were used as the fuel. The density of wood used for the cribs was approximately 580 kg/m^3, the heat of combustion was ΔH_c = 22.5 MJ/kg as measured in a bomb calorimeter, and the moisture content was less than 12%. There were three different fuel loads based on the different sizes of the wood cribs used in the experiment that are listed in the respective sections. The maximum surface-controlled heat release rate of wood cribs used in this experiment series was 520 kW (21 mm × 21 mm × 450 mm × 48 Nos), 402 kW (25 mm × 25 mm × 500 mm × 30 Nos), and 536 kW (25 mm × 25 mm × 500 mm × 40 Nos). Due to the limited availability of timber fuel the member sizes slightly vary between tests (21 mm vs. 25 mm). The ventilation controlled HRR for each RSD is provided in Table 1. A timber cladding of thickness 14 mm was used in the experiment, which had a density of 542 kg/m^3 and the heat of combustion of 18.1 MJ/kg as measured in a bomb calorimeter.

3.6. Instrumentation

K-type (1.5 mm tip diameter) thermocouples were used as the primary device for measuring temperature. Each RSD had four thermocouples, two placed at 5 mm below ceiling, one at window and door soffit. The measurements of the experiments were recorded with a data logger at every 1 Hz. The instrumentation layout of the experiment is depicted in Figure 5. Videography and photography were primary evidence to trace the fire spread patterns. The video cameras were placed at multiple angles to capture all moments of fire spread.

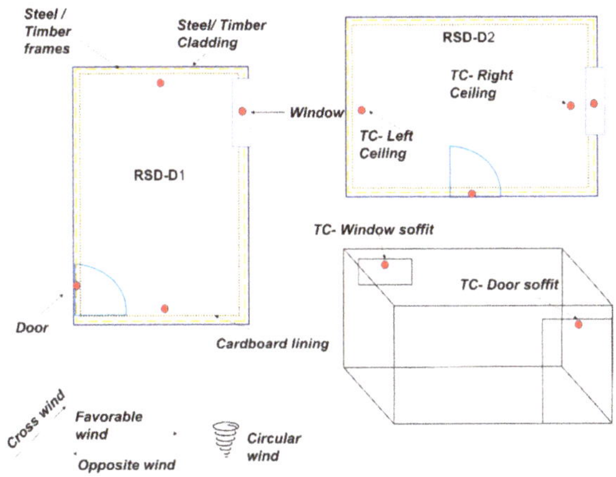

Figure 5. Instrumentation layout of an RSDs.

4. RSE Results and Discussion

The testing regime of the experiments is listed in the following sections where the experiments are discussed according to the number of dwellings tested at a time.

4.1. Two RSD Experiments

The testing matrix of double dwellings experiments shown in Table 3 and the detailed analysis has been provided in the following sections. In this two RSD test, the RSD on the left was considered as RSD-1, whereas the RSD on the right as RSD-2 and each test began with igniting the RSD-1 and fire spread to the RSD-2 was studied. The calculated fire spread rate highlights how the timber dwellings have much faster spread than steel dwellings. For the crosswind condition (Test 2) no spread occurred, whilst spread occurred very slowly with a fluctuating circular wind (Test 3).

Table 3. Test matrix of two RSD experiments.

Test No	RSD Id	RSD Configuration	Wind Speed	Fire Spread Rate (m/min)
1	1-2D-S-250-/	Steel-clad RSDs with timber cross member placed laterally.	Negligible	0.137
2	2-2D-S-250-K	Steel-clad RSDs with timber cross member placed longitudinally.	13 kmph cross	No spread
3	3-2D-S-250-C		15 kmph circular	0.008
4	4-2D-S-250-F		17 kmph favourable	0.07
5	5-2D-T-250-C	Timber-clad RSDs placed longitudinally.	11 kmph circular	0.123
6	6-2D-T-500-O		14 kmph opposing	0.121

4.1.1. Longitudinal Fire Spread between Two Steel-Clad RSDs (Test 1)

Test 1 (negligible wind): The influence of wind was minimal throughout the duration of the experiment. An initial peak in temperature was recorded at 200 s of RSD-2 came from smoke entrainment inside the dwelling from RSD-1 through the gaps between backwall and roof panel. At 561 s, flames started impinging on the wall of RSD-2 that heated the timber cross member and ignited the cardboard linings. The cardboard then allowed the fire to spread throughout the compartment leading to the ignition of the wood crib, and the RSD-2 was fully involved in fire. The fire in RSD-2 lasted for 800 s from ignition till collapse. The increase in fuel load in Test 11 prolonged the burning time relative to the tests below. However, despite the slightly higher fuel load, the wind speed is also dominant in influencing spread behaviour.

The comparative results from RSEs and FSE [9] has been provided in the Table 4. The peak roof temperatures of the RSEs in RSD-1 was 11% lower than in RSD-2, whereas this difference in peak roof temperature was 8% in FSE. This temperature difference between these dwelling was a result of limited ventilation condition. The roof temperature difference in RSEs were 28% lower than the FSEs as a result of higher heat losses and lower smoke layer thickness.

Table 4. Comparative results from longitudinal fire spread between two steel-clad dwellings.

Steel-Clad ISDs	Full-Scale ISD [9]		Reduced-Scale ISD	
	ISD-1	**ISD-2**	**RSD-1**	**RSD-2**
Fire spread mechanism	Dwelling of fire origin	Flame impingement from RSD-1 on exposed timber cross member of RSD-2	Dwelling of fire origin	Flame impingement from RSD-1 on exposed timber cross member of RSD-2
Maximum ceiling temperature	963 °C	1040 °C	685 °C	763 °C
Ventilation	Door	Door and windows	Door	Door and windows
Time from the start to flashover	360 s	160 s	210 s	480 s
Scaled time from the start to flashover	-	-	420 s	960 s
Time from start of ignition to collapse	660 s	440 s	587 s	840 s
Scaled time from start of ignition to collapse	-	-	1174 s	1680 s
Time to fire spread to RSD-2 from flashover of RSD-1	-	80 s	-	110 s
Scaled time to fire spread to RSD-2 from flashover of RSD-1	-	-	-	220 s

The time from ignition to flashover was 360 s and 160 s vs. 210 s and 480 s for the FSE and RSE, respectively, where the latter scaled to 420 s and 960 s ($t \sim s^{\frac{1}{2}}$). The spread time for the FSE was 80 s from flashover of RSD-1, whereas it was 110 s and 220 s for the unscaled and scaled times for the RSE. In case of RSD-2, the growth phase was largely smouldering after the ignition of cardboard; thus, time to flashover was substantially longer in the experiment time. The fire spread rate in Test 1 was 0.167 m/min which after scaling was approximately three times larger than in FSE at 0.84 m/min. Hence, overall temperatures, flame behaviour, spread mechanisms and fire development trends were reasonably well captured, but spread rates and experiment times were not. This is a consistent theme that follows through the experiments below.

4.1.2. Lateral Fire Spread between Two Steel-Clad RSDs (Tests 2–4)

In Test 2 (2-2D-S-250-K), Test 3 (3-2D-S-250-C) and Test 4 (4-2D-S-250-F), both RSDs were steel-clad with timber cross members. The window was introduced in RSD-1 such that the dwelling is identical to RSD-2. This updated test was proposed with a slightly lower fuel load and the RSDs were positioned laterally at 250 mm from each other. These tests were conducted outdoors with varying wind conditions to assess the impact of wind conditions. Figure 6 shows a test setup and the temperature-time results of the tests.

Test 2 (cross wind): The flames started to emerge from the openings of RSD-1 within 2 min after ignition and the flashover was observed shortly thereafter. The cross wind swept the hot gases and flames away from the adjacent RSD-2 and no fire spread was observed to the RSD-2.

Test 3 (circular wind): Flashover was observed within 2 min and flames from the openings of RSD-1 started impinging on the cladding of RSD-2. The movement of hot gases and flames in the open environment was dependent on the direction of wind, which kept changing throughout the experiment. There was a continuous heat flux (radiative and convective) from windows of RSD-1 to the steel cladding that initiated the smouldering of cardboard linings of RSD-2. The flame impingement was seen for a short span that ignited the smouldering cardboard through gaps in the RSDs. After the ignition of cardboard, the adjacent RSD was in a smouldering phase for 1550 s and later the adjacent RSD was involved in the fire.

Test 4 (favourable wind): This test was unique where a wind of 24 kmph was observed from RSD-1 towards RSD-2. After the ignition of RSD-1, all the hot gases were driven towards the direction of RSD-2. RSD-1 reached a flashover in less than 90 s and the flames emerging out of RSD-1 started impinging onto the steel cladding of RSD-2. A smouldering of cardboard in RSD-2 was seen at 185 s after ignition in RSD-1 and within a few seconds the cardboard was ignited which spread the fire to the other regions of the RSD including the wood cribs. Despite being a successful experiment in terms of fire spread to the adjacent RSD, the data could not be extracted due to technical failure, hence only videographic information was used for analysis.

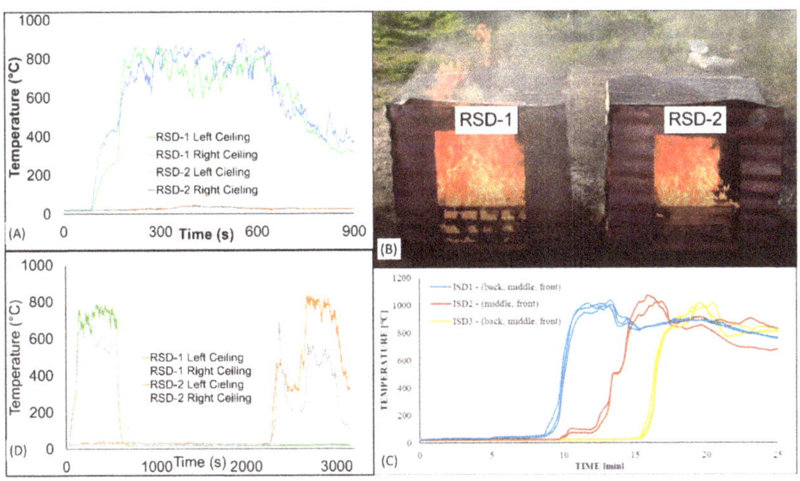

Figure 6. Longitudinal fire spread results. Left top clockwise—(**A**) Test 2: ceiling temperature on cross wind condition, (**B**) RSEs on steel-clad dwellings, (**C**) FSE results on steel-clad dwelling [7], (**D**) test 3: ceiling temperature on circular wind condition.

Figure 6 shows the results of ceiling temperature profiles of various RSEs and FSEs from multiple steel-clad dwelling tests. The experimental results from RSDs are compared with full scale experiment [7] and provided in Table 5. The temperature time curve in RSE correlate well with the FSE. Theoretically, the temperatures should be scaled for the RSEs. The ceiling temperatures in the RSEs from available data were ranging from 830 to 900 °C and found to be around 20% lower than the FSEs. A similar drop in ceiling temperatures was recorded in past work [17].

Table 5. Comparison on RSEs and FSEs results on fire spread between multiple steel-clad reduced scale dwelling.

Steel-Clad ISDs	Full-Scale ISD [7]		Reduced-Scale ISD					
	Favourable Wind		Test 2 (Cross Wind)		Test 3 (Circular Wind)		Test 4 (Favourable Wind)	
	ISD-1	ISD-2	RSD-1	RSD-1	RSD-1	RSD-2	RSD-1	RSD-2
Maximum ceiling temperature	1037 °C	1070 °C	900 °C	No spread	838 °C	840 °C	No Data	No Data
Time from start of ignition to collapse	1080 s	1140 s	537 s	No Ignition	570 s	574 s	553 s	637 s
Scaled time from start of ignition to collapse	-	-	1074 s	-	1140 s	1148 s	1106 s	1274 s
Time from the flashover to collapse	510 s	378 s	311 s	No Ignition	485 s	497 s	445 s	524 s
Scaled time from the flashover to collapse	-	-	622 s	-	970 s	994 s	890 s	1048 s
Time to fire spread to RSD-2 from flashover of RSD-1	-	240 s	-	No Spread	-	2259 s	-	319 s
Scaled time to fire spread to RSD-2 from flashover of RSD-1	-	-	-	No Spread	-	4518 s	-	638 s

It was observed that the experiment time in all RSEs followed scaled correlation ($t \sim s^{1/2}$) in all wind conditions with a maximum error margin of 11% in test 4. The time to flashover in each case was different and deviated up to 100% from the full-scale experimental results. This shows the limit of scaling hypothesis in which all phenomena cannot preserved. The time for spread from flashover in the FSE was 240 s which is not comparable to the unscaled RSE values (No spread, 2259 s, 319 s) and the scaled values (No spread, 4518 s, 638 s).

In both FSE and RSE, the fire spread from the dwelling on fire adjacent RSE was due to ignition of cardboard because of flame impingement through the gaps and constant heat flux that led to ignition. This burning cardboard then spread fire to entire dwelling. The fire spread in Test 4 (favourable) was faster than in other tests due to favourable wind currents. In Test 3 (circular), as seen in Figure 6D, although fire spread was delayed it was interesting to see an evolution of smouldering fire that subsequently led to the collapse of the RSD and spread only occurred after that. This confirms the possibility of restarting IS fires despite the original dwelling on fire is attended. In all cases, the fire spread time in the FSE was significantly lower than the RSEs by 95 and 62%.

4.1.3. Lateral Fire Spread between Two Timber-Clad RSDs (Tests 5–6)

Tests 5 to 6 were carried out to understand the influence of combustible cladding, namely 14 mm thick timber planks. Since combustible cladding is known to cause increased fire spread in settlements a large separation distance of 500 mm was also used. Low wind speeds that were circular or opposing wind spread occurred during the testing.

Test 5 (250 mm spacing): Upon ignition RSD-1 was engulfed in fire within 100 s and the hot smoke and flames emitted convective and radiative heat fluxes on to the adjacent RSD that aids in pyrolysing the timber cladding of RSD-2. Although the wind was in circular motion throughout the experiment, the adjacent RSD took less than 121 s to ignite. The entire experiment was completed within a period of 800 s that completely burned down the RSDs ending the experiment.

Test 6 (500 mm spacing): Spread from RSD to RSD took 427 s from ignition, which is approximately double that of Test 5. The total duration of an experiment was little over 1100 s.

Figure 7 shows the comparison between FSEs and RSEs. The experimental results from RSDs are compared with full scale experiment [7] and provided in Table 6. The experimental time temperature curve for FSE and RSEs had similar pattern indicating that general behaviour was captured. However, the peak temperature in both RSEs tests was lower by around 10%. The temperature profile followed in the FSD graph, as shown in Figure 7A, and is similar to the timber-clad RSEs temperature results for both Test 5 and 6. The adjacent RSDs in both the tests became involved in fire once the burning FSD reached the fully developed phase. As shown in Figure 7E, in RSEs with 250 mm separation distance experiment, RSD-2 ignited during the final stages of fully developed phase of RSD-1 closely representing full scale experiment. However, ignition of RSDs with 500 mm separation distance was delayed where the ignition phase of RSD-2 was coinciding with decay phase of fire in RSD-1 as seen in Figure 7D as the wind was opposing the fire spread. Hence, in this instance geometric scaling was suitable for capturing spread behaviour.

Table 6. Results from fire spread between timber-clad dwelling in full scale experiment and reduced scale experiment.

Timber-Clad RSD Informal Settlement Dwelling	Full-Scale ISD [7]		Reduced Scale ISD			
	Mild Favourable Wind		Favourable Wind, Spacing—250 mm		Favourable Wind, Spacing—500 mm	
	ISD-1	ISD-2	RSD-1	RSD-2	RSD-1	RSD-2
Maximum ceiling temperature	1104 °C	1176 °C	1016 °C	957 °C	1000 °C	974 °C
Time from start of ignition to collapse	798 s	858 s	580 s	420 s	780 s	410 s
Scaled time from start of ignition to collapse	-	-	1160 s	840 s	1560 s	820 s
Time from the start to flashover to collapse	306 s	318 s	480 s	390 s	600 s	359 s
Scaled time from the start to flashover to collapse	-	-	960 s	780 s	1200 s	718 s
Time to fire spread to RSD-2 from flashover of RSD-1	-	82 s	-	121 s	-	247 s
Scaled time to fire spread to RSD-2 from flashover of RSD-1	-	-	-	242 s	-	494 s

In the FSEs, the wind current was favourable to fire spread, whereas in the RSEs, they were either circular or opposing the fire spread, potentially leading to the discrepancy in spread times. The duration in each RSD from ignition to collapse was recorded as twice that of FSEs in the unfavourable wind condition. The fire spread to and adjacent RSD from a timber-clad RSEs is largely due to flame impingement and heat fluxes that ignite cardboard and also pyrolyse the timber cladding. At this scale, the fire spread time is approximately linearly dependent on separation distance between the dwellings, but this is likely to be influenced by multiple factors. The time to spread from flashover was 82 s for the FSE which is faster than the RSE unscaled (121 s, 242 s) and scaled (247 s, 494 s) values.

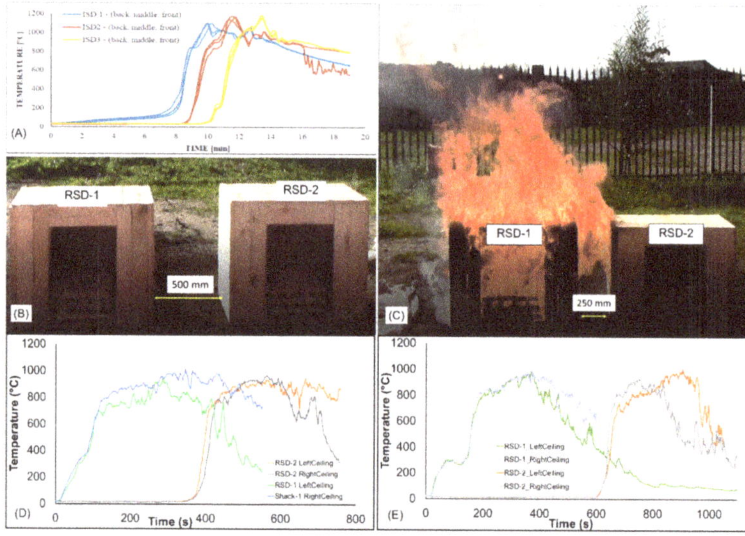

Figure 7. Results from various muti dwelling tests. Clockwise—(**A**) FSE results on timber-clad dwelling [7], (**B**) test 6: RSEs on timber-clad dwellings at 500 mm, (**C**) test 5: RSEs on timber-clad dwellings at 250 mm, (**D**) test 6: ceiling temperature in timber-clad dwelling placed at 500 mm, (**E**) test 5: ceiling temperature in timber-clad dwelling placed at 250 mm.

4.2. Three RSD Experiments

In continuation to the two RSD experiments, a third dwelling was added on the adjacent side, see Figure 8, such that the fire spread from the door and window can be analysed. In these experiments, the impact of combustible cross members with varying wind conditions has been analysed. In this series three sets of experiments were conducted that are listed in the following section and the testing matrix has been provided in Table 7. The wind speed in all tests was low at around less than 5 kmph, which led to only wind direction being captured accurately.

Table 7. Testing matrix of three RSD experiment.

Test No	RSD Id	RSD Configuration	Wind	Fire Spread Rate (m/min)
7	7-3D-S-H-300-F	Steel-clad RSDs with steel cross member spaced at 300 mm from each other, in 'L' shape.	Favourable <5 kmph	0.15
8	8-3D-S-H-250-O	Repeated above test with timber cross member. Dummy RSDs added.	Opposing <5 kmph	0.136
9	9-3D-S-H-300-O	Steel-clad RSDs 3 with timber cross member spaced laterally	Opposing <5 kmph	No spread
10	10-3D-S-H-300-R	Repeated above test in restricted wind condition	/	0.15

Test 7 (steel frame): Three steel-clad RSDs were located equidistant of 300 mm from each other in a 'L' shape. The 300 mm spacing is based on geometrically scaling the 1.2 m distance between dwellings in the FSE used for comparison. RSD-A1 was ignited and the fire spread was observed in RSD-A2 and RSD B1. The flames from door and windows of RSD-A1 started impinging on adjacent RSDs at 270 s. The constant flame impingement on backwall of RSD-B1 led to the smouldering of its cardboard lining for more than 100 s then subsequently ignited the cardboard at 450 s as seen in Figure 8. The fire quickly spread throughout the RSD-B1 including the wood cribs. The fire in RSD-A1 burned away before the fire in the adjacent RSD-B1 could establish and the fire development inside the RSD-B1 was aborted.

In addition, there was no combustible cross members which in real fires contribute to faster fire spread. The flames from the window of RSD-A1 was not steady while impinging on the window of RSD-A2. The flames were pushed away by air movement in the alley between the RSDs; thus, the flames were not able to ignite the cardboard of RSD-A2; however, smouldering gases from the RSD were seen throughout the experiment. The intensity of flame impingement represented as a temperature profile in adjacent dwellings has been provided in Figure 9A where the roof temperature of RSD-B1 indicated the ignition of the shack and window temperature of RSD-A2 indicates the temperature of pyrolysis gases.

Based on the results from Test 7, Test 8 was then used to study the influence of reduced spacing and timber frames; Test 9 considered the same spacing and timber frames whilst Test 10 was a repeat of Test 9 but was carried out with negligible air movement. Tests 9 and 10 were carried out as a single line of dwellings to ensure one-dimensional spread and effects could be ensured.

Test 8 (timber frame, 250 mm spacing): The same experimental set-up was repeated by reducing the separation distance between the RSDs from 300 mm to 250 mm. One dummy RSD added ahead of RSD-A2 such that it becomes 4-RSD arrangement that will restrict the wind near the windows. The unfavourable wind condition pushed flames away from the RSD-A2 and no significant difference was found in fire spread rates from Test 7.

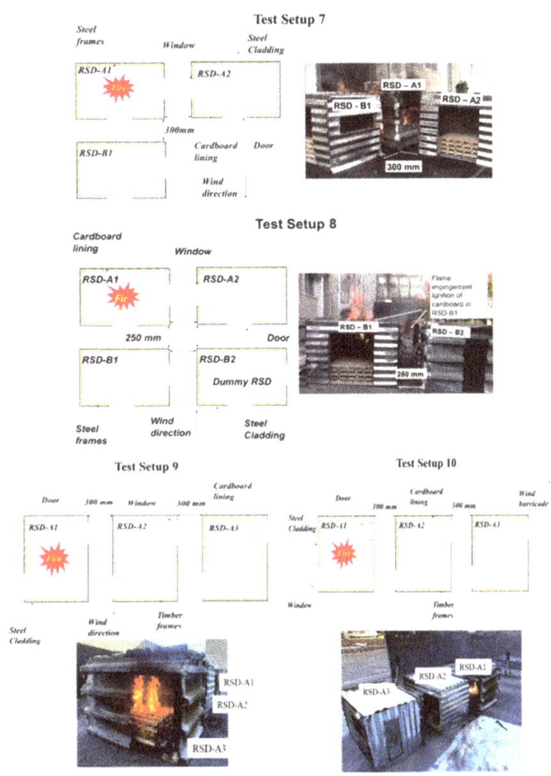

Figure 8. Tests 7 to 10 showing layouts and images during testing.

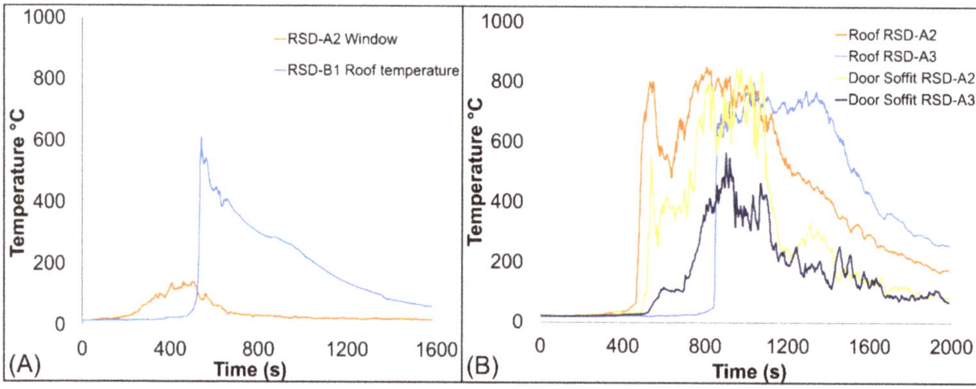

Figure 9. Three RSD experiment: (**A**) test 7: fire spread by RSD-01; (**B**) test 10: roof temperatures and door soffit temperatures.

Test 9 (timber frame, 300 mm spacing): There was no fire spread observed in Test 9 despite the presence of combustible cross members due a light unfavourable wind. Hence, depending on wind direction spread is unlikely to occur with a spacing of 300 to 400 mm based on the geometry of RSD-A1.

Test 10 (as Test 9 with no air movement): The same test was repeated as Test 10 but with barricading the experimental area with porous screen to considerably reduce the influence of wind. Fire spread was observed in all the RSDs and a fire development pattern observed with one RSD igniting the other in every 400 s as shown in Figure 9B. This test again highlights both the importance of combustible cross member and influence of wind conditions in ISDs fire spread.

From Table 7 it can be observed that the spread rate was surprisingly consistent between the different tests at around 0.136–0.15 m/min, considering the variable conditions and parameters, but of the same order as the tests with more rapid spread for the double dwelling RSDs above.

4.3. Four RSD Experiments

The four RSD experiments were conducted with varying distances and staggered arrangements of RSDs. The details of the test setups have been provided in Table 8. In these experiments, the impact of distance, cladding, and positioning of RSDs with varying wind conditions have been analysed. Setups have been designed to mimic a section of the 20-dwelling experiment presented in Section 5.2.3 [10], and the results will be analysed and presented further in the following sections.

Table 8. Testing matrix of four RSD experiment.

Test No	RSD Id	RSD Configuration	Wind Speed	Fire Spread Rate (m/min)
11	11-4D-S-550-/	Steel-clad RSDs in staggered arrangement spaced equidistant from each other such that there are 2 rows of 2 Nos RSDs. The wind condition was restricted	/	0.225
12	12-4D-2S/2T-250/550-F	Steel-clad RSDs spaced equidistant from each other such that there are 2 rows of 2 RSDs.	9 kmph favourable	0.194

Test 11 (negligible wind): The steel-clad RSDs were arranged in staggered positions with two rows of two numbers of RSDs as shown in Figure 10. The adjacent dwellings were positioned 250 mm part and the next row was placed at 550 mm away from the first row. The influence of wind during the test was restricted by a porous barricade on the test site perimeter.

RSD-A1 and RSD-A2 were ignited simultaneously, and the fire spread to adjacent rows was observed. At 400 s after ignition, the flames were seen projecting outside the RSD-A1 and were directly impinging onto backwall of RSD-B1. Then RSD-A2 became fully involved in fire at 485 s and the flames from its door opening were now impinging on the back wall of RSD-B2 and initiated the pyrolysis of the cardboard lining near its windows. The cardboard lining near the timber cross member of the backwall of RSD-B2 was ignited by the impinging flames from RSD-A2 and within 60 s the fire was spread inside the dwelling. The entire test lasted for approximately 20 min but there was no ignition or fire spread observed to RSD-B1 despite a constant flame impingement for 600 s. It is also interesting to observe that the door opening in RSD-A1 and the window opening of RSD-B1 were located on opposite sides and could be the reason for no fire spread in the test as flames were not able to trigger the ignition of the cardboard linings. In addition, the RSDs with different dwelling profile such as RSD-A1 (Type D2) and RSD-A2 (Type D1) in this section provided different results in fire spread under the same environmental conditions.

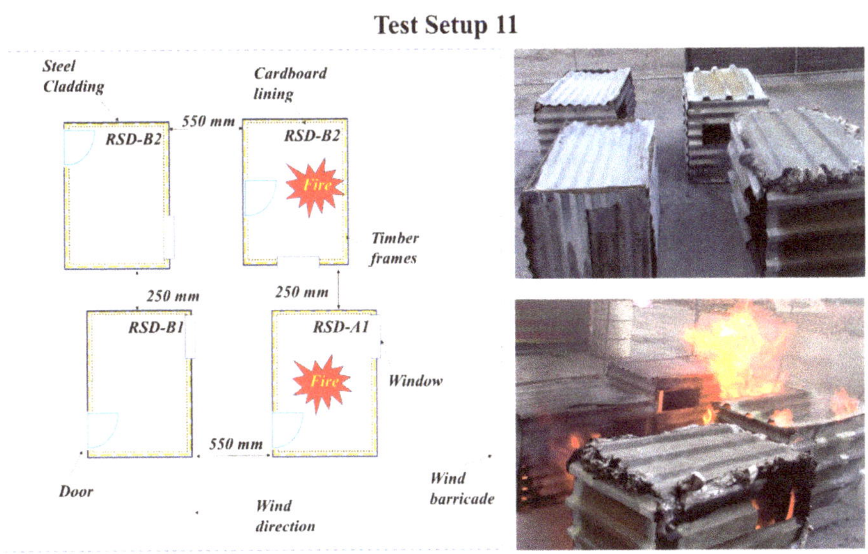

Figure 10. Four RSD experiment: test setup 12; clockwise—schematic view; test setup-12 @ 0 min; flame impingement from RSD-A2 at 11 min.

Test 12: The RSDs in each row was separated by 250 mm and RSDs in the next row were placed at 300 mm and 550 mm as shown in Figure 11. Each row consisted of one steel-clad RSD and one timber-clad. In the beginning, RSDs of row B was ignited and within 40 s the RSDs reached the fully developed stage. The ignition of combustible cladding of timber in RSD-B2 resulted in larger influence of convective and radiative heat flux in the vicinity.

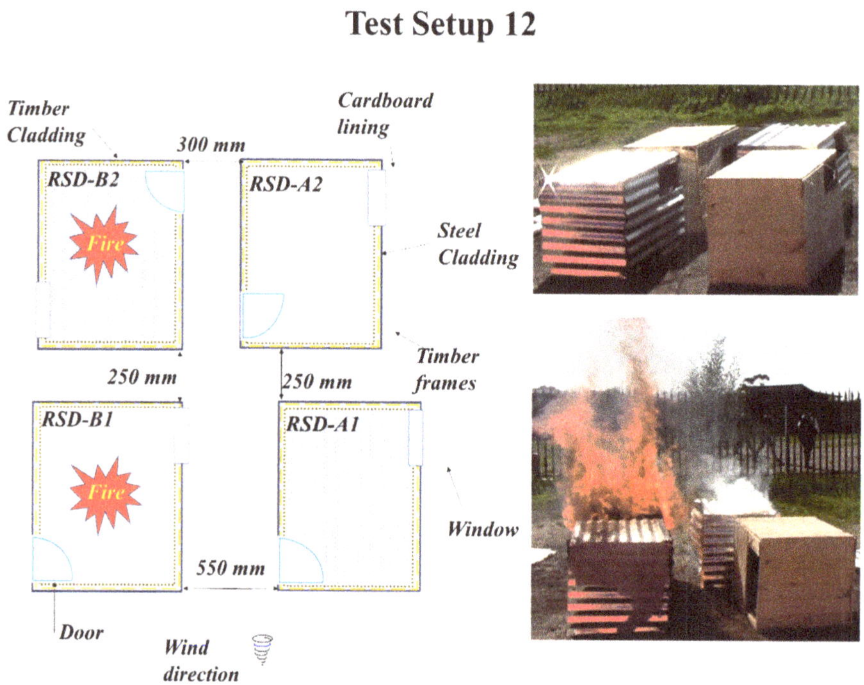

Figure 11. Four RSD experiment: test setup 12; clockwise—schematic view; test setup-11 @ 0 min; @ 6 min.

This heat flux triggered the ignition of steel-clad RSD-A2 in the next row as seen in Figure 11. The heat flux and flame impingement from the door of RSD-B2 aided in faster burning of combustibles in the dwelling RSD-A2. As a result, the total burn time of RSD-A2 reduced to 400 s. The RSD-B2 also influenced the timber RSD-A1 that was on the opposite side. The flame impingement and heat flux from the steel-clad RSD-B1 of first row, heated the timber cladding of RSD-A1 in second row which resulted in pyrolysed and charred cladding surface on the RSD. However, there was no ignition seen in the timber dwelling.

Discussion on Fire Spread Comparative Results in 4 Nos RSD Test (Tests 12) and FSE

A section of the 20 dwelling large scale experiments has been represented by the 4 RSDs in Test 11 and 12. The experimental setup in full scale experiment and reduced scale experiment with fire spread patterns are shown in Figure 12. The chosen section in the full-scale experiment was surrounded by burning dwellings which enhanced burning rate due to continuous flame impingement onto the RSD and higher heat transfer from dwellings on fire to adjacent dwelling. This phenomenon has enhanced the time to flashover and supported faster fire spread in the IS dwellings.

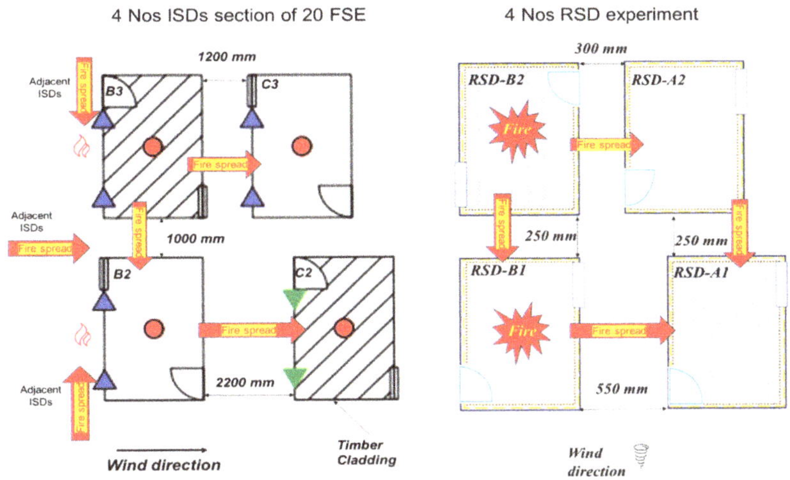

Figure 12. Fire spread in FSEs and 4 Nos RSD test.

In contrast, the reduced scale experiments had no influence of added heat transfer or flame impingement from adjacent RSD and unlike in full-scale experiment where wind direction was favourable to fire spread, in the RSEs the wind direction was changing throughout the experiment. This condition diverted the flames and heat away from the RSDs in row A. However, the effect of timber dwelling RSD-B2 dominated the wind effect and ignited the RSD-A2 within 200 s as seen in the graph B of below Figure 13. The graphs represent the ceiling temperature of dwellings in full scale experiment (top) and reduced scale experiment (bottom). Table 9 lists the difference in time to flashover, duration of dwelling to collapse, and maximum ceiling temperature.

The ceiling temperatures in full scale experiments were over 25% higher than the reduced scale experiments due to higher smoke layer build-up. In both type of experiments, the ignition to flashover to fully developed was achieved in short duration with little over a minute. However, this transition was achieved more quickly due to the impact of burning adjacent RSD. In the four RSD experiment the influence of burning adjacent RSD are clearly visible in burning time. The fire spread to adjacent row is evident in the graphs of reduced scale experiment but in the section of full-scale experiment, the burning behaviour was a collective phenomenon involving fires at multiple FSDs.

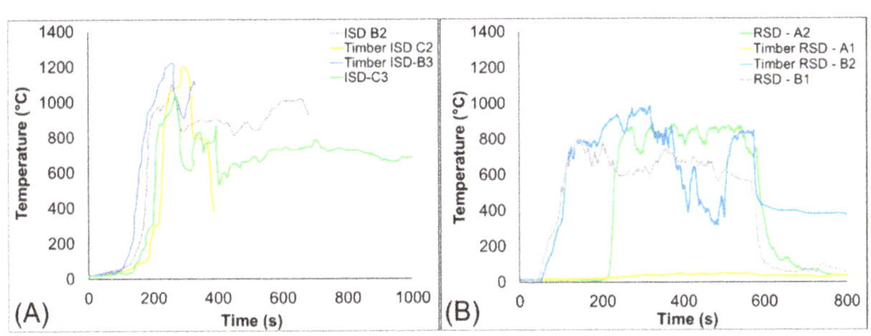

Figure 13. Ceiling temperature of 4 Nos RSDs in (A) FSEs, and (B) RSEs.

Table 9. Comparative results from 4 Nos RSEs (Test-12) and FSEs.

RSD Nos		Maximum Ceiling Temperature (°C)	Time from the Start to Flashover (s)	Scaled Time from the Start to Flashover (s)	Time from Start of Ignition to Collapse (s)	Scaled Time from Start of Ignition to Collapse (s)	Time to Fire Spread to Adjacent RSD from Flashover of RSD in 1st Row (s)	Scaled time to Fire Spread to Adjacent RSD from Flashover of RSD in 1st Row (s)
Full-Scale ISD	B2	1203	60	-	600	-	-	-
	B3	1213	60	-	240	-	-	-
	C2	1170	60	-	250	-	100	-
	C3	1039	40	-	940	-	90	-
Reduced Scale ISD	B2	986	55	110	550	1100	-	-
	B1	801	60	120	513	1026	-	-
	A2	881	87	174	490	980	677	1354
	A1	50	Nil	-	Nil	-	-	-

4.4. Six RSD Experiments

The six RSD experiment tests below provide an understanding on behaviour of fire spread between multiple mixed-clad dwellings. The influence of cladding material, fire load, wind conditions, and separating distance between RSDs on fire spread among multiple dwellings has been recorded. The testing matrix has been provided in the Table 10 and the detailed analysis has been provided in the following sections, with layouts provided in Figures 14–19. After the results from all the six dwelling experiments have been presented, they will be compared with the FSE results below.

Table 10. Testing matrix of six RSD experiment.

Test No	RSD Id	RSD Configuration	Wind Speed	Fire Spread Rate (m/min)
13	13-6D-S-300/550-O	Longitudinally placed 2 Nos Steel-clad RSDs in 3 rows spaced at 300 mm and 550 mm.	30 kmph	No spread
14	14-6D-S-300/550-F		28 kmph	
15	15-6D-S-300/550-K	Repeated above test with increased fuel load	24 kmph	0.15
16	16-6D-4S/2T-300/550-O	Repeated above test but steel RSDs were replaced with timber-clad RSD in the either side of middle and last row.	11 kmph	0.13/0.04
17	17-6D-4S/2T-300/550-O		13 kmph	0.058/0.015
18	18-6D-3S/3T-550-/	The same test was repeated but RSDs rows spaced at equal distance. The fire was ignited in the second row to avoid the influence of wind.	/	0.273/0.34
19	19-6D-4S/2T-550-/		/	0.114

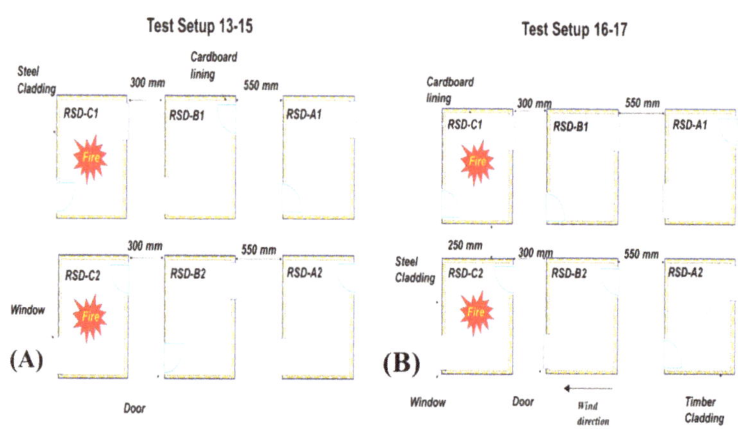

Figure 14. Six RSD experiment—(**A**) test setup 13–15, (**B**) test setup 16–17.

4.4.1. Tests 13–15—Steel Only Dwellings

Test-13 (opposed wind): In the first test, RSD-C1 and RSD-C2 of first row were ignited. The fire quickly reached flashover state and the fire became fully developed within 90 s. The wind was blowing strongly in the opposite direction of expected flame spread which drove away all the hot gases and flames away from second row RSDs. This led to loss of convective and radiative heat flux that play a crucial role in pyrolysis of available combustibles. Furthermore, the impinging flames were also diverted away from second row RSDs that would ignite the already pyrolysing combustibles inside them. No fire spread occurred to the adjacent rows.

Test 14 (favourable wind): The same test was repeated with favourable wind conditions blowing strongly at 28 kmph that directs all the hot gases towards the second row of RSDs. The flames from the ignited RSD-C1 and RSD-C2 were also seen to be carried over by the wind. The strong favourable wind current also caused the burning of the fuel at a faster rate. On the other hand, the heat losses in the burning RSDs increased due to mixing of cold air with hot gases. As a result of losing essential heat fluxes in this way, there was no effective fire spread to adjacent rows of RSDs. Though substantial flame impingement was seen towards second row RSDs, but it was inadequate to ignite the combustible materials. This is interesting given that the previous test with unfavourable wind conditions had spread across the 300 mm space to the second row.

Test 15 (crosswind and increased fuel load): There was no modifications made in geometry of the RSDs, but the fuel load was increased by two layers. There was heavy cross wind which blew the hot gases partially towards the second row of RSDs. There was a sustained flame impingement due to continuous favourable wind current and increased fire load. This aided in the fire spread onto the RSDs in the second row which was ignited after 4 min. The flames were also seen going over the RSDs and not impinging on the steel cladding. The temperature development for this test setup are shown in Figure 15 in which each RSD after its ignition, takes around 100 s to reach ceiling temperature of 600 °C indicating the flashover stage (in addition flames appear outside the RSD openings) and the total duration of each burning RSD is around 500 s. Additionally, ignition of RSDs in each row can be clearly seen in graphs where second row was ignited at around 300 s. Although from the Figure 15, it can be seen that the third row has recorded elevated temperatures in the openings but the flames from second row were inadequate to ignite the combustibles in the third row of RSDs. Thus, wind current with an optimum speed and favourable direction, and an increased fuel load, possess a higher possibility of fire spread through flame impingement and heat fluxes than from the hot gases.

Overall, only Test 15 resulted in spread through the 6 dwellings, with a rate of 0.15 m/min, which is similar to spread rates above. Both the increased fuel load and effect of wind influenced this result.

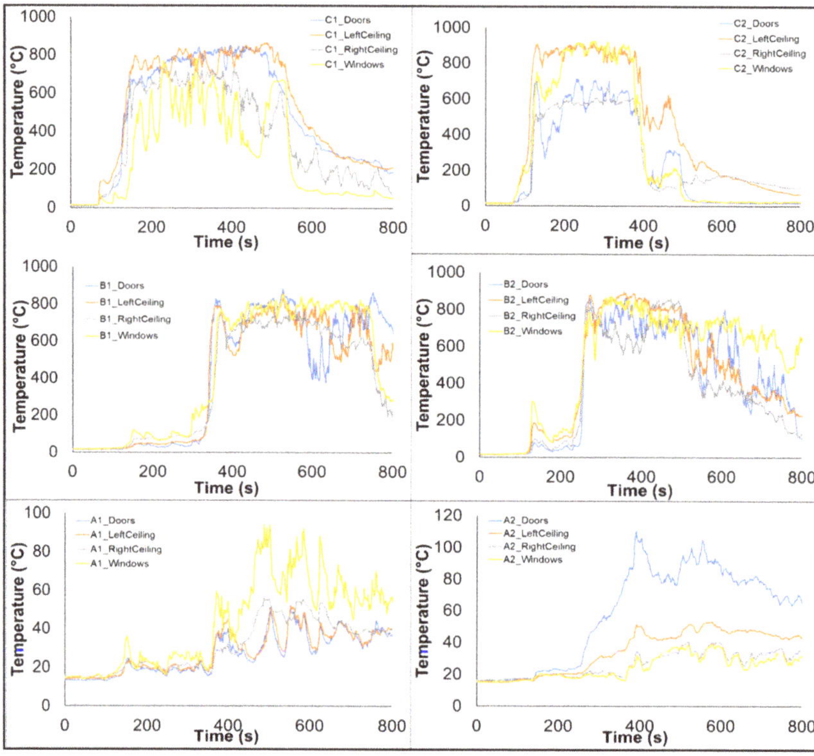

Figure 15. Six RSD experiment: temperature profiles of each RSD in Test 15.

4.4.2. Tests 16–17—Mixed Steel and Timber Dwellings

Test-16 (16-6D-4S/2T-300/550-O): The previous tests were modified with the inclusion of two timber-clad RSDs placed in second and third rows. The test was initiated by igniting two RSDs of first (C) row and within 90 s, the fire quickly reached fully developed stage. There was an opposing wind current driving all the hot gases away from the second row of RSDs. The flames were also diminished and fluctuated by the wind. Despite this factor, the timber dwelling RSD–B1 in the second row ignited at 6th minute and the heat flux from the flaming timber cladding led to steel ISD in the second row, RSD–B2 to ignite. The heat flux and flames from RSD—B1 and B2 together started pyrolysing the timber cladding of RSD–A2 in the third row. However, due to insufficient flame lengths owing to opposing wind condition, the third row was not involved in fire. This test was therefore repeated to confirm the repeatability of the results of the reduced scale experiments.

Test 17 (repeat): In this test the fire in the two dwellings of the first row namely RSD-C1 and RSD-C2 became fully developed within 60 s which is earlier than Test 16. The wind current flowing at 13 kmph was opposing the fire spread to the second row RSDs. The results such as fire spread, flashover time and test duration between from Tests 16 and 17 were similar as shown in Table 11. In both cases, the timber RSD–B2 was first to ignite in the second row due to continuous heat flux on timber cladding and an ignition by a flame from RSDs the first row. Similarly, in the test-17, the RSD–B1 took longer to ignite due to opposing wind conditions. In this test spread did occur to the third line of dwellings.

Table 11. Comparative results of test-16 and 17 with FSE.

RSD Nos		Maximum Ceiling Temperature (°C)	Time from the Start to Flashover (s)	Scaled Time from the Start to Flashover (s)	Time from Start of Ignition to Collapse (s)	Scaled Time from Start of Ignition to Collapse (s)	Time to Fire Spread to Adjacent RSD from Flashover of RSD in 1st Row (s)	Scaled Time to Fire Spread to Adjacent RSD from Flashover of RSD in 1st Row (s)
Test Setup-16	A1	51	Nil	-	Nil	-	-	-
	A2	52	Nil	-	Nil	-	-	-
	B1	967	45	90	379	758	206	412
	B2	873	42	84	400	800	390	780
	C1	899	41	82	473	946	-	-
	C2	902	45	90	414	828	-	-
Test Setup-17	A1	51	Nil	-	Nil	-	-	-
	A2	52	Nil	-	Nil	-	-	-
	B1	902	40	80	368	736	309	618
	B2	880	56	112	404	808	992	1984
	C1	817	66	132	492	984	-	-
	C2	979	67	134	489	978	-	-
Full-Scale ISD	A1	1240	100	-	500	-	-	-
	A2	1220	40	-	600	-	-	-
	B1	1170	70	-	300	-	90	-
	B2	1100	100	-	600	-	100	-
	C1	1150	90	-	720	-	130	-
	C2	1205	100	-	250	-	120	-

This similarity in temperature trends is graphically represented in Figure 16 where ceiling temperatures of ignited RSDs are presented. Despite this set of tests having a lower fuel load and opposing wind, the addition of timber-clad RSD changed the entire dynamics of fire spread. The test was initially planned to be favourable wind condition but due to sudden change in wind direction, the same test was repeated with opposing wind. However, it was interesting to see comparable results of Test 16 and Test 17, but difference in steel-clad dwelling (RSD-B2) of second row where time to fire spread was different. For the FSE the spread times were 0.8 m/min, whereas they were on average 0.05 m/min for the RSEs.

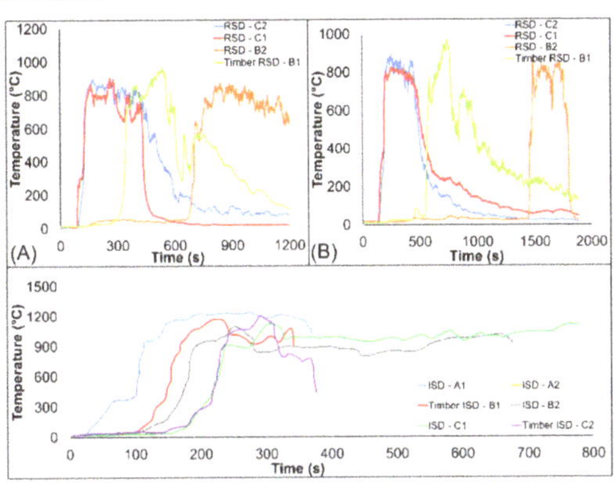

Figure 16. Six RSD experiments: ceiling temperature: top—temperature profiles of each RSD in (**A**) test 16, and (**B**) test 17, bottom—6 Nos ISDs from section of 20 FSE.

4.4.3. Comparative Study with FSE and RSE (Test-16, 17) in Six RSD Experiments

In this comparison study, six ISDs from twenty full scale dwelling experiment was considered and the relevant section is highlighted in Figure 17. The selected section consists of four steel-clad dwellings and two timber-clad dwellings. There were three rows of dwellings and each row had two dwellings. The timber-clad dwellings were placed on opposite side of second and third rows. The study will be focussing on parameters such as temperatures inside the dwellings, timelines of burning, fire spread patterns and so on.

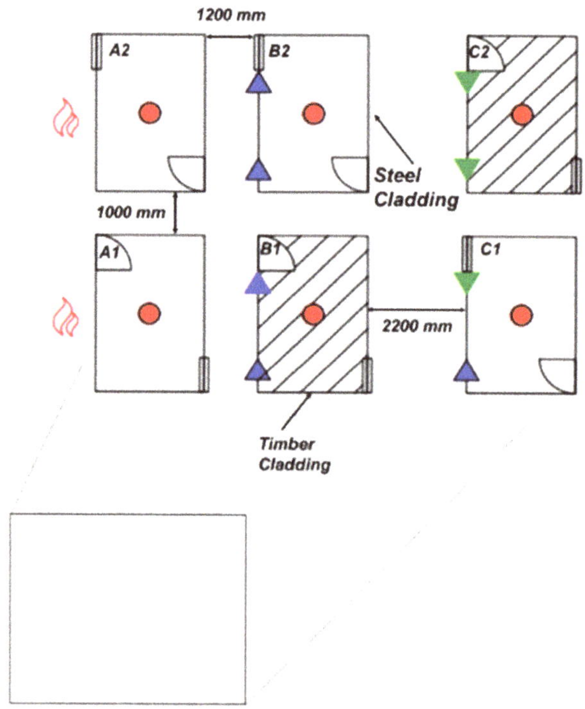

Figure 17. Section of 20 large-scale dwelling FSE on fire spread in ISDs considered for 6 Nos RSEs.

In the full-scale experiment, first row ('A' row) of dwellings were ignited and the fire spread behaviour to other rows were studied. Soon after the ISD-A1 reached flashover, the timber ISD–B1 was ignited and reached flashover even before igniting ISD–A2. The ignition of timber ISD–B1 changed the dynamics of fire spread and within 100 s other ISDs were involved in the fire. The fire spread rate as observed in FSE was 3.6 m/min. It should be noted that the adjacent section with burning ISDs also influenced the fire spread. The graph for the full-scale experiment as shown in Figure 16 (Bottom) shows that each dwelling was involved in the fire individually and there was no clear distinction of fire spread in different rows.

A section of the full-scale experimental setup as shown in Figure 17 was replicated in the reduced scale experiment. Table 11 provides comparative results on burning time and ceiling temperatures from FSEs and six RSDs. Figure 16 provides a graphical representation of ceiling temperatures in FSEs and RSDs. The first row of steel-clad RSDs was ignited, and they reached flashover stage around 50 s of ignition. At the same time, the timber-clad dwelling RSD–B1 was ignited, and it became fully developed in next 52 s. The dwelling continued to burn for the next 520 s. Meanwhile RSD–B2 was smouldering throughout the burning of the timber-clad RSD-B1 and started burning during decay phase of RSD–B1.

Unlike in the FSE, there was a clear distinction seen in burning pattern of the RSEs. Figure 18 shows the comparative pictorial representation of fire spread in-between the dwellings of FSEs and RSEs. The temperatures in the RSEs were 30% lower than that of FSEs and the total burning time in both FSEs and RSEs were almost similar for each dwelling. The higher temperatures and reduction in burning time could be a result of thermal influence from other section of the FSEs and higher smoke build-up that was clearly missing in the RSEs. The fire spread in RSE was ranging from 0.05 to 0.09 m/min in longitudinal direction and 0.01 to 0.05 m/min in lateral direction, whereas fire spread rate as observed in FSE was 3.6 m/min. The fire spread rate obtained in RSEs were negligible when compared with FSE. This could be a result of favourable wind current in FSEs, whereas in RSEs the wind was opposing fire spread, and thus, the third row in RSEs was not involved in fire and only charring was seen in the timber cladding of RSD–A2.

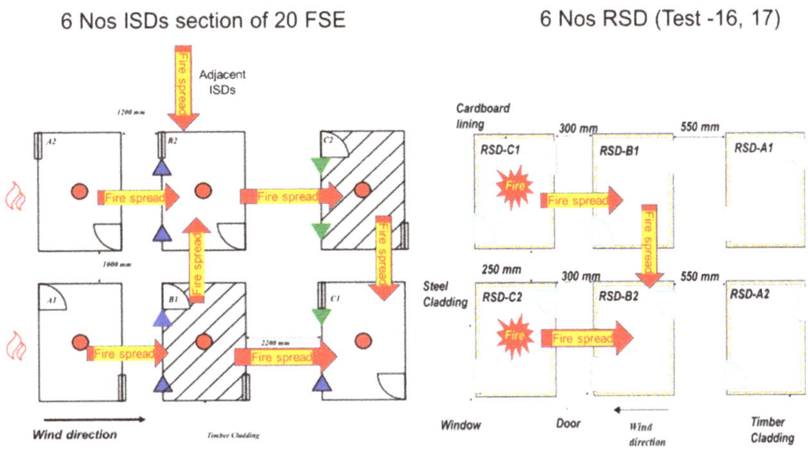

Figure 18. Fire spread in FSEs and 6 Nos RSD test (Test 16 and 17).

4.4.4. Test Setups—18 and 19—Bi-Directional Fire Spread

Test 18 (bi-directional spread, 3S/3T): From above tests it was seen that wind direction was crucial for fire spread, to reduce the dependency of that parameter, a test was setup as shown in Figure 19. The RSDs were arranged in three rows with two RSDs per row where each row was placed 550 mm apart. The steel and a timber-clad RSDs in the test setup were arranged in an alternating pattern. The fire was set-up in the second row to limit the dependency of wind direction and fire spread in both directions can be studied. From the test it was seen that timber cladding in the second row played a crucial role in the fire spread. It was seen that strong wind was driving the hot gases and the flames in the direction of wind current. This individual timber-clad RSD-B2 in the middle row triggered the ignition of timber-clad RSDs, RSD-C1 and RSD-A1 on either side of the adjacent rows.

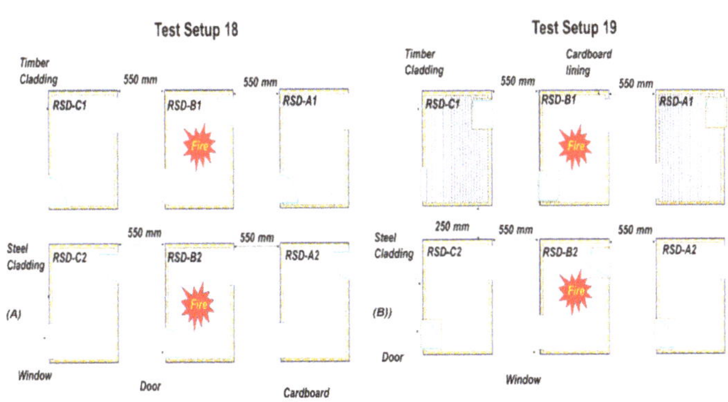

Figure 19. Six RSD experiment—(**A**) test setup 18, (**B**) test setup 19.

The temperature curve for the Test 18 is shown in Figure 20. As seen in the graph, middle row RSDs were ignited and, in that timber-clad RSD was burning at faster rate due to higher fuel load. Surprisingly, the timber-clad RSDs on either side of the adjacent rows were the first to ignite rather than the nearest steel-clad dwellings, demonstrating the impact of timber RSDs in fire spread in the test arrangement. The overall time of each RSD burning was approximately 400–450 s, and so no difference in burn duration was seen between steel-clad and timber-clad RSDs.

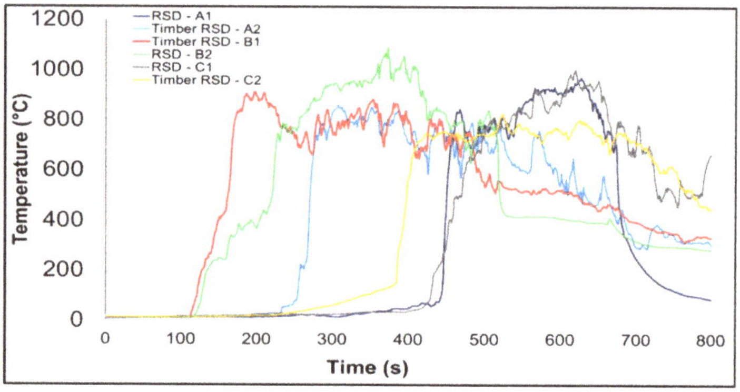

Figure 20. Ceiling temperature of each RSD in test setup 18.

Therefore, it can be said that the timber-clad RSDs influence fire spread fire in the ISD community due to high fuel load that has the potential to emit heat fluxes in all directions and project the continuous flames as compared to steel-clad RSDs. While burning in a fully developed stage, it has a potential to ignite all the combustible in its vicinity even at 550 mm (2200 m in FSE) and can alter the dynamics of fire spreading.

Test 19 (bi-directional spread, 4S/2T): In previous sections, it has been shown that timber cladding has a big influence on fire spread. However, to assess the importance of timber-clad RSD in the middle row, the wood-clad RSD was replaced by steel-clad RSD in the repeat test as shown in Figure 20.

As in Test 19, the RSDs in the middle row were ignited. The heat fluxes from the hot gases and flames of the steel-clad RSDs were continuously pyrolysing the timber cladding of RSD in first and third row. After 8 min of ignition, the back wall of the RSD-B1 collapsed near the timber-clad RSD–A1 with some of its residual cross members still burning as shown in Figure 21. The pyrolysing of timber cladding in RSD–A1 was increased due to sudden exposure to higher heat fluxes from burning RSD–B1 owing to the collapse of the back wall. At the 9th minute of the test, pyrolysis gas was ignited by residual cross members of the back wall, such that RSD–A1 started burning and it continued burning for the next 10 min. Despite exposure to higher heat fluxes in the vicinity, the steel-clad RSD–A2 of third row was smouldering for 15 min and then with a sudden gust of wind it started burning and after 5 min the dwelling collapsed. Throughout the test, the first row remained unaffected as the heat fluxes and the flames were driven by wind currents towards the third row of dwellings.

Thus, absence of timber-clad dwelling from the middle row, reduced the fuel load and heat release rate consequently reducing the radiative and convective heat flux that enhances the fire spread. A timber-clad RSD (or other combustible cladding) in the settlement is a key contributor to fire spread. In the absence of timber-clad dwellings, fire spread is primarily dependent on the direction of wind current. It was also interesting to see that despite the fire being suppressed in dwellings, the remnants of burning RSD can trigger smouldering of combustible in its vicinity that can start another informal settlement dwelling fire.

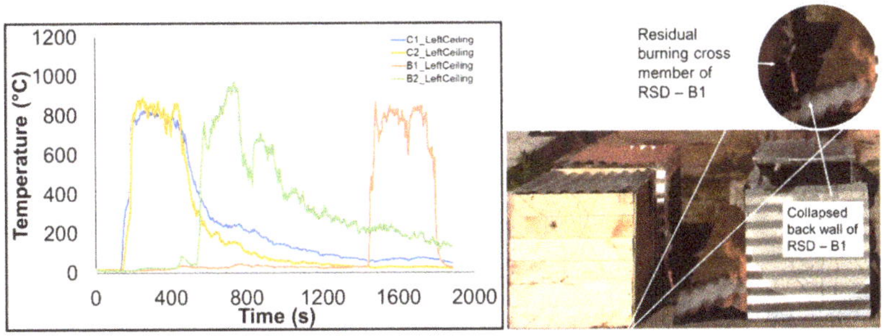

Figure 21. Six RSD experiments—left: ceiling temperatures of RSDs in test 19; right: burning of residual cross members.

5. Conclusions

This paper investigated fire spread using reduced scale dwellings at 1/4th of full-scale informal settlement experiments. The objective of the study was to identify the impact of factors such as cladding material, wind current and separation distance. The study also investigated the relevance of previous multiple-dwelling full-scale experiments in reduced scaling of informal settlement dwellings. The study involved fire spread between two, three, four, and six sets of reduced scaled dwellings with various orientations, separation distances, and cladding materials. Although primarily flame impingement, and secondarily heat flux, are responsible for fire spread to a greater extent, burning remnants from a collapsed dwelling have potential to ignite a new dwelling.

The direction of wind current plays a crucial role in directing flames and driving heat fluxes in both FSEs and RSEs. An optimum wind speed can aid in faster fire spread between the dwellings. However, a strong favourable wind current may also increase the heat losses in the burning RSDs, leading to no effective fire spread to adjacent rows (Test 14), as discussed in Section 3.2. In some instances, the flames were also seen going over the RSDs instead of hitting the roof panel. Wind can become a dominant factor to fire spread for inter dwelling spacing above 300 mm. Although, wind direction and speed measured at the beginning of each test fluctuations in wind current during most of the test was visible. Since this factor is so dominant, it will be beneficial to conduct such experiments in a wind tunnel in the future where the wind speed can be regulated. Additionally, capturing localised wind effects in between rows of dwellings can support the predication of fire spread and the pulsating behaviour of flame length and flame heights emerging from the openings.

The possibility of ignition is further increased due to the availability of timber cross members (Test 8) and combustible lining inside the dwelling walls. The exposed combustible cross members in the event of fire become involved in fire in the early stage, and they also contribute to fire spread to other combustibles inside the dwelling. The dwelling with higher fuel tends to have longer burning time. A dwelling is likely to be ignited as the increased burning time enhances the devolatilisation process and flame impingement due to longer exposure time. In addition, timber-cladded dwellings increase the risk of ignition in the vicinity to substantial level (Tests 5–6). The timber-clad RSDs have a significant impact on spread fire in ISs. They have the potential to ignite all the combustibles in its vicinity even at 550 mm (2.2 m in FSE) and can alter the dynamics of fire spreading despite having opposing wind currents.

The time to flashover in each case was different and deviated up to 100% from the full-scale experimental results, with RSEs typically ranging between 30 s and 180 s depending upon type of test and location of the RSD. The fire spread time between the dwellings increased with an increase in their separation distances, as would be expected. It was seen from the test that the fire spread due to flame impingement through the door opening contributes more than the window openings. The dwelling with windows on shorter side walls has greater fire spread prospects than the dwelling windows on longer back wall.

The addition of an opening has negligible impact on fire spread to the adjacent dwelling, but fire spread in type-1,2 RSDs provided different results under the same environmental conditions, proving the influence of the location of openings in a dwelling (test 11).

The average fire spread rate from all 19 RSEs was 0.092 m/min, ranging from 0.007 m/min in Test 3 to 0.27 m/min in Test 17 (FSE—0.137 m/min), which is significantly lower than the range of 1.2–3.6 m/min as found for real incidents and large-scale experiments. The scaled spread rates varied between 0.14 m/min and 0.55 m/min between rows of dwellings, in comparison to FSEs where spread times ranged between 0.167 m/min and 3.6 m/min. This highlights that even though trends, temperatures, and general fire dynamic behaviour can be captured, RSEs cannot currently be used for quantifying spread rates for informal settlement fires, although the occurrence of spread can be captured to a certain degree.

It was observed that the temperature profile for both FSE and RSEs were comparable with similar profiles, but the peak temperature was lower in RSEs than FSE, by around 10–20%. The experiment time from ignition to collapse in all wind conditions and for all RSDs was similar to FSE, but the flashover time in each case was different and deviated from the full-scale experimental results. The time variable in the RSEs was predominantly influenced by wind currents that affected the scaling of experimental time, flashover time, and fire spread rate. However, in few instances where wind current was similar to FSE, time variable was scaled well. Heat flux data in RSEs were not captured which would have further provided details to understand the FSEs and RSEs quantitatively. The results from individual RSEs (test-1–6) provided more comparable results with FSEs than the test with multiple rows, as greater numbers of steps in the flame paths leads to higher levels of uncertainty. It was observed that thermal influence from adjacent section with burning ISDs had great impact on the fire spread, and the same was missing in RSEs. In the experiments with six RSDs, each dwelling was involved in the fire individually and there was no clear distinction of fire spread in different rows.

To sum up, (a) wind, combustible cladding, and separation distance between dwellings significantly influence results. However, (b) the influence of cross member, fuel load inside the dwelling, and type of dwellings also had an impact on fire spread. (c) The comparative results from FSEs and RSEs demonstrate good fire dynamic correlations within the dwellings and in between the dwellings with comparable profiles but (d) have limited correlation to the fire spread rate. The database presented in this work provides a useful basis for enhancing scaling methodologies, but further data are needed before the results can be consistently applied due to the variability encountered. A study on the flame length and flame height emerging from the RSDs with various wind conditions would be beneficial to understand the impact of these parameters on fire spread. By quantifying flame length and height, and developing analytical equations for scaled dwellings, it is hypothesised that the behaviour observed in this paper can be more accurately defined, especially if equations can account for (albeit approximate) wind conditions. It is envisaged that separation distances between dwellings could be specified based on FSEs and RSEs causing the same level of flame impingement on the target dwellings.

Based on the results above, and observations of real informal settlement incidents, more research and testing are needed before large-scale multi-dwelling spread will be predicted accurately. Nevertheless, in future, these experiments can further be extended by changing selected parameters and studying the relative change in the test results using an initial test as a validated benchmark (i.e., all tests conducted at $\frac{1}{4}$ scale but the relative change in spread rate can be readily quantified by adjusting parameters). Such results can assist in quantifying empirical correlations to predict fire spread mechanisms in relation to multiple informal dwellings with reduced scale experiments considering various parameters. In addition, a numerical simulation can be developed that can predict the influence of numerous variables that can influence fire spread involving a large number of dwellings. As noted above, tests in a wind tunnel would be ideal for more accurately quantifying spread rates and comparing the influence of parameters. This work would serve as preliminary guidance for such a study.

Author Contributions: Investigation, Methodology—V.N.; Testing—V.N., A.O.; Writing and Editing—V.N., R.S.W.; Review—R.S.W. and A.O. All authors have read and agreed to the published version of the manuscript.

Funding: The authors would like to acknowledge the financial support of (a) the Lloyd's Register Foundation under the "Fire Engineering Education for Africa" project (Grant GA 100093), and (b) the Royal Academy of Engineering/Lloyd's Register Foundation under the "Engineering Skills Where They are Most Needed" grant (Grant ESMN 192-1-141).

Institutional Review Board Statement: Not applicable.

Informed Consent Statement: Not applicable.

Data Availability Statement: Not applicable.

Acknowledgments: The authors are grateful to Klapmuts Fire Department for their extended support to conduct the experiments on their premises.

Conflicts of Interest: The authors declare no conflict of interest.

References

1. United Nations (UN). 'Population 2030: Demographic Challenges and Opportunities for Sustainable Development Planning', New York. 2015. Available online: http://www.un.org/en/development/desa/population/publications/pdf/trends/Population2030.pdf (accessed on 11 July 2022).
2. Statistics South Africa (Stats SA). Concepts and Definition. Available online: www.statssa.gov.za (accessed on 11 July 2022).
3. Walls, R.; Zweig, P. Towards Sustainable Slums: Understanding Fire. In *Advanced Technologies for Sustainable Systems*; Springer: Berlin, Germany, 2017; pp. 93–98. [CrossRef]
4. Walls, R.; Olivier, G.; Eksteen, R. Informal settlement fires in South Africa: Fire engineering overview and full-scale tests on "shacks". *Fire Saf. J.* **2017**, *91*, 997–1006. [CrossRef]
5. Moradi, A. *A Physics-Based Model for Fire Spreading in Low Cost Housing in South African*; Stellenbosch University: Stellenbosch, South Africa, 2016.
6. Cicione, A.; Beshir, M.; Walls, R.S.; Rush, D. Full-Scale Informal Settlement Dwelling Fire Experiments and Development of Numerical Models. *Fire Technol.* **2019**, *56*, 639–672. [CrossRef]
7. Cicione, A.; Walls, R.; Kahanji, C. Experimental study of fire spread between multiple full scale informal settlement dwellings. *Fire Saf. J.* **2019**, *105*, 19–27. [CrossRef]
8. Cicione, A.; Walls, R.; Stevens, S.; Sander, Z.; Flores, N.; Narayanan, V.; Rush, D. An Experimental and Numerical Study on the Effects of Leakages and Ventilation Conditions on Informal Settlement Fire Dynamics. *Fire Technol.* **2021**, *58*, 217–250. [CrossRef]
9. Cicione, A.; Walls, R.; Sander, Z.; Flores, N.; Narayanan, V.; Stevens, S.; Rush, D. The Effect of Separation Distance Between Informal Dwellings on Fire Spread Rates Based on Experimental Data and Analytical Equations. *Fire Technol.* **2021**, *57*, 873–909. [CrossRef]
10. De Koker, N.; Walls, R.S.; Cicione, A.; Sander, Z.R.; Löffel, S.; Claasen, J.J.; Fourie, S.J.; Croukamp, L.; Rush, D. 20 Dwelling Large-Scale Experiment of Fire Spread in Informal Settlements. *Fire Technol.* **2020**, *56*, 1599–1620. [CrossRef]
11. Wang, Y.; Gibson, L.; Beshir, M.; Rush, D. Determination of Critical Separation Distance Between Dwellings in Informal Settlements Fire. *Fire Technol.* **2021**, *57*, 987–1014. [CrossRef]
12. Beshir, M.; Wang, Y.; Centeno, F.; Hadden, R.; Welch, S.; Rush, D. Semi-empirical model for estimating the heat release rate required for flashover in compartments with thermally-thin boundaries and ultra-fast fires. *Fire Saf. J.* **2021**, *120*, 103124. [CrossRef]
13. Walls, R.; Kahanji, C.; Cicione, A.; Vuuren, M.J.V. Fire dynamics in informal settlement "shacks": Lessons learnt and appraisal of fire behavior based on full-scale testing. In *Asia-Oceania Symposium on Fire Science and Technology*; Springer: Singapore, 2018; pp. 15–27.
14. Cicione, A.; Walls, R.S. Towards a simplified fire dynamic simulator model to analyse fire spread between multiple informal settlement dwellings based on full-scale experiments. *Fire Mater.* **2020**, *45*, 720–736. [CrossRef]
15. Cicione, A.; Wade, C.; Spearpoint, M.; Gibson, L.; Walls, R.; Rush, D. A preliminary investigation to develop a semi-probabilistic model of informal settlement fire spread using B-RISK. *Fire Saf. J.* **2020**, *120*, 103115. [CrossRef]
16. Quiroz, N.F.; Walls, R.; Cicione, A.; Smith, M. Fire incident analysis of a large-scale informal settlement fire based on video imagery. *Int. J. Disaster Risk Reduct.* **2021**, *55*, 102107. [CrossRef]
17. Narayanan, V.; Cicicone, A.; Botha, A.D.; Walls, R.S. Reduced-scale experiments and numerical simulations of informal settlement dwelling fires. *Int. J. Prog. Scale Model. Int. J.* **2022**, *3*, 3023. [CrossRef]
18. Beshir, M.; Omar, K.; Centeno, F.; Stevens, S.; Gibson, L.; Rush, D. Experimental and Numerical Study for the Effect of Horizontal Openings on the External Plume and Potential Fire Spread in Informal Settlements. *Appl. Sci.* **2021**, *11*, 2380. [CrossRef]
19. Quintiere, J.G. *Fundamentals of Fire Phenomena*; John Wiley & Sons, Ltd.: Hoboken, NJ, USA, 2006.

20. Bryner, N.; Johnsson, E.L.; Pitts, W. Carbon Monoxide Fires—Production in Compartment Reduced-Scale Enclosure Test Facility, NIST, NSTIR 5568. 1994. Available online: https://nvlpubs.nist.gov/nistpubs/Legacy/IR/nistir5568.pdf (accessed on 11 July 2022).
21. Babrauskas, V. Heat release rates. In *SFPE Handbook of Fire Protection Engineering, 5th ed*; Springer: New York, NY, USA, 2016; pp. 799–904. [CrossRef]

fire

Article

Fire Risk Assessment on Wildland–Urban Interface and Adjoined Urban Areas: Estimation Vegetation Ignitability by Artificial Neural Network

Maria Mahamed (Polinova) [1], Lea Wittenberg [2], Haim Kutiel [3] and Anna Brook [1,*]

[1] Spectroscopy & Remote Sensing Laboratory, Spatial Analysis Research Center (UHCSISR), Department of Geography and Environmental Studies, University of Haifa, Mount Carmel, Haifa 3498838, Israel
[2] Geomorphology Laboratory, Department of Geography and Environmental Studies, University of Haifa, Mount Carmel, Haifa 3498838, Israel
[3] Climatology Laboratory, Department of Geography and Environmental Studies, University of Haifa, Mount Carmel, Haifa 3498838, Israel
* Correspondence: abrook@geo.haifa.ac.il

Abstract: Fire risk assessment on the wildland–urban interface (WUI) and adjoined urban areas is crucial to prevent human losses and structural damages. One of many interacting and dynamic factors influencing the structure and function of fire-prone ecosystems is vegetation ignitability, which plays a significant role in spreading fire. This study sought to identify areas with a high-level probability of ignition from time series multispectral images by designing a pattern recognition neural network (PRNN). The temporal behavior of six vegetation indices (VIs) before the considered wildfire event provided the input data for the PRNN. In total, we tested eight combinations of inputs for PRNN: the temporal behavior of each chosen VI, the temporal behavior of all indices together, and the values of VIs at specific dates selected based on factor analysis. The reference output data for training was a map of areas ignited in the wildfire. Among the considered inputs, the MSAVI dataset, which reflects changes in vegetation biomass and canopy cover, showed the best performance. The precision of the presented PRNN (RMSE = 0.85) in identification areas with a high potential of ignitability gives ground for the application of the proposed method in risk assessment and fuel treatment planning on WUI and adjoined urban areas.

Keywords: wildland–urban interface; vegetation ignitability; fire risk assessment; artificial neural network

Citation: Mahamed, M.; Wittenberg, L.; Kutiel, H.; Brook, A. Fire Risk Assessment on Wildland–Urban Interface and Adjoined Urban Areas: Estimation Vegetation Ignitability by Artificial Neural Network. *Fire* **2022**, 5, 184. https://doi.org/10.3390/fire5060184

Academic Editor: Tiago Miguel Ferreira

Received: 1 October 2022
Accepted: 2 November 2022
Published: 3 November 2022

1. Introduction

The concept of Wildland–Urban Interface (WUI) is a transition zone between the natural landscape and the build-up environment, officially proposed in February 1987 by the U.S. Department of Agriculture [1]. The basis for allocating territories to WUI with a specific fire management approach was reasoned by evidence that protecting structures from wildland fires is challenging, and human-caused fire ignitions are the most common, which became the basis for a specific legislative framework for WUI management [2]. Further, considering that anthropogenic factors increase the risk of a wildfire [3], the management of WUI decided to create buffer (sanitary) zones and fuel breaks to protect urban territories from fire [4]. A particular concern was that compact city planning is more resistant to fire [5] and that urban areas were not considered in fire management. The concept of non-flammable cities has worked for a long time until the climate changes observed in recent times led to an increase in the frequency and intensity of fire weather [6].

Nevertheless, wildfires that hit the city have become frequent in the last decade. Among numerous examples, wildfires have been reported in Greece (Athens, 2009 and 2015, Thasos, 2016, Mati, 2018), France (Marseilles, 2009 and 2016), Spain (Javea, 2012 and 2016), Italy (Palermo, 2022), Israel, 2016 and 2021, and the United States (California, 2018).

The catastrophic consequences of the wildfire spread in cities have made professional societies reconsider the existing approach to fire management and include urban vegetation adjoin to WUI in the risk assessments system [7].

Fire spread in forests and WUI is a combination of two main strategies: direct propagation from adjacent vegetation and spotting fire through ember attacks [8,9]. The specificity of urban areas limits direct propagation due to interspersing vegetation with fire-resistant structures, while firebrands are the primary fire spread strategy on the built-in part of the WUI [10–12]. In laboratory studies where fuel is subjected to contact with a lightning source, e.g., firebrands, the ignitability is defined as 100% due to the experiment conditions excluding external limiting factors, while the time of ignition and flaming duration vary [13,14]. Studies that have assessed actual wildfires, however, show that not all firebrands drive new ignitions; rather, fire spotting propagation depends on the meteorological conditions, the amount of fine fuel, and species composition on the specific patches [15]. Thus, the ability to identify areas with high ignition probability allows a better estimate fire connectivity network in the specific area, which in turn supports risk assessment and fuel treatment planning [16,17]. Although the probability of ignition depends on various environmental factors, there now exists a wealth of evidence that the main factor affecting the likelihood of the fire from firebrands is vegetation and its characteristics: moisture content, biomass amount, and biofuel type [18–20].

The three main approaches used to evaluate fire risks in vegetated areas are biophysical models, statistical models, and fire behavior models [21]

- Biophysical models estimate fire risk based on the scientifically validated weights of terrain physical parameters: vegetation, elevation, slope and aspect, roads, and settlements [22,23];
- Statistical models use GIS-based historical summaries to estimate the correlation between fire-affecting parameters and observed fire frequency at specific locations [24];
- Fire behavior models use mathematical models that predict fire spread based on biophysical parameters that simulate a fire dynamic in particular conditions [25]

Currently, the most advanced methodologies consider continuous risk assessment, learning from past events, and using dedicated techniques to process relevant data, support decisions, and enable risk management. Recent studies, therefore, propose a risk assessment approach based on machine learning [26,27]. Practice shows a good performance of various machine learning methods: support vector machines, decision trees, random forests, artificial neural networks (ANN), and k-nearest neighbors [28–30]. The advantage of ANN for fire studies is the ability to solve complex non-linear relationships between multiple inputs and the probability of ignition that allows for achieving predictive accuracy higher than in traditional statistical approaches [31]. ANNs are already implemented to predict forest fire probability based on common parameters for biophysical models. However, when considering the accuracy of estimated relationships to predict fire risks, both approaches have comparable difficulties in practical application due to spatial and temporal site specifics.

Although the vegetation characteristics are typical for fire risk, the specific dataset of vegetation parameters varies among studies, e.g., tree height, canopy cover, and vegetation type. The idea for monitoring vegetation state by spectral data obtained from remote sensing satellite missions appeared in 1970 [32]. Firstly, the studies were focused on the direct effect of plan biophysical properties, e.g., chlorophyll content, green biomass, and leaf area index, on vegetation reflectance in the different spectral ranges. As a result, a dozen vegetation indices (Vis) and statistical models were proposed for the estimation of vegetation characteristics. Further, the implementation of modern technologies and time-series spectral data allows for the estimation of indirect parameters such as plant phenology and forest overstorey fuel attributes and supports fuel model classification [33]. VIs allow for the detection of forest degradation [34], discriminating vegetation covers [35], and mapping vegetation according to the fuel type [36]. In practice, fire risk assessment uses datasets of time-series VIs for better fuel classification performance [37,38].

The present study focuses on supporting fire risk assessment on WUI and adjoined urban areas with the ANN application, which estimates the probability of ignition based on temporal VIs behavior. The study is conducted on empirical knowledge of Haifa's 2016 wildfire; we test the hypothesis that time series multispectral images provide sufficient information to predict vegetation ignitability using a pattern recognition neural network (PRNN).

2. Materials and Methods

2.1. Case Study

Haifa is a coastal Mediterranean city in Israel on Mt. Carmel (32°48′56″N, 34°59′21″E). The local topography includes steep mountain slopes and dry riverbeds (wadis) that frame the sprawling city with "green fingers", which leads to a considerable length of the WUI [39]. The urban area includes native greenery, flora, and many planted trees [40]. Within the city, the vegetation consists primarily of decorative plantings, low-growing trees, conifers, and maquis shrubland: *Pinus halepensis*, *Quercus* spp., *Quercus calliprinos*, *Ceratonia siliqua*, *Pistacia* spp., *Pistacia letiscus*, *Cistus salviifolius*, *Cistus criticus*, *Sarcopoterium spinosum*, *Calicotome villosa*, *Genista fasselata Decne* [41–43]. In the study region, the vegetation tends to be extremely flammable because of its short time-to-ignition and long flame duration. Previous ecological studies also indicate extensive connectivity among open spaces in Haifa; backyards and other urban in-between areas complement the semi-natural landscape and ensure wildlife movement between habitat patches [44].

Like the Mediterranean region, Haifa is prone to fires due to the "Fire Bioclimates" climate, characterized by dry and hot summers and wet and mild winters [45]. Easterly winds from deserts called "Sharav" aggravate the fire situation, which intensifies in the transitional seasons and brings high temperatures of nearly 40 °C and low humidity below 30% [46]. Wildfires in the areas surrounding Haifa are frequent and well-studied in the context of fire management; however, likely due to the non-flammable cities concept [39,47], urban areas are often excluded.

The wildfire considered in this study occurred in Haifa on 24 November 2016. Meteorological stations reported low humidity above 30% and strong southwest wind 10–15 m s^{-1}. The start of the fire was the ignition of wildland adjoined to the urban development [48]. Bypassing non-flammable constructions and burning nearly 9 ha of vegetation, the wildfire rapidly crossed the city in the first hour (Figure 1). The surrounding wildlands' total burned area was 120 ha [49]. According to the assessment, the total damage and loss amounted to 130,000,000 USD [50].

Figure 1. Map of the actual burned area at the first hour of the wildfire event in Haifa on 24 November 2016 [48].

2.2. Data Collection and Pre-Processing

The source for vegetation data collection is the LANDSAT 8 satellite which consists of two sensors: Operational Land Imager and Thermal Infrared Sensor. In the presented study, we use data from Operational Land Imager 8 (OLI8), which consists of 9 bands. Band 1 and band 9 are supporting bands for image correction according to environmental conditions (atmosphere and clouds). Bands 2–7 present visible and infrared spectral data. Bands 2–6 have spatial resolution of 30 m, and band 7 has resolution of 60 m. Band 8 is panchromatic channel proposed for data fusion and improvement of spatial resolution of the spectral bands to 15 m. We acquired OLI8 images with radiometric and geometric correction (Level-2 Data Product) during 2014–2016, as provided by the United States Geological Survey (USGS). Images have undergone atmospheric correction by the FLAASH algorithm [51] and Gram–Schmidt pan-sharpening [52] in the ENVI environment (L3Harris Technologies, Exelis Inc., Broomfield, CO, USA). Images with cloudiness of less than 5% on the region of interest were selected from the obtained data set. To reach a better spatial resolution that is important for urban studies, the collection of selected OLI8 images was downscaled to achieve 1 m resolution and to reconstruct pre-fire vegetation conditions [53]. The downscaling method is based on machine learning technique that estimates spatial distribution of vegetation from low-resolution spectra data by discovering dependencies between 1 m resolution aerial imagery and 15 m resolution satellite-acquired pixels.

The reference output data to train the proposed PRNN was a map of the actual burned area at the first hour of the wildfire event in Haifa on 24 November 2016, reconstructed based on crowd knowledge and firefighters' data by Polinova [48].

2.3. Pattern Recognition Neural Network

2.3.1. Neural Network Design

PRNN with the maximum likelihood principle was designed to estimate the relationship between temporal VIs and the probability of ignition on WUI and adjoined urban areas (Figure 2). Neural Network toolbox developed in MATLAB2020b environment (The MathWorks, Inc., Natick, MA, USA) that provides ready-use modules was chosen for PRNN designing. The backpropagation method with the sigmoid activation function was used for PRNN training. This approach allowed for the optimization of the weights and minimized a combination of squared errors so that the neural network learned how to estimate relationships between inputs and outputs correctly. To provide random data distribution for test-training-validation and to generalize the network by determining the correct combination of squared errors and weights, Bayesian regularization that eliminates the need for lengthy cross-validation process was applied for the designed PRNN. The number of hidden layers of PRNN was defined as 10 for all input datasets. The PRNN input data is temporal VIs behavior maps, and output data was classified into 'Ignited' and 'Not ignited' pixels.

2.3.2. PRNN Input

The collected and downscaled OLI8 images were processed to produce VIs maps. The VIs considered in the study were selected according to their relevance to the main fire-related vegetation characteristics: canopy cover, leaf area index, biomass amount, and moisture (Table 1). In total, six indices were considered in the study: Enhanced Vegetation Index (EVI), Normalized Difference Vegetation Index (NDVI), Modified Simple Ratio (MSR), Modified Soil Adjusted Vegetation Index (MSAVI), Transformed Difference Vegetation Index (TDVI), Normalized Multi-band Drought Index (NMDI).

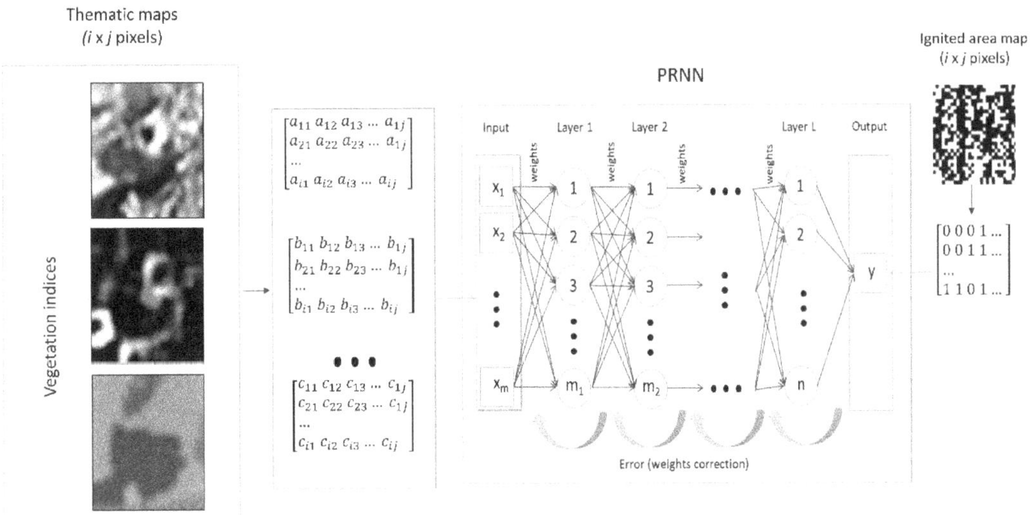

Figure 2. PRNN for estimating the relationship between temporal VIs behavior and probability of ignition.

Table 1. VIs considered in the study.

VIs	Equation	Related Vegetation Characteristics	Reference
EVI	$2.5 * \frac{(NIR - Red)}{(NIR + 6 * Red - 7.5 * Blue + 1)}$	Biomass, canopy cover	[54,55]
NDVI	$\frac{NIR - Red}{NIR + Red}$	Biomass, tree productivity, leaf area index	[56–58]
MSR	$\frac{\left(\frac{NIR}{RED}\right) - 1}{\left(\sqrt{\frac{NIR}{Red}}\right) + 1}$	Leaf area index, fraction of photosynthetically active radiation, biomass	[59,60]
MSAVI	$\frac{2 * NIR + 1 - \sqrt{(2 * NIr + 1)^2 - 8(NIR - Red)}}{2}$	Biomass, canopy cover	[60–62]
TDVI	$1.5 * \left[\frac{(NIR - Red)}{\sqrt{NIR^2 + Red + 0.5}}\right]$	Canopy cover	[63]
NMDI	$\frac{NIR - (SWIR1 - SWIR2)}{NIR + (SWIR1 - SWIR2)}$	Vegetation moisture	[64]

The PRNN input dataset had several configurations: all indices for all considered dates, each index for all dates, and a set of indices at the specific dates chosen based on the statistical analysis named the 'PCA dataset'. Factor analysis with the principal component method was used to reduce the number of variables by a multivariate technique that analyzes a matrix of numerous inter-correlated quantitative dependent variables [65,66]. 'PCA dataset' was proposed to evaluate factor-based combinations (different VIs at different dates) for better ignitability prediction.

2.3.3. PRNN Output

The map of the actual burned area at the first hour of the wildfire event was used to prepare the common output dataset for PRNNs with various inputs. Pixels matching the actual burned area at the first hour were marked to class 'Ignited'; pixels near the burned patches and staying resistant to the fire were mapped to the class 'Not ignited'. In total, 1500 pixels of each class were selected for the neural network training.

2.3.4. RNN Accuracy Measures

Trained PRNNs with the best performance of estimated ignition probability were subjected to accuracy assessment. The precision of the PRNNs was assessed by the percentage of pixels within the boundaries of the actual ignited area that is classified as 'Ignited' (i.e., with a probability threshold of 0.5). Pixels with an ignition probability of more than 0.5 are mapped together with polygons of the actual ignited area for visual inspection [67]. In addition, the quantitative assessment of accuracy was performed by calculating the average value and standard deviation of ignition probability, estimated by PRNNs in pixels falling into the actual ignited area.

3. Results

3.1. Data Preparation

After sorting satellite images and excluding images with a high level of cloudiness, the result was fourteen OLI8 images captured from November 2014 to October 2016. Due to the cloudiness, two large gaps occurred in the temporal data sequence; data was absent for the period from May to October 2015 and from March to July 2016. The resulting images were used to produce six VI's temporal maps—one temporal map contained all declared indices named 'All indices'.

The factor analysis was performed to reduce the number of indices from the 'All indices' dataset. The resulting two first principal components produced by the factor analysis explained 81.8% of the variance: 60.3% explained by the first component, and 21.5% explained by the second component. The strength of correlation between considered VIs and the first two principal components was used to select 39 VIs obtained at the specific dates and introduced in a new map named 'PCA dataset'.

3.2. PRNN Training

The predefined samples of pixels from the input maps were introduced in PRNN for training. The results of the confusion matrix for each prepared input dataset are presented in Table 2. All of the eight considered datasets showed good performance in identifying both ignited and not ignited pixels. When considering the prediction accuracy in each individual group—training, test, validation—we see a chaotic distribution of estimated accuracy caused by the random dividing of data into groups supported by Bayesian regularization. Therefore, to assess the PRNN performance regarding the input dataset, the total accuracy considering together results of training, test and validation was chosen as a key indicator. The lowest prediction accuracy was observed in the NMDI data set: 85.9% for ignited pixels and 87.5 for not ignited pixels. The 'All indices' dataset' accuracy was also relatively low—92.9% for ignited and 88.4% for not ignited. One more dataset with relatively low precision was PRNN trained on the NDVI dataset: 91.3% for ignited pixels and 83% for not ignited. The accuracy of PRNN trained by the TDVI dataset was 90.9% for ignited and 89.1% for not ignited pixels. The PRNN trained by the MSR dataset predicted ignited areas with an accuracy of 95% and not ignited with an accuracy of 86%. The neural network trained on the 'PCA dataset' and EVI had similar results in accuracy: 96.3% for ignited pixels in both datasets, 90.6% in the 'PCA dataset', and 87% in the EVI dataset for not ignited pixels. PRNN trained by the MSAVI dataset showed the best performance: 96.4% for ignited and 90.6% for not ignited areas. The best performance of trained PRNN was observed in MSAVI, EVI, and 'PCA dataset'.

3.3. PRNN Accuracy Assessment

Based on the training results, three ignition probability maps were reconstructed using PRNNs with the best performance: the EVI, MSAVI, and PCA datasets. A total of 94,043 pixels from the reconstructed PRNN map fell within the boundaries of the actual ignited area. Among them, a probability of ignition of more than 0.5 was 77.8% of the pixels estimated by PRNN trained on the EVI dataset, 85.6% of the pixels estimated by PRNN trained on the MSAVI dataset, and 83.6% of the pixels estimated by PRNN trained on PCA dataset (Figure 3).

Table 2. The results of the Confusion Matrix of trained PRNN.

Dataset		Correspondence of Predicted and Actual Targets (%)			
		Training	Validation	Test	Total (Training, Validation, and Test)
EVI	Ignited	96.8	100	90	96.3
	Not ignited	82.4	93.8	100	87
NDVI	Ignited	90.6	100	86.7	91.3
	Not ignited	80.6	86.7	92.3	83
MSR	Ignited	96.3	90	93.8	95
	Not ignited	89	76.9	78.6	86
MSAVI	Ignited	96.9	90	100	96.4
	Not ignited	92.5	87.5	84.6	90.6
TDVI	Ignited	90.3	84.6	100	90.9
	Not ignited	91.5	77.8	93.3	89.1
NMDI	Ignited	84.5	88.9	91.7	85.9
	Not ignited	88.1	85.7	86.7	87.5
All indices	Ignited	94.1	90	85.7	92.9
	Not ignited	87.3	87.5	93.8	88.4
PCA dataset	Ignited	95.1	100	100	96.3
	Not ignited	84.5	100	92.9	87.9

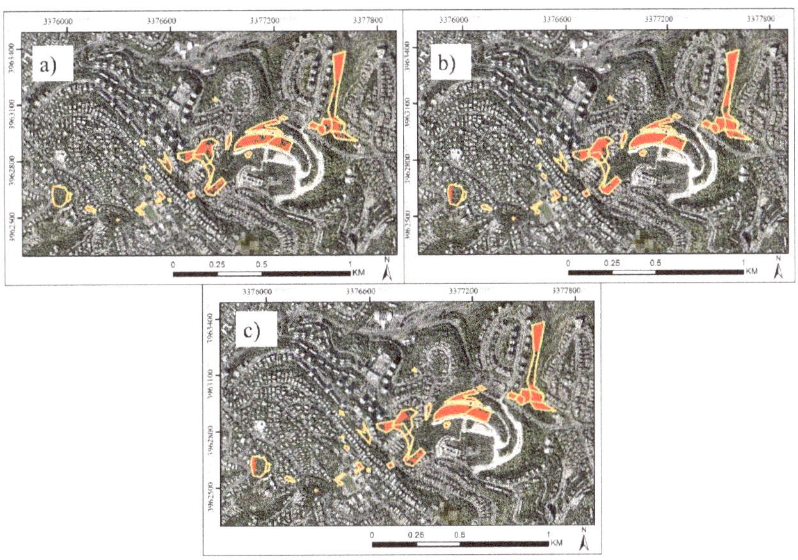

Figure 3. Maps of actual ignited area (orange line) via estimated ignited area with a probability more than 0.5 (red color): (**a**) by PRNN trained on EVI dataset; (**b**) by PRNN trained on MSAVI dataset; (**c**) by PRNN trained on 'PCA dataset'.

According to the methodology, the quantitative assessment of the predictive accuracy provided by the trained PRNNs with the best performance was calculated by the average value and standard deviation of ignition probability in pixels falling into the ignited area. Due to the ability of MSAVI to separate between vegetation and soil and sensitivity to plant dryness and phenology [68], the best precision for detecting ignited areas were observed in the map created by PRNN trained on this index: the average value of ignition probability was 80.21%, and the standard deviation was 24.9%. The PRNN trained on the PCA dataset also showed high precision in estimating ignition probability in the considered area: the average value was 80.06%, and the standard deviation was 26.28%. The lowest accuracy in prediction ignited areas was shown by PRNN trained on the EVI dataset: the average value was 73.57%, and the standard deviation was 31.17%.

4. Discussion

By utilizing biophysical vegetation characteristics expressed by VIs and calculated from OLS8 time series multispectral imagery, we effectively identified and mapped areas with a high-level probability of ignition within the WUI and adjoined urban areas. Whereas other methods based on the temporal deviation of vegetation characteristics obtain an RMSE $\approx$ 0.6–0.80 [69–71], the method used in this study was based on vegetation behavior patterns expressed by VIs and estimated to be more accurate (RMSE $\approx$ 0.85). It is important to note that obtained results, similar to other studies i.e., [69–71], refer to the informativeness of the input dataset and suitability of analytical technique, while the ability of generalization correlative approaches for fire risk assessment is still challenging [72]. The precision of the vegetation-only method presented in this paper is comparable to statistical approaches which consider complex environmental data [73–75] due to the advantage of machine learning techniques over traditional analyses [76,77].

The prediction accuracy of ignition probability by neural networks trained on complex environmental parameters depends highly on the number and informativeness of variables: the precision is directly correlated with the completeness of the input data [78]. The analysis of these networks highlights that among the considered environmental parameters, fuel moisture and the amount of precipitation are the main factors for ignitability prediction [79–81]. As noted in our introduction, vegetation is the main factor predicting ignitability and adds information about landscape and weather to increase the accuracy of the neural network analysis by a few percent [31]. The increased accuracy is because the information on vegetation state dynamics allows for obtaining indirect information about precipitation and anthropogenic activity by the plant growth rate [64,82,83].

The study area investigated in this paper represents a variety of vegetation types under different water regimes, including trees, shrubs, grass, and ornamental plants. The advantage of this work is the ability to predict the ignitability in a diversity of vegetation species with different moisture contents typical to urban areas, which has previously been a challenge for many researchers [74,84,85]. In contrast to the generally accepted approach of live fuel moisture content analysis as the primary ignitability estimator [86,87], spectral remote sending data and VIs, in particular, allow monitoring phenological status as relevant drivers of leaf biomass and moisture contents [88].

The feature of multispectral satellite systems such as OLS8 is collected by spectral signals together with the biochemical and physiological characteristics of vegetation [89]. The advantage of this feature for fire risk assessment is the ability to capture the changes both in water content [90] and phenology [91], which allows for the estimation of fuel flammability and supports ignitability prediction [13,14,92]. In the conducted study, the most appropriate VI for vegetation ignitability prediction on WUI and adjoint urban areas was MSAVI. Correlating with green biomass and vegetation cover, MSAVI makes the index a powerful tool for estimating vegetation vitality [93]. In recently published works, MSAVI has been established to predict land use and land cover classes such as native forest, shrublands, grassland, and vegetation adjoint to the built-up areas [94,95].

The high predictive accuracy of MSAVI observed in this study exceeded the results obtained with the input data configuration based on PCA analysis, indicating that the informativeness of this index in fire risk assessment is underestimated and has the potential for ignitability mapping. While NDVI is wildly used for fire risk assessment in vegetation [71], PRNN trained on the NDVI dataset is not among the top three methods in terms of accuracy due to the low sensitivity of the index to vegetation moisture in shrubs and trees [96].

The precision of the presented PRNN gives ground for applying this approach to estimate vegetation ignitability and can be implemented in fire risk assessment as input data that describes fuel [97,98]. MSAVI reflects vegetation characteristics relevant to flammability and, together with other environmental data such as topography and climate, can support fire management and decision-making on WUI and adjoined urban areas [92,99].

5. Conclusions

The present study proposes to support fire risk assessment on WUI and adjoined urban areas by estimating the probability of vegetation ignition by ANN. The PRNN was designed to predict ignitability based on temporal VIs behavior and assess its performance in comparison to the actual ignited area observed in a wildfire that occurred in Haifa, Israel, in 2016. The results of the study confirm that time series multispectral images provide sufficient information to classify vegetation according to its probability of ignition. Among the considered indices, the best performance in identifying areas with a high potential of ignitability was MSAVI, which reflects changes in vegetation biomass and canopy cover. The precision of more than 85% of the presented PRNN gives ground for applying this approach to assess vegetation ignitability and to support fire management and decision-making on WUI and adjoined urban areas.

Author Contributions: Conceptualization, M.M.; methodology, M.M.; software, M.M.; validation, M.M.; formal analysis, M.M.; investigation, M.M.; resources, M.M.; data curation, M.M.; writing—original draft preparation, M.M.; writing—review and editing, H.K., A.B. and L.W.; visualization, M.M.; supervision, H.K., A.B. and L.W.; project administration, A.B.; funding acquisition, A.B. All authors have read and agreed to the published version of the manuscript.

Funding: The authors declare that no funds, grants, or other support were received during the preparation of this manuscript.

Institutional Review Board Statement: Not applicable.

Informed Consent Statement: Not applicable.

Data Availability Statement: Not applicable.

Acknowledgments: Outstanding Doctoral Program "IDIT" at the Faculty of Social Science University of Haifa, Israel. We gratefully acknowledge the European Cooperation in Science and Technology (COST) action CA18135 "FIRElinks" (Fire in the Earth System: Science & Society) and CA16219 "HARMONIOUS" (Harmonization of UAS techniques for agricultural and natural ecosystems monitoring) for technical support.

Conflicts of Interest: The authors declare no conflict of interest.

References

1. Sommers, W.T. The emergence of the Wildland-urban interface concept. *FOREST* **2008**, 13–18.
2. Radeloff, V.C.; Hammer, R.B.; Stewart, S.I.; Fried, J.S.; Holcomb, S.S.; McKeefry, J.F. The wildland–urban interface in the United States. *Ecol. Appl.* **2005**, *15*, 799–805. [CrossRef]
3. Assaker, A.; Darwish, T.; Faour, G.; Noun, M. Use of remote sensing and GIS to assess the anthropogenic impact on forest fires in Nahr Ibrahim watershed, Lebanon. *Leban. Sci. J.* **2012**, *13*, 15–28.
4. Rigolot, E.; Castelli, L.; Cohen, M.; Costa, M.; Duché, Y. Recommendations for fuel-break design and fuel management at the wildland urban interface: An empirical approach in south eastern France. In Proceedings of the Institute of Mediterranean Forest Ecosystems and Forest Products Warm International Workshop, Athens, Greece, 2004; pp. 131–142. Available online: http://www.fria.gr/WARM/chapters/warmCh16Rigolot.pdf (accessed on 14 August 2022).

5. Neuman, M. The compact city fallacy. *J. Plan. Educ. Res.* **2005**, *25*, 11–26. [CrossRef]
6. Pörtner, H.O.; Roberts, D.C.; Adams, H.; Adler, C.; Aldunce, P.; Ali, E.; Begum, R.A.; Betts, R.; Kerr, R.B.; Biesbroek, R.; et al. *Climate Change 2022: Impacts, Adaptation and Vulnerability*; IPCC Sixth Assessment Report; Cambridge University Press: Cambridge, UK; New York, NY, USA, 2022.
7. Kochilakis, G.; Proussalidi, D.; Chrysoulakis, N.; Varella, V.; Kotroni, V.; Eftychidis, G.; Mimikou, M. A web based DSS for the management of floods and wildfires (FLIRE) in urban and periurban areas. *Environ. Model. Softw.* **2016**, *86*, 111–115. [CrossRef]
8. Storey Michael, A.; Price Owen, F.; Sharples Jason, J.; Bradstock Ross, A. Drivers of long-distance spotting during wildfires in south-eastern Australia. *Int. J. Wildland Fire* **2020**, *29*, 459–472. [CrossRef]
9. Penman Sandra, H.; Price Owen, F.; Penman Trent, D.; Bradstock Ross, A. The role of defensible space on the likelihood of house impact from wildfires in forested landscapes of south eastern Australia. *Int. J. Wildland Fire* **2019**, *28*, 4–14. [CrossRef]
10. Martin, J.; Hillen, T. The spotting distribution of wildfires. *Appl. Sci.* **2016**, *6*, 177. [CrossRef]
11. Manzello, S.L.; Foote, E.I. Characterizing firebrand exposure from wildland–urban interface (WUI) fires: Results from the 2007 Angora Fire. *Fire Technol.* **2014**, *50*, 105–124. [CrossRef]
12. Manzello, S.L.; Suzuki, S.; Gollner, M.J.; Fernandez-Pello, A.C. Role of firebrand combustion in large outdoor fire spread. *Prog. Energy Combust. Sci.* **2020**, *76*, 100801. [CrossRef]
13. Ganteaume, A.; Guijarro, M.; Jappiot, M.; Hernando, C.; Lampin-Maillet, C.; Pérez-Gorostiaga, P.; Vega, J.A. Laboratory characterization of firebrands involved in spot fires. *Ann. For. Sci.* **2011**, *68*, 531–541. [CrossRef]
14. Ganteaume, A.; Jappiot, M.; Lampin, C.; Guijarro, M.; Hernando, C. Flammability of some ornamental species in wildland–urban interfaces in southeastern France: Laboratory assessment at particle level. *Environ. Manag.* **2013**, *52*, 467–480. [CrossRef]
15. Molina, J.R.; Lora, A.; Prades, C.; Silva, F.R.Y. Roadside vegetation planning and conservation: New approach to prevent and mitigate wildfires based on fire ignition potential. *For. Ecol. Manag.* **2019**, *444*, 163–173. [CrossRef]
16. Gray, M.E.; Dickson, B.G. Applying fire connectivity and centrality measures to mitigate the cheatgrass-fire cycle in the arid West, USA. *Landsc. Ecol.* **2016**, *31*, 1681–1696. [CrossRef]
17. Kaur, I.; Pagnini, G. Fire-spotting modelling and parametrisation for wild-land fires. *Int. Congr. Environ. Model. Softw.* **2016**, *55*.
18. Manzello, S.L.; Cleary, T.G.; Shields, J.R.; Maranghides, A.; Mell, W.; Yang, J.C. Experimental investigation of firebrands: Generation and ignition of fuel beds. *Fire Saf. J.* **2008**, *43*, 226–233. [CrossRef]
19. Dahanayake, K.C.; Chow, C.L. Moisture content, ignitability, and fire risk of vegetation in vertical greenery systems. *Fire Ecol.* **2018**, *14*, 125–142. [CrossRef]
20. Cawson, J.G.; Pickering, B.J.; Filkov, A.I.; Burton, J.E.; Kilinc, M.; Penman, T.D. Predicting ignitability from firebrands in mature wet eucalypt forests. *For. Ecol. Manag.* **2022**, *519*, 120315. [CrossRef]
21. Rauscher, H.M.; Sands, Y.; Lee, D.C.; Beatty, J.S. *Advances in Threat Assessment and Their Application to Forest and Rangeland Management—Volume 2*; Gen. Tech. Rep. PNW-GTR-802; US Department of Agriculture, Forest Service, Pacific Northwest and Southern Research Stations: Portland, OR, USA, 2010; pp. 249–708.
22. Cáceres, C.F. Using GIS in hotspots analysis and for forest fire risk zones mapping in the Yeguare Region, Southeastern Honduras. *Pap. Resour. Anal.* **2011**, *13*, 1–14.
23. Adab, H.; Kanniah, D.; Solaimani, K. GIS-based probability assessment of fire risk in grassland and forested landscapes of Golestan Province, Iran. In Proceedings of the International Conference on Environmental and Computer Science IPCBEE, Singapore, 16–18 September 2011; Volume 19, p. 2011.
24. Gerdzheva, A.A. A comparative analysis of different wildfire risk assessment models (a case study for Smolyan district, Bulgaria). *Eur. J. Geogr.* **2014**, *5*, 22–36.
25. Rothermel, R.C. *A Mathematical Model for Predicting Fire Spread in Wildland Fuels*; Intermountain Forest & Range Experiment Station, Forest Service, US Department of Agriculture, 1972; Volume 115.
26. Rodrigues, M.; Jiménez, A.; de la Riva, J. Analysis of recent spatial–temporal evolution of human driving factors of wildfires in Spain. *Nat. Hazards* **2016**, *84*, 2049–2070. [CrossRef]
27. Paltrinieri, N.; Comfort, L.; Reniers, G. Learning about risk: Machine learning for risk assessment. *Saf. Sci.* **2019**, *118*, 475–486. [CrossRef]
28. Mountrakis, G.; Im, J.; Ogole, C. Support vector machines in remote sensing: A review. *ISPRS J. Photogramm. Remote Sens.* **2011**, *66*, 247–259. [CrossRef]
29. Belgiu, M.; Drăguţ, L. Random forest in remote sensing: A review of applications and future directions. *ISPRS J. Photogramm. Remote Sens.* **2016**, *114*, 24–31. [CrossRef]
30. Maxwell, A.E.; Warner, T.A.; Fang, F. Implementation of machine-learning classification in remote sensing: An applied review. *Int. J. Remote Sens.* **2018**, *39*, 2784–2817. [CrossRef]
31. Goldarag, Y.J.; Mohammadzadeh, A.; Ardakani, A.S. Fire risk assessment using neural network and logistic regression. *J. Indian Soc. Remote Sens.* **2016**, *44*, 885–894. [CrossRef]
32. Gates, D.M. Physical and Physiological Properties of Plants. In *Remote Sensing with Special Reference to Agriculture and Forestry*; National Academy of Sciences: Washington, DC, USA, 1970; pp. 224–252.
33. Gale, M.G.; Cary, G.J.; Van Dijk, A.I.; Yebra, M. Forest fire fuel through the lens of remote sensing: Review of approaches, challenges and future directions in the remote sensing of biotic determinants of fire behaviour. *Remote Sens. Environ.* **2021**, *255*, 112–282. [CrossRef]

34. Hasanah, A.; Indrawan, M. Assessment of tropical forest degradation on a small island using the enhanced vegetation index. *IOP Conf. Ser. Earth Environ. Sci.* **2020**, *481*, 012061. [CrossRef]
35. Rhyma, P.P.; Norizah, K.; Hamdan, O.; Faridah-Hanum, I.; Zulfa, A.W. Integration of normalised different vegetation index and soil-adjusted vegetation index for mangrove vegetation delineation. *Remote Sens. Appl. Soc. Environ.* **2020**, *17*, 100280. [CrossRef]
36. Stefanidou, A.; Gitas, I.Z.; Katagis, T. A national fuel type mapping method improvement using sentinel-2 satellite data. *Geocarto Int.* **2022**, *37*, 1022–1042. [CrossRef]
37. Bajocco, S.; Dragoz, E.; Gitas, I.; Smiraglia, D.; Salvati, L.; Ricotta, C. Mapping forest fuels through vegetation phenology: The role of coarse-resolution satellite time-series. *PLoS ONE* **2015**, *10*, e0119811.
38. Michael, Y.; Helman, D.; Glickman, O.; Gabay, D.; Brenner, S.; Lensky, I.M. Forecasting fire risk with machine learning and dynamic information derived from satellite vegetation index time-series. *Sci. Total Environ.* **2021**, *764*, 142844. [CrossRef]
39. Kopel, D.; Malkinson, D.; Wittenberg, L. Characterization of vegetation community dynamics in areas affected by construction waste along the urban fringe. *Urban Ecosyst.* **2015**, *18*, 133–150. [CrossRef]
40. Kolodney, Z.; Kallus, R. The Politics of Landscape (Re) Production Haifa Between Colonialism and Nation Building. *Landsc. J.* **2008**, *27*, 173–189. [CrossRef]
41. Wittenberg, L.; Malkinson, D. Spatio-temporal perspectives of forest fires regimes in a maturing Mediterranean mixed pine landscape. *Eur. J. For. Res.* **2009**, *128*, 297. [CrossRef]
42. Carmel, Y.; Paz, S.; Jahashan, F.; Shoshany, M. Assessing fire risk using Monte Carlo simulations of fire spread. *For. Ecol. Manag.* **2009**, *257*, 370–377. [CrossRef]
43. Tessler, N.; Sapir, Y.; Wittenberg, L.; Greenbaum, N. Recovery of Mediterranean vegetation after recurrent forest fires: Insight from the 2010 forest fire on Mount Carmel, Israel. *Land Degrad. Dev.* **2016**, *27*, 1424–1431. [CrossRef]
44. Toger, M.; Malkinson, D.; Benenson, I.; Czamanski, D. The connectivity of Haifa urban open space network. *Environ. Plan. B Plan. Des.* **2016**, *43*, 848–870. [CrossRef]
45. Naveh, Z. The evolutionary significance of fire in the Mediterranean region. *Vegetatio* **1975**, *29*, 199–208. [CrossRef]
46. Winstanley, D. Sharav. *Weather* **1972**, *27*, 146–160. [CrossRef]
47. Paz, S.; Carmel, Y.; Jahshan, F.; Shoshany, M. Post-fire analysis of pre-fire mapping of fire-risk: A recent case study from Mt. Carmel (Israel). *For. Ecol. Manag.* **2011**, *262*, 1184–1188. [CrossRef]
48. Polinova, M.; Wittenberg, L.; Kutiel, H.; Brook, A. A novel urban vegetation mapping approach for fire risk assessment: A Mediterranean case study. *J. Urban Ecol.* **2022**, in press.
49. Tessler, N.; Borger, H.; Rave, E.; Argaman, E.; Kopel, D.; Brook, A.; Wittenberg, L. Haifa fire restoration project–urban forest management: A case study. *Int. J. Wildland Fire* **2019**, *28*, 485–494. [CrossRef]
50. Wikipedia. November 2016 Israel fires. Wikimedia Foundation. Available online: https://en.wikipedia.org/wiki/November_2016_Israel_fires (accessed on 14 August 2022).
51. Perkins, T.; Adler-Golden, S.M.; Matthew, M.W.; Berk, A.; Bernstein, L.S.; Lee, J.; Fox, M. Speed and accuracy improvements in FLAASH atmospheric correction of hyperspectral imagery. *Opt. Eng.* **2012**, *51*, 111707. [CrossRef]
52. Laben, C.A.; Brower, B.V. Process for Enhancing the Spatial Resolution of Multispectral Imagery Using Pan-Sharpening. US6011875A, 29 April 1998.
53. Polinova, M.; Wittenberg, L.; Kutiel, H.; Brook, A. Reconstructing pre-fire vegetation condition in the wildland urban interface (WUI) using artificial neural network. *J. Environ. Manag.* **2019**, *238*, 224–234. [CrossRef] [PubMed]
54. Huete, A.; Didan, K.; Miura, T.; Rodriguez, E.P.; Gao, X.; Ferreira, L.G. Overview of the radiometric and biophysical performance of the MODIS vegetation indices. *Remote Sens. Environ.* **2002**, *83*, 195–213. [CrossRef]
55. Nepita-Villanueva, M.R.; Berlanga-Robles, C.A.; Ruiz-Luna, A.; Barcenas, J.H.M. Spatio-temporal mangrove canopy variation (2001–2016) assessed using the MODIS enhanced vegetation index (EVI). *J. Coast. Conserv.* **2019**, *23*, 589–597. [CrossRef]
56. Rouse, J.W.; Haas, R.H.; Schell, J.A.; Deering, D.W. Monitoring vegetation systems in the Great Plains with ERTS. *NASA Spec. Publ.* **1974**, *351*, 309.
57. Wang, J.; Rich, P.M.; Price, K.P.; Kettle, W.D. Relations between NDVI and tree productivity in the central Great Plains. *Int. J. Remote Sens.* **2004**, *25*, 3127–3138. [CrossRef]
58. Wang, Q.; Adiku, S.; Tenhunen, J.; Granier, A. On the relationship of NDVI with leaf area index in a deciduous forest site. *Remote Sens. Environ.* **2005**, *94*, 244–255. [CrossRef]
59. Chen, J.M. Evaluation of vegetation indices and a modified simple ratio for boreal applications. *Can. J. Remote Sens.* **1996**, *22*, 229–242. [CrossRef]
60. Das, S.; Singh, T.P. Correlation analysis between biomass and spectral vegetation indices of forest ecosystem. *Int. J. Eng. Res. Technol* **2012**, *1*, 1–13.
61. Qi, J.; Chehbouni, A.; Huete, A.R.; Kerr, Y.H.; Sorooshian, S. A modified soil adjusted vegetation index. *Remote Sens. Environ.* **1994**, *48*, 119–126. [CrossRef]
62. Herbei, M.V.; Herbei, R.; Smuleac, L.; Salagean, T. Using Remote Sensing Techniques in Environmental Management. *Bull. Univ. Agric. Sci. Vet. Med. Cluj-Napoca Agric.* **2016**, *73*, 230–237. [CrossRef]
63. Bannari, A.; Asalhi, H.; Teillet, P.M. Transformed difference vegetation index (TDVI) for vegetation cover mapping. In Proceedings of the IEEE International Geoscience and Remote Sensing Symposium, Toronto, ON, Canada, 24–28 June 2002; IEEE: Piscataway, NJ, USA, 2002; Volume 5, pp. 3053–3055.

64. Wang, L.; Qu, J.J. NMDI: A normalized multi-band drought index for monitoring soil and vegetation moisture with satellite remote sensing. *Geophys. Res. Lett.* **2007**, *34*, L20405. [CrossRef]
65. Wold, S.; Esbensen, K.; Geladi, P. Principal component analysis. *Chemom. Intell. Lab. Syst.* **1987**, *2*, 37–52. [CrossRef]
66. Abdi, H.; Williams, L.J. Principal component analysis. *Wiley Interdiscip. Rev. Comput. Stat.* **2010**, *2*, 433–459. [CrossRef]
67. Congalton, R.G. Accuracy assessment and validation of remotely sensed and other spatial information. *Int. J. Wildland Fire* **2001**, *10*, 321–328. [CrossRef]
68. Guerra, F.; Puig, H.; Chaume, R. The forest-savanna dynamics from multi-date Landsat-TM data in Sierra Parima, Venezuela. *Int. J. Remote Sens.* **1998**, *19*, 2061–2075. [CrossRef]
69. Jurdao, S.; Chuvieco, E.; Arevalillo, J.M. Modelling fire ignition probability from satellite estimates of live fuel moisture content. *Fire Ecol.* **2012**, *8*, 77–97. [CrossRef]
70. Argañaraz, J.P.; Landi, M.A.; Bravo, S.J.; Gavier-Pizarro, G.I.; Scavuzzo, C.M.; Bellis, L.M. Estimation of live fuel moisture content from MODIS images for fire danger assessment in Southern Gran Chaco. *IEEE J. Sel. Top. Appl. Earth Obs. Remote Sens.* **2016**, *9*, 5339–5349. [CrossRef]
71. Xie, L.; Zhang, R.; Zhan, J.; Li, S.; Shama, A.; Zhan, R.; Wu, R. Wildfire Risk Assessment in Liangshan Prefecture, China Based on An Integration Machine Learning Algorithm. *Remote Sens.* **2022**, *14*, 4592. [CrossRef]
72. Vega-Nieva, D.J.; Nava-Miranda, M.G.; Calleros-Flores, E.; López-Serrano, P.M.; Briseño-Reyes, J.; Corral-Rivas, J.J.; Alvarado-Celestino, E. Temporal patterns of fire density by vegetation type and region in Mexico and its temporal relationships with a monthly satellite fuel greenness index. *Fire Ecol.* **2019**, *15*, 28. [CrossRef]
73. Guo, Z.; Fang, W.; Tan, J.; Shi, X. A time-dependent stochastic grassland fire ignition probability model for Hulun Buir Grassland of China. *Chin. Geogr. Sci.* **2013**, *23*, 445–459. [CrossRef]
74. D'Este, M.; Ganga, A.; Elia, M.; Lovreglio, R.; Giannico, V.; Spano, G.; Sanesi, G. Modeling fire ignition probability and frequency using Hurdle models: A cross-regional study in Southern Europe. *Ecol. Process.* **2020**, *9*, 1–14. [CrossRef]
75. Smith, J.T.; Allred, B.W.; Boyd, C.S.; Davies, K.W.; Jones, M.O.; Kleinhesselink, A.R.; Maestas, J.D.; Naugle, D.E. Where there's smoke, there's fuel: Dynamic vegetation data improve predictions of wildfire hazard in the Great Basin. *bioRxiv* **2022**, arXiv:2021.06.25.449963. [CrossRef]
76. Zhang, J.; Li, Z.; Pu, Z.; Xu, C. Comparing prediction performance for crash injury severity among various machine learning and statistical methods. *IEEE Access* **2018**, *6*, 60079–60087. [CrossRef]
77. Devi, S.S.; Radhika, Y. A survey on machine learning and statistical techniques in bankruptcy prediction. *Int. J. Mach. Learn. Comput.* **2018**, *8*, 133–139. [CrossRef]
78. Fernandes, L.C.; Cintra, R.S.; Nero, M.A.; da Costa Temba, P. Fire Risk Modeling Using Artificial Neural Networks. In *International Conference on Engineering Optimizationi*; Springer: Cham, Switzerland, 2018; pp. 938–948.
79. Vasilakos, C.; Kalabokidis, K.; Hatzopoulos, J.; Matsinos, I. Identifying wildland fire ignition factors through sensitivity analysis of a neural network. *Nat. Hazards* **2009**, *50*, 125–143. [CrossRef]
80. Satir, O.; Berberoglu, S.; Donmez, C. Mapping regional forest fire probability using artificial neural network model in a Mediterranean forest ecosystem. *Geomat. Nat. Hazards Risk* **2016**, *7*, 1645–1658. [CrossRef]
81. Joshi, J.; Sukumar, R. Improving prediction and assessment of global fires using multilayer neural networks. *Sci. Rep.* **2021**, *11*, 1–14.
82. Meng, M.; Huang, N.; Wu, M.; Pei, J.; Wang, J.; Niu, Z. Vegetation change in response to climate factors and human activities on the Mongolian Plateau. *PeerJ* **2019**, *7*, e7735. [CrossRef] [PubMed]
83. Bashir, B.; Cao, C.; Naeem, S.; Zamani Joharestani, M.; Bo, X.; Afzal, H.; Mumtaz, F. Spatio-temporal vegetation dynamic and persistence under climatic and anthropogenic factors. *Remote Sens.* **2020**, *12*, 2612. [CrossRef]
84. Yebra, M.; Dennison, P.E.; Chuvieco, E.; Riaño, D.; Zylstra, P.; Hunt Jr, E.R.; Jurdao, S. A global review of remote sensing of live fuel moisture content for fire danger assessment: Moving towards operational products. *Remote Sens. Environ.* **2013**, *136*, 455–468. [CrossRef]
85. Marino, E.; Yebra, M.; Guillén-Climent, M.; Algeet, N.; Tomé, J.L.; Madrigal, J.; Hernando, C. Investigating live fuel moisture content estimation in fire-prone shrubland from remote sensing using empirical modelling and RTM simulations. *Remote Sens.* **2020**, *12*, 2251. [CrossRef]
86. Alvarado, S.T.; Fornazari, T.; Cóstola, A.; Morellato, L.P.C.; Silva, T.S.F. Drivers of fire occurrence in a mountainous Brazilian cerrado savanna: Tracking long-term fire regimes using remote sensing. *Ecol. Indic.* **2017**, *78*, 270–281. [CrossRef]
87. Yebra, M.; Quan, X.; Riaño, D.; Larraondo, P.R.; van Dijk, A.I.; Cary, G.J. A fuel moisture content and flammability monitoring methodology for continental Australia based on optical remote sensing. *Remote Sens. Environ.* **2018**, *212*, 260–272. [CrossRef]
88. Fares, S.; Bajocco, S.; Salvati, L.; Camarretta, N.; Dupuy, J.L.; Xanthopoulos, G.; Corona, P. Characterizing potential wildland fire fuel in live vegetation in the Mediterranean region. *Ann. For. Sci.* **2017**, *74*, 1–14. [CrossRef]
89. Brosinsky, A.; Lausch, A.; Doktor, D.; Salbach, C.; Merbach, I.; Gwillym-Margianto, S.; Pause, M. Analysis of spectral vegetation signal characteristics as a function of soil moisture conditions using hyperspectral remote sensing. *J. Indian Soc. Remote Sens.* **2014**, *42*, 311–324. [CrossRef]
90. Yebra, M.; Scortechini, G.; Badi, A.; Beget, M.E.; Boer, M.M.; Bradstock, R.; Ustin, S. Globe-LFMC, a global plant water status database for vegetation ecophysiology and wildfire applications. *Sci. Data* **2019**, *6*, 1–8.

91. Landi, M.A.; Di Bella, C.M.; Bravo, S.J.; Bellis, L.M. Structural resistance and functional resilience of the Chaco forest to wildland fires: An approach with MODIS time series. *Austral Ecol.* **2021**, *46*, 277–289. [CrossRef]
92. Molina, J.R.; Martín, T.; Silva, F.R.Y.; Herrera, M.Á. The ignition index based on flammability of vegetation improves planning in the wildland-urban interface: A case study in Southern Spain. *Landsc. Urban Plan.* **2017**, *158*, 129–138. [CrossRef]
93. Cho, S.H.; Lee, G.S.; Hwang, J.W. Drone-based vegetation index analysis considering vegetation vitality. *J. Korean Assoc. Geogr. Inf. Stud.* **2020**, *23*, 21–35.
94. Younes, N.; Joyce, K.E.; Northfield, T.D.; Maier, S.W. The effects of water depth on estimating Fractional Vegetation Cover in mangrove forests. *Int. J. Appl. Earth Obs. Geoinf.* **2019**, *83*, 101924. [CrossRef]
95. da Silva, V.S.; Salami, G.; da Silva, M.I.O.; Silva, E.A.; Monteiro Junior, J.J.; Alba, E. Methodological evaluation of vegetation indexes in land use and land cover (LULC) classification. *Geol. Ecol. Landsc.* **2020**, *4*, 159–169. [CrossRef]
96. Hardy, C.C.; Burgan, R.E. Evaluation of NDVI for monitoring live moisture in three vegetation types of the western US. *Photogramm. Eng. Remote Sens.* **1999**, *65*, 603–610.
97. Blackhall, M.; Raffaele, E. Flammability of Patagonian invaders and natives: When exotic plant species affect live fine fuel ignitability in wildland-urban interfaces. *Landsc. Urban Plan.* **2019**, *189*, 1–10. [CrossRef]
98. Farahmand, A.; Reager, J.T.; Behrangi, A.; Stavros, E.N.; Randerson, J.T. Using NASA satellite observations to map wildfire risk in the United States for allocation of fire management resources. In *AGU Fall Meeting Abstracts*; American Geophysical Union: Washington, DC, USA, 2017; p. NH21E-02.
99. Galiana-Martin, L.; Herrero, G.; Solana, J. A wildland–urban interface typology for forest fire risk management in Mediterranean areas. *Landsc. Res.* **2011**, *36*, 151–171. [CrossRef]

fire

MDPI

Article

Assessment and Mitigation of the Fire Vulnerability and Risk in the Historic City Centre of Aveiro, Portugal

Dener Silva [1], Hugo Rodrigues [1,*] and Tiago Miguel Ferreira [2]

[1] RISCO, Department of Civil Engineering, University of Aveiro, 3810-193 Aveiro, Portugal
[2] Department of Geography and Environmental Management, University of the West of England, Bristol BS16 1QY, UK
* Correspondence: hrodrigues@ua.pt

Abstract: Identifying fire risk in urban centres is instrumental for supporting informed decision-making and outlining efficient vulnerability mitigation strategies. Historic centres are particularly complex in this regard due to the high density of combustible materials in these areas, the favourable fire propagation conditions between buildings, and the complex urban morphology, which makes the evacuation of inhabitants difficult in case of a fire emergency. Recent safety regulations tend not to be fully applicable to historic city centres, where the specificities of the buildings, together with the need to safeguard their heritage value, make the rules for new buildings incompatible. For that reason, an adaptation of current evaluation methods is required to assure the safety of these places. The present paper aims to contribute to this topic by presenting and discussing the results obtained from the application of a simplified fire risk assessment methodology to a representative part of the historic city centre of Aveiro, Portugal. Data were collected through fieldwork building inspections and the results were mapped using a Geographic Information System tool. The study reveals that around 63% of the assessed buildings have a level of fire risk greater than the level of risk which is acceptable for buildings with this type of use and value. Based on the work developed, different mitigation strategies are suggested and compared. Finally, the results obtained in this work are compared with results published for historic urban areas with similar characteristics.

Keywords: fire risk; risk assessment; urban risk; historic city centres; fire safety

check for updates

Citation: Silva, D.; Rodrigues, H.; Ferreira, T.M. Assessment and Mitigation of the Fire Vulnerability and Risk in the Historic City Centre of Aveiro, Portugal. *Fire* **2022**, 5, 173. https://doi.org/10.3390/fire5050173

Academic Editor: Katherine Cashell

Received: 4 September 2022
Accepted: 19 October 2022
Published: 21 October 2022

1. Introduction

Due to their material and morphological characteristics, traditional buildings are not usually prepared for current comfort and safety standards [1–4]. Fire is the most significant potential hazard in urban areas. Although the safeguarding of occupants' lives is certainly the most important point, the safeguarding of the building itself is also very relevant, since most of these buildings have heritage significance and value. This is, however, a very challenging task, not only due to the material and construction characteristics of these buildings but also due to the significant amount of fire loads involved, such as wooden ceilings and floors, textiles, and paintings, and oftentimes, the impossibility of installing fire protection devices in these buildings, such as sprinklers or smoke detectors [5].

These issues, together with a lack of adequate maintenance practices, have been responsible for the loss of buildings of inestimable value, such as the Notre Dame Cathedral in France [6] or the National Museum in Brazil [7,8]. With lesser patrimonial value, but potentially more significant consequences in terms of human losses, buildings located in Historic City centres do not easily comply with the generalities of current fire safety requirements, as those in the Portuguese Code [9], it thus being necessary to evaluate each case separately and accept conditions that, in normal circumstances, would be inadequate, such as narrow emergency paths and more combustible construction materials. This situation, alongside difficult firefighting settings, and favourable propagation conditions due to the proximity of buildings, poses a significant fire risk to these areas.

The number of urban fires has grown worldwide in recent years, particularly in Portugal [10], where areas of high fire risk are relatively well identified. Among a number of other historic areas, some of them already well-studied and characterised (such as the historic urban areas of Lisbon and Oporto), is the historic city centre of Aveiro, for which the level of knowledge regarding fire risk and safety is very limited, despite its high heritage and socio-cultural value [11]. Some studies have been conducted around the world [12–16] and have yielded very interesting results about how alternative methodologies work and how they can help historic sites with respect to fire risk. In addition, these methods have been significant in confirming the ANEQP [11] risk maps and specifying which areas are more likely to be affected, as well as determining the state of the buildings in those areas.

In Portugal, the regulations regarding the building fires prevention [9,17] was until 2019 applicable to new and existing buildings, prescribing for example minimal dimensions of evacuation paths, ventilation, and materials. This fact made the use of the rules and the preservation of heritage building very difficult. In 2019, identifying heritage buildings as particular cases, the "Decreto-Lei no. 95/2019" was implemented [18], allowing the exemption of some requirements since they were justified and analysed by the governmental commission. In sequence, the ARICA methodology developed by The Civil Engineering National Laboratory (LNEC) was approved as a legal tool to help engineers to justify their exemption choices regarding projects.

From this perspective, it is considered that the fundamental importance of the preservation of historic sites is to give the authorities precise information about their vulnerabilities. Based on the background described above, this work aimed to assess the fire risk of a part of the historic city centre of Aveiro through the application of a simplified fire risk assessment approach, which has already been applied in the evaluation of some historic urban areas in Portugal; see [16]. As comprehensively presented and discussed herein, this work involved extensive fieldwork to collect data about the specific characteristics of the buildings included in the study area, which were then used to get a fire risk index (FRI) for each building. These individual results were subsequently mapped using a Geographic Information System (GIS) tool, and the buildings identified as most critical were analysed with a view to suggesting possible risk mitigation strategies.

2. Methodological Framework

As mentioned above, the first step of this research involved the delimitation of the study area, which should be simultaneously wide enough to be representative of the whole building stock of the historic city centre of Aveiro, but still a reasonable size considering the amount of time and human resources allocated to the fieldwork. Once the data collection stage was finished, individual fire risk indices were calculated, integrated, and mapped using a GIS tool. Each one of these steps, from the identification of the study area and the description of the data gathering procedure to the explanation of the simplified fire risk assessment methodology used in this work, is reported in the following subsections.

2.1. Study Area and Field Survey

As a small medieval city, Aveiro grew inside its walls for defence against external invasions. Later, with the end of territorial disputes, the city started expanding beyond the walls. In the next centuries, due to the main source of local income being salt extraction, the city faced growth in the direction of the Ria de Aveiro lagoon, which provided salt water and agricultural organic fertiliser [19], as well as acting as a delimiter throughout the flooded areas near the city.

The expansion of the urbanisation process was started in the mid-16th century and, by the 19th century, it had already consolidated numerous buildings [20]. As a result of this growth, the area of the "Praça do Peixe" was created, surrounded by canals, and mainly composed of residential buildings and small businesses.

Good quality stone for construction is scarce in this region due to its geomorphologic characteristics and proximity to the sea. Because of this, only the most important buildings, such as churches, governmental buildings, or those belonging to the wealthiest families, were built with stone coming from other regions. Ordinary buildings, on the other hand, were typically built using adobe and timber [21,22] as the main construction materials, with ceramic tiles covering the façades [23].

Combining the lower values usually assigned to traditional buildings and the absence of maintenance, as well as the noncompliance of the buildings with applicable standards (and, later, touristic pressure [24]), led to the substitution or conversion of several buildings. With the spread of reinforced concrete structures associated with a more efficient construction, the mixing of reinforced concrete structures with traditional masonry and timber elements started to appear and, in other cases, total replacement by new reinforced concrete structures. This created a variety of different building typologies in the area.

Today, the area that shelters bars, hotels and residences is now one of the busiest and central zones of the city, gathering a great number of old and heritage buildings together. As a place conditioned by the morphology of its buildings and streets, resulting from its growth (Figure 1), the area is considered a key zone for applying alternative methodologies to the ones applied for the evaluation of new buildings.

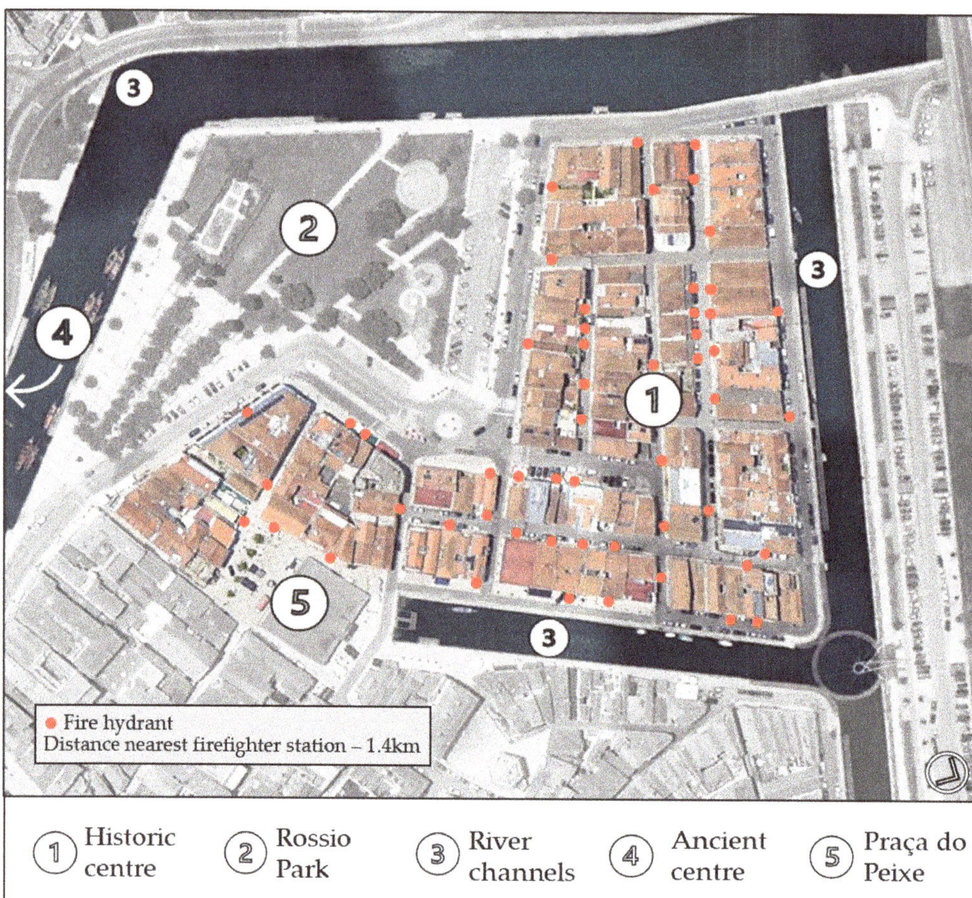

Figure 1. The historic city centre of Aveiro—Study area.

Based on the data acquired and conditioned by the great heritage value that highlights the local history and culture, a zone covering all the most recurrent building typologies in use was defined. The borders created by the canals were used as a divisor of the area (Figure 1). The total number of buildings placed in the zone was codified and catalogued after definition; the application of this work was directed through visual inspection of the group of facades and the collection of relevant information.

Several fire hydrants can be identified in the study area (Figure 1); however, it should be noticed that they are mostly located on the façade walls of the oldest buildings without periodical maintenance. In this sense, it could not confirm the reliability of these hydrants (posteriorly considered in the method as low reliability), exposing the probability that most of them may not work properly.

In total, the study area encompasses 153 buildings divided across 16 blocks. A significant percentage of these buildings (about 70%) have a residential or a mixed residential/commercial use. The remaining buildings have commercial, industrial, or cultural uses, mainly linked to the tourism industry (Figure 2). There were also three buildings undergoing major renovations at the time of the survey, which were left out of this analysis since the building works imply that they will comply with current fire safety requirements.

Figure 2. Distribution of the buildings regarding their type of use.

It is worth adding that 33 out of the 150 buildings included in this study (about 22%) were unoccupied or in a state of abandonment, which is a relevant aspect from a fire safety point of view. Another important fire risk parameter is related to the height and the number of floors of the buildings: in this case, it was found that 18% of the buildings considered in this work are 1-storey buildings, 34% are 2-storey buildings, and 41% are 3-storey buildings. The tallest buildings, representing 7% of the building stock across the study area, have 4-storeys.

2.2. Description of the Fire Risk Assessment Method

The fire risk assessment methods for buildings and public spaces result from the compilation of data related to the built environment, construction elements, surrounding conditions and the capacity of warning and fire extinguishing capacity. It is statistically weighted through the importance of each of the items considered.

Aiming to adapt and simplify the regulations for new buildings by the specificities of different construction typologies, exemplified by Gholitabar et al. [24], some methods developed for existing buildings stand out for their applicability and have been used by several research teams to validate the parameters and assess the risk to historic centres. In Portugal alone, some of these methods have been successfully applied to assess fire risk in several cities, including Guimarães [13,25], Coimbra [14], Castelo Branco et al. [15], Seixal [16,26] and Oporto [27]. These works not only allowed the consolidation of several risk assessment methodologies, making them more compatible and adjusted to the specificities of the Portuguese building stock, but they also generated significant amounts of data regarding the fire safety of these areas.

Most of the methods correlate data on the morphological characteristics of the buildings with information related to the existence and level of operationality of fire protection and firefighting equipment, such as extinguishers, smoke alarms and hydrants, differing essentially in the scope and scale of the evaluation. The ARICA method, for example, focuses more on the expeditious evaluation of the general aspects that most influence the ignition, propagation, combat, and evacuation processes, whereas the Gretener and FRAME methods [15] comprehensively address the internal characteristics of the buildings.

Because it was more adequate for the data set (mainly collected in an expeditious way), in visually evaluating the content of the area, the ARICA method (mentioned in the introduction section) was applied throughout the Fire Risk Index (FRI) simplification for the definition of fire risk in the urban context of the neighbourhood. This method, which is explained in more detail in [12], is based on the collection of information about the buildings and public space by correlating two large groups of data (Table 1): data related to the risk resulting from the characteristics of the building itself, which are used to calculate the Global Risk Factor (GRF), and data related to all the aspects affecting firefighting capacity, which are used to work out the Global Efficiency Factor (GEF).

Table 1. Determination of the fire risk index (FRI) global factors.

Global Risk Factor (GRF)			Global Efficiency Factor (GEF)
Sub-Factors (SF)			
SF_I (Ignition)	SF_P (Propagation)	SF_E (Evacuation)	SF_C (Combat)
Partial Factors (PF)			
PF_{A1} (Building conservation) PF_{A2} (Conditions of electric installations) PF_{A3} (Conditions of gas installations) PF_{A4} (Fire load nature)	PF_{B1} (Distance between parallel openings) PF_{B2} (Safety teams in the building) PF_{B3} (Existence of fire alarm) PF_{B4} (Intern compartmentalisation) PF_{B5} (Fire loads)	PF_{C1} (Conditions of the evacuation routes) PF_{C2} (Building properties) PF_{C3} (Correction factor for evacuation)	PF_{D1} (Conditions of external firefighting equipment) PF_{D2} (Conditions of internal firefighting equipment) PF_{D3} (Preparation of safety teams in the building)

The diagram shown in Figure 3 highlights the link between the data collected on-site as components of the partial factors (PF), which are converted into sub-factors (SF) and applied to the fire risk FRI methodology. The four SF groups are then weighted by their importance in the overall risk composition, with SF_I (Ignition) being increased by 20%, SF_p (Propagation) increased by 10%, and SF_E and SF_C (Evacuation and Combat, respectively) not being increased, according to the method.

The result of the weighted average of the sub-factors results in the Global Risk (GR) (Figure 3), and this result needs to be corrected according to the height and the use typology of the building through the Reference Risk (RR) equations. The product of this process is the Global Fire Risk (GFR), presented in Equation (1), which is a dimensionless factor which measures how far the GR factor is from the RR factor and makes it possible to quantitatively determine whether buildings are suitable for habitation or not:

$$GFR = \frac{\frac{(1.2 \times SF_I + 1.1 \times SF_P + SF_E + SF_C)}{4}}{RR} \quad (1)$$

Divided by RR, the GFR must be less than 1.00, which means the same as the maximum acceptable reference risk for that specific building's typology. When this condition is not met, vulnerability mitigation strategies should, potentially, be put in place to reduce the fire risk of the building. This is addressed in more detail in Section 4 of this manuscript.

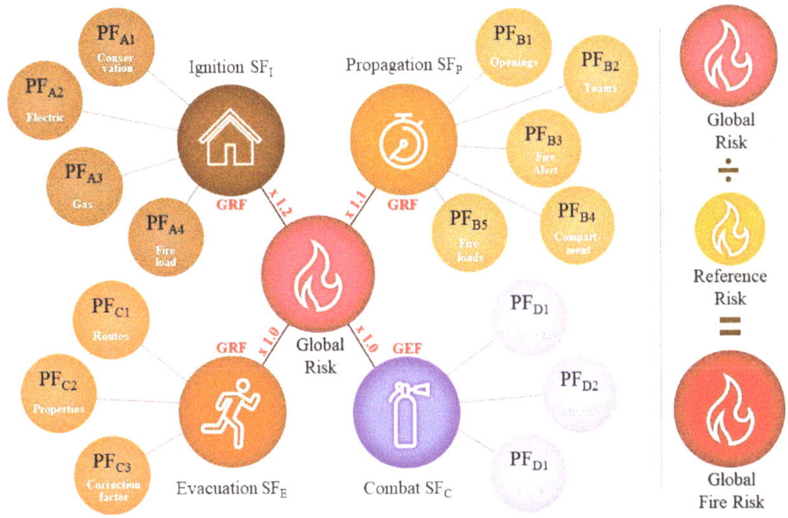

Figure 3. Application diagram of the Fire Risk Index methodology.

For the sake of a better understanding and mapping of the results, the GFR value was divided into four categories of increasing risk: 1. Low, 2. Medium, 3. High, and 4. Very High (Table 2). These four categories were derived and adapted from the ARICA methodology which determines that the risk is 'just acceptable' when GFR is lower than 1.0 and 'not acceptable' when GFR is greater than 1.0. Thus, in this case, it can be considered that, until the medium category, the building has an acceptable fire risk and, up to this, it is important to develop some actions to mitigate the risk [16].

Table 2. Classifications of GFR.

Value of GFR	Classification
$GFR \leq 0.9$	Low Risk
$0.9 > GFR \geq 1.0$	Medium Risk
$1.0 > GFR \geq 1.1$	High Risk
$GFR > 1.1$	Very High Risk

With the production of quantitative data, there is the possibility of punctually finding the aggravating factors of the risk, thus enabling decision-making and appropriate actions to be taken to mitigate the determining factors and contributing more succinctly to the overall improvement of the buildings and the area.

3. Fire Risk in the Historic City Centre of Aveiro

Based on the FRI method, a visual survey was carried out to acquire the required information. The site data collection was performed in two days combining facade observation and randomly getting complementary information with residents. Jointly with the authors, a group of 10 master students were evolved to perform the collection of the information. Once the surveys were filled, the site collected data was complemented with online data sources, such as Google Maps and Google Earth, and provided by the municipality.

The gathered information was allocated in a spreadsheet which performs the necessary application of the equations and classification establishing, subsequently, the method results. All the existing buildings were analysed through the survey and the method was applied to each of them. Based on the results, the buildings were divided into four categories, according to the Global Fire Risk and typology of use (Table 3). These data allow the use of the GIS system to visualise the fire risk and intuitively assess the most compromised areas and buildings.

Table 3. Global Fire Risk per category/typology of use.

Typology of Use	Global Fire Risk Categorisation				Total Σ (Use)
	Low	Medium	High	Very High	
Residential	1.7%	21.0%	30.6%	16.1%	69.4%
Theatre/auditorium	0.0%	0.6%	0.0%	0.0%	0.6%
Hotel/restaurant	0.6%	11.3%	5.9%	0.8%	18.6%
Commercial	0.0%	1.3%	4.7%	0.7%	6.7%
Museum/art gallery	0.0%	0.6%	0.0%	0.0%	0.6%
Industrial/storage	0.0%	0.0%	0.0%	4.2%	4.2%
Total Σ (Categories)	2.3%	34.8%	41.2%	21.8%	100.0%

According to Table 3, only 37% of the buildings present a fire risk classification which is lower than the maximum risk acceptable, according to the FRI methodology; this result is also shown in Figure 4. When it focuses on residential use, which represents more than two-thirds of the building typologies, this number decreases to less than 23%, and this high number of residential buildings at fire risk is recurrent and has already been highlighted by [28]. It is important to note that a great part of the buildings is used over long periods of time, so they need special attention when it comes to people's permanent exposure to the fire risk. Compared to the historic city centre of Guimarães [25], with similar conditions using a similar methodology, a lower risk was observed; just 6% of the buildings showed a low-medium risk of fire. The main reason for the differences found may be related, with the narrow streets and characteristics based on building evacuation directly pointing to the reason for that difference.

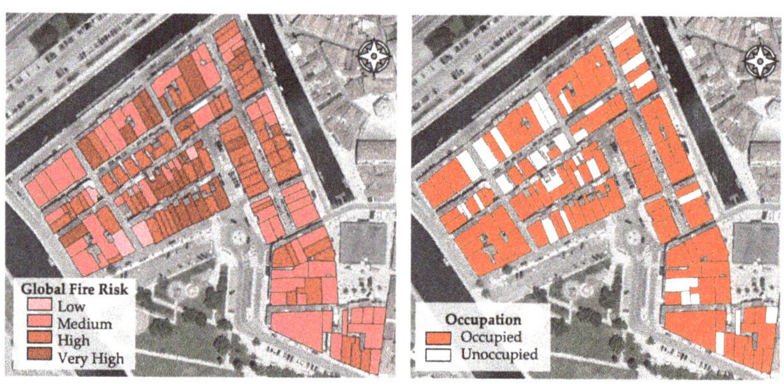

Figure 4. Global Risk Factor and unoccupied buildings, respectively.

On the other hand, besides residential buildings, the historic city centre of Aveiro presents better results in the other typologies; in this case, low-medium fire risk corresponds to half of the constructions and, when it comes to high frequented places (excluding industrial/storage), this number increases a bit more. This situation is mainly justified by the national regulations for building uses with public concentration, but a considerable number of buildings are still exposed to greater risks than what is deemed to be acceptable (Figure 4).

As a result of the high exposure to fire risk, the probability of urban fires remains very high, as observed by the Portuguese National Agency in the Aveiro region [10]. Moreover, the problem is aggravated by the uninhabited buildings (Figure 4), which make a special contribution to the global risk, justified by late warnings (after the start of a fire). It is noticeable that about 22% of the buildings are currently in that state and the pattern of use is repeated by 74% of those characterised as being residential. In the case of Seixal and Guimarães, the empty properties were also identified, and it can be seen as a recurring reality in historic city centres [16,24].

In Figure 4, in the central blocks of the area, there is a higher concentration of buildings containing a high fire risk. This effect is explained by the higher number of residential buildings. As already observed in Table 3, the category of residential typology is related to the increase in the fire risk; for example, in the Very High Fire Risk category, the residential buildings represent 70% of the buildings in the category.

Fire Risk per Sub-Factor

Separately evaluating the sub-factors helps define and design the mitigation solutions.

These different hazard sources have already been identified by different authors [25,29] and largely exemplified in the case of the historic city centre of Coimbra [14]. According to the FRI methodology, the most important ones are related to the beginning of the fire (SF_I), and its spreading (SF_P) (increased factors of 20% and 10%, respectively); such a case is explained by the city morphology which promotes a bigger probability of global risk.

Figure 5 presents the four sub-factors individually, showing some heterogenicity between the buildings. As mentioned before, the first two sub-factors are the most important, particularly in the case of SF_P, where the fire exposure directly affects the neighbouring set. In this group, the risk is aggravated, mainly due to the lack of fire alarms and detection systems, as well as the nature of traditional buildings that do not present very efficient cut-fire compartmentation.

Figure 5. Sub-factors of fire risk, SF_I, SF_P, SF_E, and SF_C, respectively.

Still referring to the beginning of the fire, by analysing SF_I in Figure 5, it is possible to infer that the sub-factor related to the fire ignition is very heterogeneous, and it is the one that most contributes to the GFR. In this case, old infrastructure, like exposed gas pipes and unprotected electric installations, is a major influence and is increased by generally poor maintenance or conservation. Hence, this confirms that higher risk is usually related to a lack of conservation or maintenance and failure to renew existing infrastructure.

When evaluating the last two sub-factors (SF_E and SF_C), buildings present more homogeneous indicators. In the SF_E case, there is notably higher risk located in the middle part of the area that corresponds to buildings with two or three floors, especially those where their exits are partly blocked, or they have narrow stairs and doors. These are important issues when there is a fire because people need to have a fast and easy evacuation route and those with reduced mobility will have even greater difficulties.

Reviewing the last sub-factor, most of the buildings have a low combat sub-factor, and more than 85% of the buildings (around 130 buildings) have a SFc value lower than 1.81 and 25% (37 buildings) lower than 1.30, by combining the preparation of the firefighting crews and the fire extinguishing capacity of the buildings, considering the water availability and accessibility of the external help. However, an area with higher risk is identified close to the corner of the river, which is mainly due to the lack of firefighting facilities in the area.

Considering all the values, the high deviation of SF_I and SF_P, compared to SF_E and SF_C, concords with the statement that the first two display more heterogeneous behaviour. This situation is relevant from the perspective that the buildings with high fire risk also expose the neighbouring buildings.

4. Mitigation Strategies

With the aim of enhancing some characteristics that have a strong contribution to the severity of the fire risk, as evidenced in the previous section, in 63% of the cases, the GFR is deemed to be acceptable; therefore, some preventive solutions to help the mitigation and increase security in the urban core are proposed.

The selected method for developing the mitigation strategies for the area being studied was to look at each sub-factor and assess its weight in the contribution to the SF risk in the group. Although, considering that each individual solution can be hard to execute, depending on the engagement of the residents in the area, the first option is to develop public policies and facilities to offer a reasonable security improvement to the site.

4.1. Public Mitigation Measures with Strong Impact

During the visual inspection, the lack of reliability of urban fire hydrants (or even their absence from some streets) was indicated as one of the problems with a relatively easy solution. The municipality can elaborate a maintenance plan for the existing fire hydrants and proceed with the installation of the missing ones, to ensure water availability for the firefighters, in case of an emergency [16].

By adjusting the Partial Factor related to water availability in case of an emergency, which is placed in the last sub-factor (Global Efficiency Factor–Combat (SF_C)), a relevant increase was found in the number of buildings with a fire risk that was less than acceptable, increasing from 37% to 54% with the interventions.

In addition, fire alarms could be distributed (with buttons accessible to the inhabitants). This solution has already been proposed by other studies, showing effectiveness in the reduction of evacuation time [4], once the system is relatively simple and economically affordable to introduce into the buildings, facilitating an alert for the whole neighbourhood, including the firefighters. Installing this type of system is also important from a social point of view, as several houses are used by older people who may have more difficulties warning others about risky situations, as well as evacuating from the building after a fire has been initiated.

Action in this Partial Factor is particularly helpful seeing as it relates to the Global Risk Factor–Propagation (SF_P) and Evacuation (SF_E) and, in this way, it has an important influence on the GFR.

In addition, assuming that Package 1 of mitigation measures is implemented, an important decrease in the absolute risk value can be observed, of around 67% of the total building stock analysed. This presents a low/medium risk and exposure to the 'very high risk' category decreases to less than 5% of the building stock (Table 4).

Table 4. Global Fire Risk per category/typology of use—After Package 1 of Mitigation Measures.

Typology of Use	Global Fire Risk Categorisation			
	Low	Medium	High	Very High
Residential	3.0%	48.2%	17.5%	0.8%
Theatre/auditorium	0.0%	0.6%	0.0%	0.0%
Hotels/restaurants	0.6%	16.5%	1.4%	0.8%
Commercial	0.0%	2.6%	3.5%	0.8%
Museum/art gallery	0.0%	0.6%	0.0%	0.0%
Industrial/storage	0.0%	0.0%	1.4%	2.3%
Total Σ (Categories)	3.6%	63.8%	28.0%	4.7%

After the application of this first package of mitigation measures, which are mainly promoted by the public authorities, a positive effect was observed on the global risk.

4.2. Individual Mitigation Measures with Strong Impact

In a second approach, Package 2 of mitigation measures was considered, introduced by stimulating the population to make adaptations to their buildings, specifically in terms of electric and gas installations. In this case, public authorities are coadjutants in the measures, promoting their implementation through tax policies, aiming to provide incentives for people upgrading buildings to safer security requirements.

Considering the lack of electrical safety systems in old installations and the frequent neglect of cables and other components for electrical devices, which increases the exposure of fire starting, this situation is synthesised from several studies by the assessing workplace facilities [30]. Consequently, the alteration of those systems is considered for this package. The Partial Factor related to the reliability of electric installations, which is placed in the first Sub-factor (Global Risk Factor–Ignition (SFI)), was recalculated, and it was found that a reduction in the High and Very High-risk groups placed about 72% of the buildings into 'acceptable' risk.

Considering that gas networks are available in the street areas under study, the mitigation package also proposes the linking of all the buildings to the public distribution network, avoiding individual gas bottles inside houses, often with deficient installation. However, in other cases, there may be a different approach; for example, in Oporto [27], it was decided to keep gas bottles, in order to minimise the financial impact. Table 5 shows the improvement precisely in the residential group after this strategy, placing more than 90% in the Low/Medium group following the FRI method. Apart from the residential improvement, this measure did not have any other expressive effect.

4.3. Low Effort Strategies with Minor Impact

Lastly, Package 3 of mitigation measures would not take a significant effort to be implemented but would have a lower impact. Commercial places should follow the national law regarding fire safety, and public authorities must inspect those places to make sure that they are operating within the rules. This action tends to increase the safety of buildings, mainly commercial ones; however, they are frequently not adequate because of a lack of inspections or the inattention of owners, and, consequently, they only have a minor impact on the real safety of the area.

Table 5. Global Fire Risk per category/typology of use—After Package 2 of Mitigation Measures.

Typology of Use	Global Fire Risk Categorisation			
	Low	Medium	High	Very High
Residential	4.2%	55.0%	9.2%	0.0%
Theatre/auditorium	0.0%	0.7%	0.0%	0.0%
Hotels/restaurants	0.6%	16.7%	1.4%	0.8%
Commercial	0.0%	4.0%	2.2%	0.8%
Museum/art gallery	0.0%	0.6%	0.0%	0.0%
Industrial/storage	0.0%	0.0%	1.4%	2.3%
Total Σ (Categories)	4.8%	77.1%	14.2%	3.9%

The simulated impact of rigorous inspections, in accordance with the numbers of security teams, firefighters and detection, would promote better Global Fire Risk. These changes would promote a reduction of 2.3% of the Very High-risk buildings and 1.4% of the High-risk ones, as previously referred to in non-residential cases.

Another measure with some impact comes from the provision of fire extinguishers for some buildings, especially those with High/Very High risk. Assuming that the non-residential buildings were already improved by the previous step, only the residential ones with High/Very High risk would receive the equipment, resulting in a movement of about 1.4% from the High-risk group to the Medium Risk one (Table 6). This policy is also not considered a very adequate action because it involves other variables; for example, the inhabitant may not be able to use it or they may not provide proper care and maintenance of the equipment.

Table 6. Global Fire Risk per category/typology of use—After Package 3 of Mitigation Measures.

Typology of Use	Global Fire Risk Categorisation			
	Low	Medium	High	Very High
Residential	4.2%	57.3%	7.1%	0.0%
Theatre/auditorium	0.0%	0.7%	0.0%	0.0%
Hotels/restaurants	0.6%	17.5%	1.4%	0.0%
Commercial	0.0%	5.4%	1.5%	0.0%
Museum/art gallery	0.0%	0.6%	0.0%	0.0%
Industrial/storage	0.0%	0.7%	1.5%	1.6%
Total Σ (Categories)	4.9%	82.2%	11.4%	1.6%

After all the modifications of the factors, Table 6 shows that about 87% of the buildings are now between the acceptable values specified by the FRI methodology, placed in the Low/Medium group of risk. It is also possible to see that in only 1.6% of the cases (corresponding to two buildings) is the Global Fire Risk Very High; the use of these buildings is storage, and the intrinsic fire load promotes their risk.

The increment of buildings in the acceptable area is very significant.

Figure 6 shows that the group of Low/Medium risk (green area) had an important reaction to the first two adjustments: the third package only having minor impacts on the global scheme.

Even though High and Very High GFR are scattered in all the blocks, Figure 7 presents a bigger concentration of the more exposed buildings in the central zone of the study area. This effect is directly related to the fact that such buildings have bigger PF in the Evacuation Sub-factor. After the implementation of the three packages, the Evacuation Sub-factor was not widely changed due to the difficulty of implementing the measures, justifying the higher risk focus.

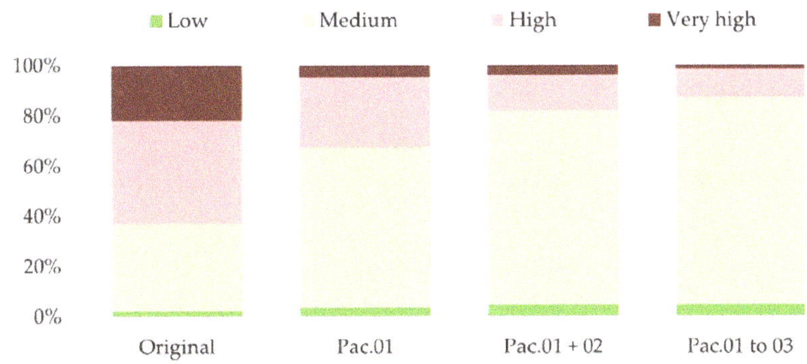

Figure 6. GFR classification at each adjustment stage.

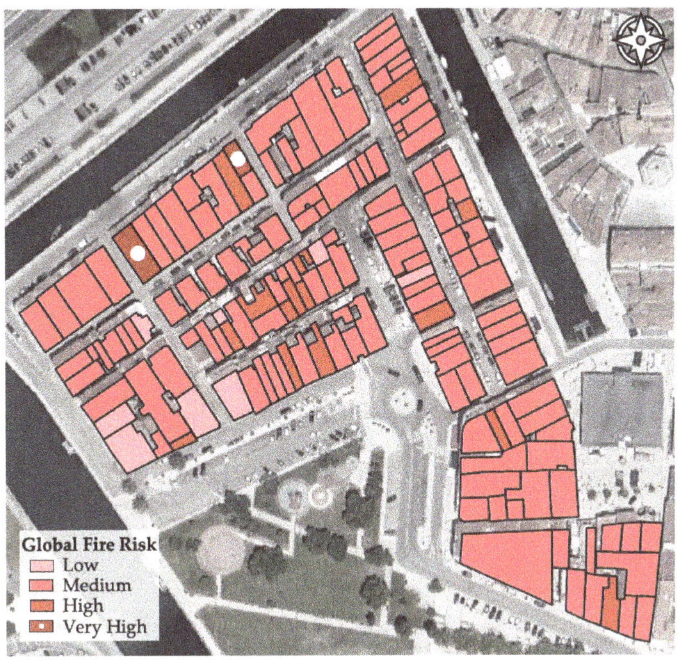

Figure 7. GRF neighbourhood map after all the adjustment packages.

With the aim of decreasing the fire risk, it might be important to implement further mitigation measures, such as enabling all the inhabitants as components of the fire teams at cultural sites, or proposing the complete rehabilitation of the buildings [31]. Those actions were not deemed to be completely effective in this case, as it was assumed that they are financially consuming or dependent on the acts of the inhabitants. Such actions would certainly decrease the risk even more than the ones chosen, but they were considered optional measures.

5. Conclusions

Historic sites, particularly the old historic city centres, need to be paid special attention regarding the application of rules and standards that are designed for new buildings. The relationship between the need to maintain traditional restrictions and the safety requirements imposed by codes for new buildings are usually mismatched. The historic city centre of Aveiro fits this case, and this justifies the option for a Fire Risk Index methodology to investigate the actual exposure of the area to fire risk and study measures to mitigate that risk.

After applying the method, some vulnerability was found as city centres show some similar patterns. As observed in other cases, the results of the method without any local intervention showed that, from a fire risk point of view, less than half of the buildings were considered safe, compared to the new ones. This number was even worse when it came to looking at residential occupation (long permanency), where about 75% of the buildings did not show an acceptable level of risk.

Using the collected data, some mitigation strategies were considered to understand each solution's impact on decreasing the fire risk in historic areas. The proposed changes were divided into three packages. The first package significantly reduced the fire risk, and the two subsequent packages promoted security growth, from 37% to 87% of the buildings now achieving acceptable values, according to the methodology. Some deeper modifications could be performed to increase the security of the site even more, but those would involve larger social and economic costs. Further studies aligned to other risk assessments might be performed.

Lastly, it is important to note that the application of a simplified method might exclude important data from the fire risk analysis and compromise the deeper results added to it. Other variables can be included to create a more complete panorama of the area. Having this in mind, the study contributes to a general overview and can be used as a first approach to critically needed interventions. Furthermore, the application of the method contributes to establishing and enlarging the fire risk assessment data of historic city centres.

Author Contributions: Conceptualization, H.R. and T.M.F.; methodology, D.S., H.R. and T.M.F.; validation, D.S., H.R. and T.M.F.; formal analysis, D.S.; writing—original draft preparation, D.S. writing—review and editing, H.R. and T.M.F. All authors have read and agreed to the published version of the manuscript.

Funding: This work was supported by the Foundation for Science and Technology (FCT)—Aveiro Research Centre for Risks and Sustainability in Construction (RISCO), Universidade de Aveiro, Portugal (FCT/UIDB/ECI/04450/2020).

Institutional Review Board Statement: Not applicable.

Informed Consent Statement: Not applicable.

Data Availability Statement: Not applicable.

Acknowledgments: The authors would like to thank the students of the Master in Built Heritage Conservation from the University of Aveiro for helping with the site data collection.

Conflicts of Interest: The authors declare no conflict of interest.

References

1. Lidelöw, S.; Örn, T.; Luciani, A.; Rizzo, A. Energy-Efficiency Measures for Heritage Buildings: A Literature Review. *Sustain. Cities Soc.* **2019**, *45*, 231–242. [CrossRef]
2. Arfa, F.H.; Zijlstra, H.; Lubelli, B.; Quist, W. Adaptive Reuse of Heritage Buildings: From a Literature Review to a Model of Practice. *Hist. Environ. Policy Pract.* **2022**, *13*, 148–170. [CrossRef]
3. Machete, R.; Silva, J.R.; Bento, R.; Falcão, A.P.; Gonçalves, A.B.; Lobo de Carvalho, J.M.; Silva, D.V. Information Transfer between Two Heritage BIMs for Reconstruction Support and Facility Management: The Case Study of the Chalet of the Countess of Edla, Sintra, Portugal. *J. Cult. Herit.* **2021**, *49*, 94–105. [CrossRef]
4. Zhang, F.; Shi, L.; Liu, S.; Shi, J.; Zhang, J. CFD-Based Framework for Fire Risk Assessment of Contiguous Wood-Frame Villages in the Western Hunan Region. *J. Build. Eng.* **2022**, *54*, 104607. [CrossRef]
5. Torero, J.L. Fire Safety of Historical Buildings: Principles and Methodological Approach. *Int. J. Archit. Herit.* **2019**, *13*, 926–940. [CrossRef]
6. Ferreira, T.M. Notre Dame Cathedral: Another Case in a Growing List of Heritage Landmarks Destroyed by Fire. *Fire* **2019**, *2*, 20. [CrossRef]
7. Barbosa, F.H.D.S.; Araújo-Júnior, H.I. de Skeletal Pathologies in the Giant Ground Sloth Eremotherium Laurillardi (Xenarthra, Folivora): New Cases from the Late Pleistocene of Brazil. *J. S. Am. Earth Sci.* **2021**, *110*, 103377. [CrossRef]

8. Motta, F.M.D.V.; Silva, R.A.R. da A Adoção de Tecnologias Digitais Na Reconstrução Do Patrimônio: Relato Da Experiência Do Museu Nacional, Brasil. *Inf. Soc. Estud.* **2020**, *30*, 1–16. [CrossRef]
9. *Decreto-Lei, n.º 224/2015 [Procede à Primeira Alteração Ao Decreto-Lei n.º 220/2008, de 16 de Dezembro, Que Estabelece o Regime Jurídico Da Urbanização e Edificação]. Diário Da República, 1.ª Série*, Portugal, 2015; pp. 8740–8774. (In Portuguese)
10. ANEPC. *Relatório de Atividades 2019; ANEPC*, Oeiras, Portugal, 2020.
11. ANEPC. *Avaliação Nacional de Risco; ANEPC*, Oeiras, Portugal, 2019; Volume 1ª Atualiz.
12. Granda, S.; Ferreira, T.M. Large-Scale Vulnerability and Fire Risk Assessment of the Historic Centre of Quito, Ecuador. *Int. J. Archit. Herit.* **2021**, *15*, 1043–1057. [CrossRef]
13. Tozo Neto, J.; Ferreira, T.M. Assessing and Mitigating Vulnerability and Fire Risk in Historic Centres: A Cost-Benefit Analysis. *J. Cult. Herit.* **2020**, *45*, 279–290. [CrossRef]
14. Mendes, P.; Correia, A.J.P.D.M. *Análise Do Risco de Incêndio Em Zonas Urbanas Antigas–Centro Histórico de Coimbra*, IPC-Instituto Politécnico de Coimbra: Coimbra, Portugal, 2015.
15. Pais, P.; Santos, C. Avaliação de Risco de Incêndio Em Centros Históricos-o Caso de Castelo Branco. *Agroforum Rev. Esc. Super. Agrária Castelo Branco* **2015**, *23*, 39–50.
16. Ferreira, T.M.; Vicente, R.; Raimundo Mendes da Silva, J.A.; Varum, H.; Costa, A.; Maio, R. Urban Fire Risk: Evaluation and Emergency Planning. *J. Cult. Herit.* **2016**, *20*, 739–745. [CrossRef]
17. *Decreto-Lei, n.º 220/2008 [Estabelece o Regime Jurídico Da Segurança Contra Incêndios Em Edifícios]. Diário Da República, 1.ª Série.* Portugal, 2008; pp. 7903–7922. (In Portuguese)
18. *Decreto-Lei, n.º 95/2019 [Estabelece o Regime Aplicável à Reabilitação de Edifícios Ou Frações Autónomas]. Diário Da República, 1.ª Série*, Portugal, 2019; pp. 35–45. (In Portuguese)
19. Guerrero-Meseguer, L.; Veiga, P.; Sampaio, L.; Rubal, M. Resurgence of Zostera Marina in the Ria de Aveiro Lagoon, Portugal. *Aquat. Bot.* **2021**, *169*, 103338. [CrossRef]
20. Camara Municipal de Aveiro. *Operação de Reabilitação Urbana (ORU)-Programa Estratégico de Reabilitação Urbana (PERU)*; Camara Municipal de Aveiro: Aveiro, Portugal, 2019.
21. Varum, H.; Costa, A.; Fonseca, J.; Furtado, A. Behaviour Characterization and Rehabilitation of Adobe Construction. *Procedia. Eng.* **2015**, *114*, 714–721. [CrossRef]
22. Martins, T.; Fernández, J.; Varum, H. Influence of Moisture on the Mechanical Properties of Load-Bearing Adobe Masonry Walls. *Int. J. Archit. Herit.* **2019**, *13*, 841–854. [CrossRef]
23. Silveira, D.; Varum, H.; Costa, A.; Neto, C. Survey of the Facade Walls of Existing Adobe Buildings. *Int. J. Archit. Herit.* **2016**, *10*, 867–886. [CrossRef]
24. Gholitabar, S.; Alipour, H.; da Costa, C.M.M. An Empirical Investigation of Architectural Heritage Management Implications for Tourism: The Case of Portugal. *Sustainability* **2018**, *10*, 93. [CrossRef]
25. Granda, S.; Ferreira, T.M. Assessing Vulnerability and Fire Risk in Old Urban Areas: Application to the Historical Centre of Guimarães. *Fire Technol.* **2019**, *55*, 105–127. [CrossRef]
26. Ferreira, A.S.R.; Vicente, R.D.S.; Rodrigues, M.F. *da S. Risco de Incêndio Em Centros Históricos: Índice de Risco*; University of Aveiro: Aveiro, Portugal, 2010.
27. Muculo, C. *Avaliação de Risco de Incêndio Pelo Método ARICA a Edifícios no Porto*; University of Porto: Oporto, Portugal, 2013.
28. Wang, K.; Yuan, Y.; Chen, M.; Wang, D. A POIs Based Method for Determining Spatial Distribution of Urban Fire Risk. *Process Saf. Environ. Prot.* **2021**, *154*, 447–457. [CrossRef]
29. Masoumi, Z.; Genderen, J.V.L.; Maleki, J. Fire Risk Assessment in Dense Urban Areas Using Information Fusion Techniques. *ISPRS Int. J. Geoinf.* **2019**, *8*, 579. [CrossRef]
30. Hassanain, M.A.; Al-Harogi, M.; Ibrahim, A.M. Fire Safety Risk Assessment of Workplace Facilities: A Case Study. *Front Built. Env.* **2022**, *8*, 23. [CrossRef]
31. Lee, J.H.; Chun, W.Y.; Choi, J.H. Weighting the Attributes of Human-Related Activities for Fire Safety Measures in Historic Villages. *Sustainability* **2021**, *13*, 3236. [CrossRef]

fire

MDPI

Article

Estimating the Suppression Performance of an Electronically Controlled Residential Water Mist System from BS 8458:2015 Fire Test Data

Charlie Hopkin [1,*], Michael Spearpoint [2], Yusuf Muhammad [3] and William Makant [3]

[1] Ashton Fire, This Is the Space, 68 Quay Street, Manchester M3 3EJ, UK
[2] OFR Consultants, Sevendale House, Lever Street, Manchester M1 1JA, UK
[3] Plumis, Unit 4, Phoenix Trading Estate, Bilton Road, London UB5 7DZ, UK
* Correspondence: charlie.hopkin@ashtonfire.com

Abstract: It is commonly assumed in fire modelling that suppression systems can control the heat release rate of a fire. However, many performance-based assumptions are derived from experimental data for sprinklers, and uncertainty remains for their application to water mist systems. In the UK, residential water mist systems are usually tested to the BS 8458:2015 standard, but the heat release rate in these tests is not quantified and focus is instead placed on thermocouple temperatures. This paper details a series of fire tests to the BS 8458:2015 standard for an electronically controlled water mist system. The paper also includes B-RISK zone modelling of these tests to estimate the suppression performance of the system, comparing model outputs to thermocouple test data. Three traditional suppression assumptions, historically derived from experimental data for sprinklers, have been adopted in the zone modelling to examine whether their application following system activation can be extended to the tested water mist system. The work indicates that applying these suppression assumptions remains reasonable in the context of the performance of the tested water mist system, noting the constraints of the test methods in representing a limited number of fire scenarios.

Keywords: water mist; suppression; performance-based design; heat release rate; residential

check for updates

Citation: Hopkin, C.; Spearpoint, M.; Muhammad, Y.; Makant, W. Estimating the Suppression Performance of an Electronically Controlled Residential Water Mist System from BS 8458:2015 Fire Test Data. *Fire* **2022**, *5*, 144. https://doi.org/10.3390/fire5050144

Academic Editor: Tiago Miguel Ferreira

Received: 26 August 2022
Accepted: 16 September 2022
Published: 21 September 2022

1. Introduction

1.1. Background

In most cases, water is used as the main suppression agent for buildings due to its relative abundance and lack of cost [1], as well as its useful fire extinguishing characteristics such as high specific heat and high latent heat of vaporisation [2]. In the fire safety design of residential buildings, there are different methods used to introduce water droplets into a fire-affected enclosure, with sprinkler systems representing the most commonly employed approach. Water sprays from sprinkler systems suppress a fire by directly wetting and cooling the combusting surface (and any surrounding surfaces), cooling the air by vaporisation (energy absorption) and diluting the air with water vapour. An alternative and more recently developed form of fire suppression system is a water mist system, described by Mawhinney and Back [3] as a system that discharge fine water sprays with droplets no larger than 1 mm (1000 μm). Water mist nozzles therefore produce sprays that have a higher fraction of very fine droplets when compared to a standard sprinkler spray. In contrast to sprinklers, water mist generally relies on the cooling and dilution mechanisms with less support from surface wetting [4].

Typically in a fire engineering assessment it is assumed that suppression systems can control, or reduce, the growth and spread of a fire in some fashion, albeit with differing methods and assumptions being adopted by practitioners. PD 7974-1:2019 [5] suggests suppression systems introduce cooling effects into the enclosure, reducing the severity of a fire. It goes on to acknowledge that this effect is difficult to quantify, but it is *"often assumed that the heat release rate* [HRR] *of the fire remains fixed at the point at which the system is first activated*". This is commonly referred to in industry as a 'sprinkler-capped' or 'sprinkler-controlled' fire [6], in which the HRR is fixed ('capped') to an appropriate steady-state HRR value to represent the control effect of the suppression system, with an example presented in Figure 1 for an idealised αt^2 fire growth HRR.

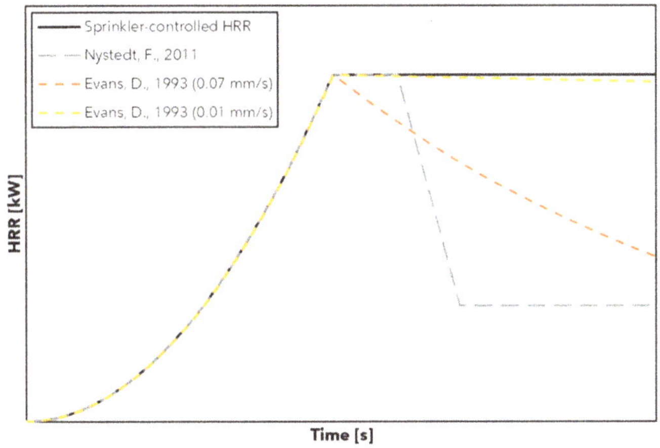

Figure 1. Example suppression models commonly adopted in fire modelling, including a sprinkler-controlled HRR [5], the Nystedt model [7], and the Evans model [8].

For a design fire impacted by sprinkler suppression, Nystedt [7] suggests that the following correlation can be used: (a) at the time of sprinkler activation, the HRR remains constant for 60 s; (b) following this, the fire decays to 1/3 of its maximum HRR (at the time of sprinkler activation) linearly over a 60 s period; and (c) for the remaining time, the HRR remains constant at 1/3 of the maximum HRR. An example of this is presented in Figure 1. Nystedt suggests that this can be applied for deterministic fire modelling where the maximum HRR is less than 5000 kW, but that "*the heat release rate should remain constant at the time of sprinkler activation*" for fires which are larger than 5000 kW.

Alternatively, Evans [8] proposed a means of quantifying the impact of sprinkler suppression on the HRR of an unshielded fire using the following exponential decay correlation:

$$\dot{Q}_{(t-t_{act})} = \dot{Q}_{(t_{act})} \times \exp\left[\frac{-(t-t_{act})}{3\left(\dot{w}''\right)^{-1.85}}\right] \tag{1}$$

where $\dot{Q}_{(t_{act})}$ is the HRR (kW) at the time of system activation (t_{act}, s), $\dot{Q}_{(t-t_{act})}$ is the decaying HRR following system activation, and $\dot{w}''$ is the water spray density of the sprinkler (mm/s). The relationship was estimated from wood crib fire experiments for commercial sprinklers. An example of the exponential decay curve is presented on Figure 1, assuming a spray density of 0.07 mm/s (4.2 mm/min), where Evans notes "*most of the data analyzed in this study were from experiments in which the sprinkler spray density was greater than 0.07 mm/s*" [8]. Evans suggests the correlation provides a "*conservative bound*" for the minimum expected reduction in the HRR from the suppression of sprinklers, but it does not account for potential variations in the water spray density. Also shown in Figure 1 is a decay curve for a much lower spray density of 0.01 mm/s (0.6 mm/min), which is broadly representative of the local discharge density achieved by the electronically controlled watermist system analysed in this paper. The system is introduced later in Section 1.4.

It is not immediately clear whether these common methods of estimating the impact of sprinklers on the HRR can be applied to alternative suppression methods like water mist, particularly as alternative systems can rely on different physical mechanisms to aid in controlling a fire. To explore this point further, the next section includes a brief review of international standards and experimental literature available on the suppression performance of water mist, with a primary focus on residential design.

1.2. Literature Review

In their review of water mist systems, Liu and Kim [9] indicated that *"due to the complex extinguishing processes, the relationship between a fire scenario and the characteristics of a water mist system is not well enough understood to apply a 'first principles' approach"*. Liu and Kim concluded that a combination of experiments and computational modelling studies are needed to support the development of water mist systems. However, performance-based design documents often provide limited information on the representation of water mist systems. For example, PD 7974-1:2019 [5] is intended to provide practicing engineers in the UK with fire safety guidance on the initiation and development of fire within an enclosure, but it makes no direct reference to water mist.

With particular reference to the effect of water discharge on the HRR, the New Zealand C/VM2 verification method [10] suggests that, for sprinklered buildings, *"the fire is expected to be controlled (i.e., with a constant HRR) after the sprinkler activates ... "*, which aligns with the concept of a sprinkler-controlled fire. However, C/VM2 makes no reference to water mist in its recommendations. VDI 6019-1 [11] details recommendations for determining HRR curves for use in smoke control calculations and, as with C/VM2, it proposes that the HRR and area of a fire can assumed to be constant from the time of activation onwards for a sprinkler system. The Swedish general recommendations on the analytical design of a building's fire protection, BBRAD [12], proposes that the effects of an automatic fire suppression system can be applied using a method consistent with that described by Nystedt [7], but only makes reference to sprinkler systems when detailing the calculation method. NFPA 92 [13] provides design fire properties to be adopted in smoke management calculation procedures, suggesting that the HRR is permitted to reach a steady HRR based on fire test data or *"engineering analysis of fire growth and sprinkler response"* and the HRR is permitted to decay based on *"analysis of the effect of sprinkler protection on the fuel at the prevailing ceiling height"*, again omitting consideration of water mist. In the absence of this information, NFPA 92 does suggest that the HRR curve can be based on *"fire test data"*, and hence practitioners may gain insight from experimental data available in technical reports and research articles.

Arvidson [14] carried out fire experiments for five different low- and high-pressure commercial water mist systems in a representative residential enclosure, with a nominal water flow range ranging from 8.2 to 36.7 L/min. A series of upholstered chairs were burned within a 3.66 m by 3.66 m by 2.5 m tall enclosure. In initial free burn experiments of the chairs, the HRR was shown to follow a trend between a medium and fast αt^2 fire growth rate, reaching a maximum HRR in the region of 400 to 500 kW. However, it appears that the HRR was not measured or estimated during the experiments where the suppression systems were incorporated (i.e., it was only determined in the free burns), and results instead focused on temperatures and carbon monoxide (CO) concentrations. In all instances, the systems were shown to substantially reduce temperatures when compared to those observed in the free burn experiments and, by visual inspection of the graphs presented by Arvidson, the commercial water mist systems generally resulted in similar or lower temperatures than the residential sprinkler systems. In previous residential sprinkler and water mist system experiments carried out by Arvidson and Larsson [15] in 2001 for living room fires, it was observed that a *"larger amount of fresh air was drawn to the fire"* for a water mist system, increasing the turbulence in the fire plume compared to sprinklers, although temperatures within the enclosure typically reduced once the system activated.

Also related to residential fires, Chow [16] suggested that water mist suppression can be used to target and suppress small fires at an early stage, such as for open kitchen fires in residential buildings. In relation to this topic, Qin et al. [17] undertook experimental studies on suppressing cooking oil fires in kitchens with water mist operation at three different operating pressures (0.2, 0.4, and 0.6 MPa, ranging from larger to finer droplets). They observed that the HRR increased rapidly after discharging the water mist. The increase in the HRR was partly attributed to water droplets colliding with the fuel surface, causing an increase in the fuel surface area. The rate of increase was shown to be inversely proportional to the operating pressure, i.e., larger droplets at a lower pressure produced a greater increase in the HRR. However, in all instances, the HRR quickly decayed following a short peak upon system activation.

There has been some work carried out to model the effect of water mist discharge using zone models, for example that of Vaari [18], and also Li and Chow [19]. The one-zone model developed by Wighus and Brandt [20] included mechanisms of heat transfer to the water spray, and the effect of the oxygen concentration in the enclosure on the HRR of the fire. The model predicts a gradual decrease in HRR as the oxygen concentration limits combustion and this results in a corresponding decrease in the average gas temperature. The decrease in HRR is similar in form to that of the Evans [8] model.

It can be seen from the above discussion that the recommended practices and correlations for suppression performance are largely derived from experimental observations for sprinkler performance and are not necessarily intended for alternative forms of suppression or, at least, it is not explicitly stated whether they can be applied to other systems. Similarly, there appears to be a limited amount of experimental research around the impact of water mist systems on the HRR of a fire when compared to sprinkler systems, particularly for residential building applications, although the research that is available does appear to indicate the effectiveness of these systems in reducing either the HRR of the fire or the gas temperatures of the affected enclosure. Uncertainty therefore remains on whether the fire suppression assumptions commonly applied to sprinklers are transferrable when assessing the performance of water mist systems. This uncertainty becomes even stronger when such systems are more conceptually 'novel' and do not necessarily operate in a similar manner to conventional systems, either in their activation methods, nozzle arrangements, and how they introduce water droplets into the fire affected enclosure [21].

1.3. Research Overview and Purpose

As a means of standardising the performance of residential water mist systems for a representative fire hazard, BS 8458:2015 provides a recommended series of fire tests (Annex C). It suggests that these tests are *"an important method of demonstrating that the water spray pattern and smaller droplet sizes produced by each specific system are capable of suppressing the test fires and reducing temperatures in the fire test room"*. The suppression performance of these tests with respect to the HRR is not quantified, and instead the tests focus on thermocouple temperatures. The thermocouple data provides valuable insight into the capability of the system to control and suppress a fire and can be supported by other observations like the post-test evaluation of fire damage. However, the tests alone do not necessarily provide direct evidence to support the application of the historical sprinkler correlations to alternative forms of suppression, where this can be a point of contention among fire safety practitioners.

To explore the above problem, this paper details a series of BS 8458:2015 fire tests which were undertaken for an electronically controlled water mist system. The fire modelling tool B-RISK [22] has then been used to represent a simplified form of the tests computationally. The aim of the modelling exercise is to attempt to quantify the suppression performance of the water mist system on the HRR of the fire. As the BS 8458:2015 fire tests do not include measurements of the HRR, the outcomes for temperature estimated in the modelling are compared to the thermocouple test data for the different HRR suppression assumptions introduced in Section 1.1. Ultimately, the intent is to provide initial suppression correlations/assumptions for the electronically controlled water mist system which can be built upon in future research. However, in undertaking this exercise, it is acknowledged that the BS 8458:2015 tests only represent a small range of fire scenarios and there will be inherent limitations to their scope of application, and further testing and experimentation is likely needed.

The research in this paper part of ongoing work to explore how electronically controlled water mist systems perform, and how the performance of these systems can be represented in simple fire calculations and models. A separate article [21] has been produced which discusses how the activation time of the system might be replicated using effective values for the response time index (RTI) and conductivity factor (C factor) for an equivalent sprinkler system. Further preliminary work has been undertaken to consider the reliability of the electronically controlled water mist system and how it performs in comparison to conventional sprinklers.

1.4. An Electronically Controlled Water Mist System

The electronically controlled water mist system used in the tests, and detailed in this paper, is designed for residential applications. The system does not operate in the same way as traditional water mist systems but instead is initiated by a wireless combined smoke and heat detector. Once initiated, the system uses an infrared (IR) thermopile sensor embedded within the nozzle head(s) to scan the room for a fire. The IR sensor measures temperature as a function of IR radiation, assessing for high temperature readings or differential increases in temperature between scans. Once the rate of change in temperature exceeds a given threshold, the head is considered to have successfully located a fire and discharges water droplets in its direction. The water is discharged by the activation of a high-pressure pump (Figure 2a), which drives mains-linked water through the nozzle unit (Figure 2b). The spray nozzles for the system are wall-mounted and positioned around light switch height, e.g., 1.45 m from floor level. The nozzle achieves a water rate discharge of around 5.6 L/min with water droplets less than 100 μm in size. Typically only a single nozzle activates within the enclosure and directs its spray towards the fire, rather than spray being distributed throughout the enclosure. As a result, the system does not necessarily achieve a fixed design discharge density (commonly expressed in mm/s or mm/min). Previous studies carried out on the system using pans to collect the water spray have shown that the discharge density varies across the affected area depending on the nozzle location within the enclosure and its discharge direction. The local discharge density can reach up to 0.015 mm/s, with a modal value in the region of 0.01 mm/s. For a given fire incident, the local discharge density will also be influenced by the location of the fire relative to the nozzle head and the interaction of the spray with the fire plume.

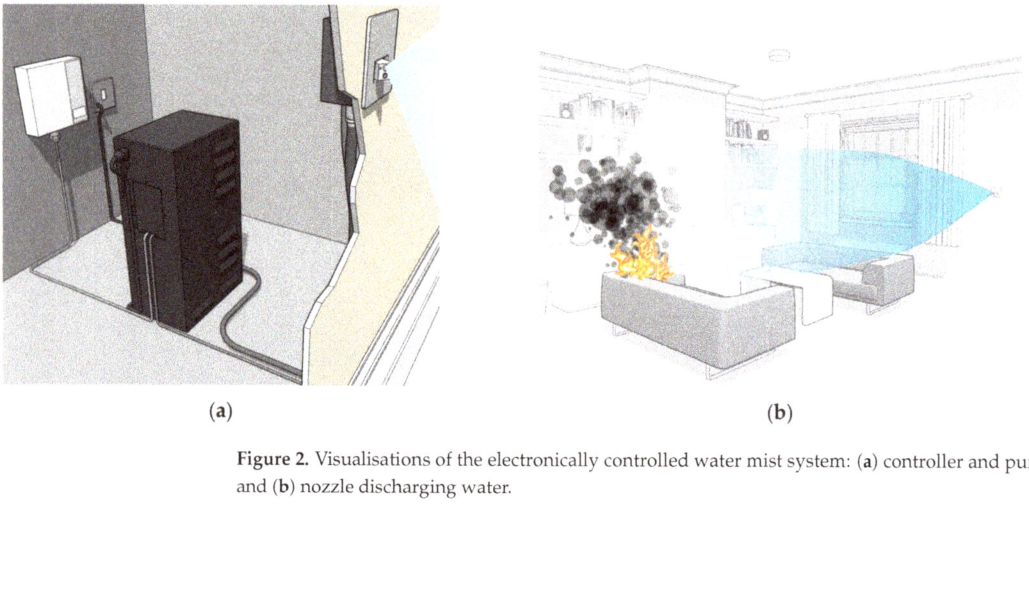

(a) (b)

Figure 2. Visualisations of the electronically controlled water mist system: (**a**) controller and pump; and (**b**) nozzle discharging water.

For the tests presented in this paper, the system used was the Automist Smartscan Hydra, and the detector was an Apollo 51000 multi-criteria optical smoke and heat wireless system. A more detailed description of the system and its design motivations can be found in Spearpoint et al. [21], with further component information available from the system supplier [23,24] as well as in the design, installation, operation, and maintenance manual [25].

2. BS 8458:2015 Fire Tests

2.1. BS 8458:2015 Annex C

The electronically controlled water mist system was independently tested to Annex C of the British Standard BS 8458:2015 [26]. Annex C of BS 8458:2015 details test procedures of fire tests for water mist systems with automatic nozzles, where the test is deemed a pass should the criteria listed in Table 1 be met. The subsequent sections summarise the various elements of the test procedures.

Table 1. BS 8458:2015 test criteria.

Thermocouple Location (Relative to the Ceiling/Floor)	Maximum Allowable Temperature [°C]
75 mm below the underside of the ceiling	320
1.6 m above the floor	95
1.6 m above the floor	55 (for not more than any 120 s interval)

Note: the maximum allowable temperature is observed up to four thermocouple locations, discussed later.

2.2. Enclosure Arrangements

The BS 5458:2015 enclosure is specified as 8 m long by 4 m wide by 2.5 m high. For the test room used in this work the enclosure ceiling and walls were covered by 12.5 mm thick Type F fire-rated plasterboard. The BS 5458:2015 tests cover three physical enclosure arrangement, either incorporating a four-walled enclosure or only two walls. There are two four-walled arrangements, dependent on whether the test incorporates a mechanical fan or not. Each arrangement incorporates full height (2.5 m) openings, as detailed in Figure 3.

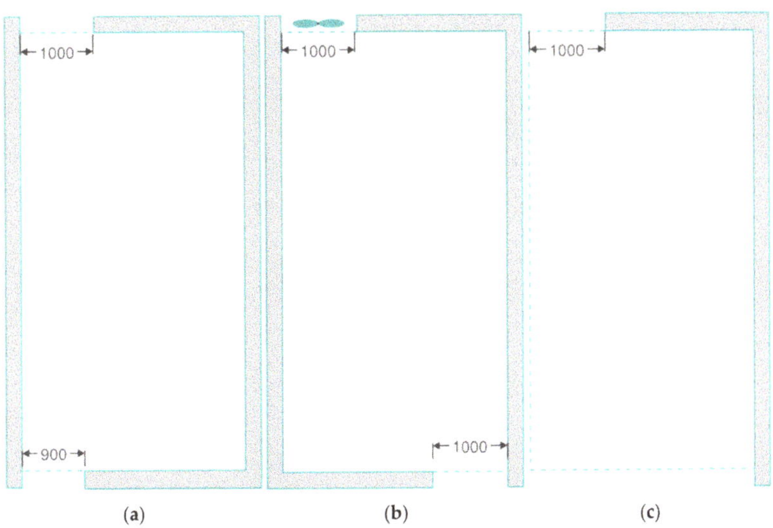

Figure 3. Enclosure arrangements for the tests: (**a**) four walls (without fan) [A-01 to A-03 and A-05 to A-06]; (**b**) four walls (with fan) [A-04 and A-08]; and (**c**) two walls [A-09 to A-11]. Solid wall obstructions are shown in grey with dashed green lines indicating openings. All dimensions in mm.

For the four-walled arrangement incorporating a fan, BS 5458:2015 specifies a ventilation test be undertaken, i.e., at least one test be repeated with the ambient air achieving a minimum velocity of 1 m/s. This velocity is measured inside the room 1 m above floor level and at a horizontal distance of 1 m from the fan. The fan is 500 mm in diameter and positioned 1 m above floor level, with airflow directed parallel to the floor. For these tests, the fan supplied ('pushed') air into the enclosure, rather than extracting from the enclosure. BS 5458:2015 notes that the purpose of this fan is to provide *"an assessment of the effect of air flows on the watermist droplets"*.

2.3. Fire Locations and Fuel Package

Depending on the test, the fire was placed in three different locations, as indicated in Figure 4. For each of the tests, a fuel package was adopted which comprises an 'ignition package' and a 'fuel package'.

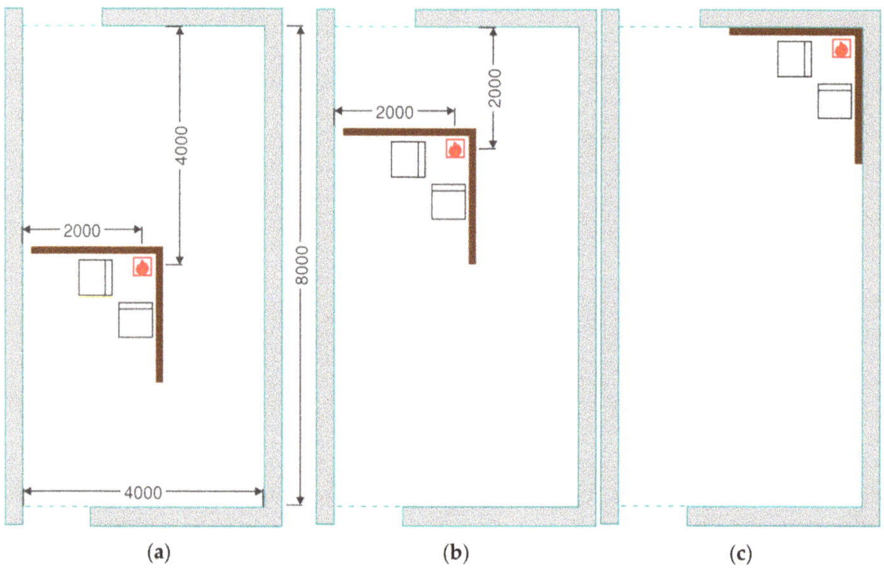

Figure 4. Fire locations for the tests: (**a**) centre 1 [A-02, A-04, A-06 and A-09]; (**b**) centre 2 [A-03, A-07, and A-10]; and (**c**) corner [A-01, A-05, A-08, and A-11]. All dimensions in mm.

The ignition package is a combination of: a 300 mm wide by 300 mm long by 100 mm high steel tray containing heptane; and a wood crib, with eight layers of wood sticks (collectively 305 mm in each direction), positioned on top of this tray. The fuel package comprises two sheets of polyether foam, 865 mm long by 775 mm wide by 75 mm thick, glued to a sacrificial backing board of the same width and length (with a 12 mm depth). This backing board is then attached to a supporting wooden frame. Two cotton wicks, soaked in heptane, are placed on a fire brick, with 150 mm laid along the edge of the foam sheets below. In addition to the ignition and fuel packages, the test arrangement incorporates 12 mm thick plywood panels, either 2.2 m or 2.4 m in length, and covering a height of either 1.2 m or 2.5 m, depending on the fire location. A more detailed description of the fuel package can be found in BS 8458:2015 [26] and Spearpoint et al. [21].

2.4. Thermocouple, Nozzle, and Detector Locations

Depending on the test arrangement, up to four thermocouple (TC) trees were included in the enclosure. These are presented in Figure 5. The thermocouples were positioned 1.6 m from floor level and 75 mm below the ceiling.

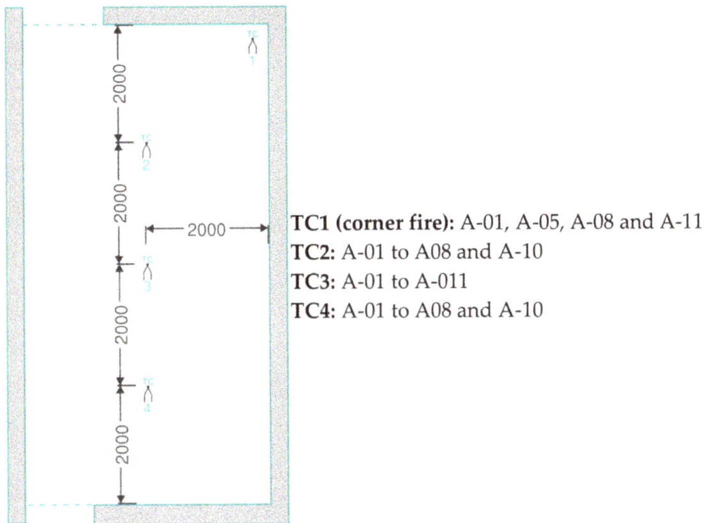

Figure 5. Test thermocouple locations. All dimensions in mm.

Figure 6 provides the three different nozzle arrangements used in the tests, with each arrangement incorporating two nozzles. The nozzles were wall-mounted and positioned along a long wall edge, located 2 m from the nearest perpendicular wall/boundary. All nozzles were positioned 1.45 m from floor level. For each arrangement, an optical smoke and heat wireless detector was located at the centre of the enclosure and used to activate the nozzle scan procedure.

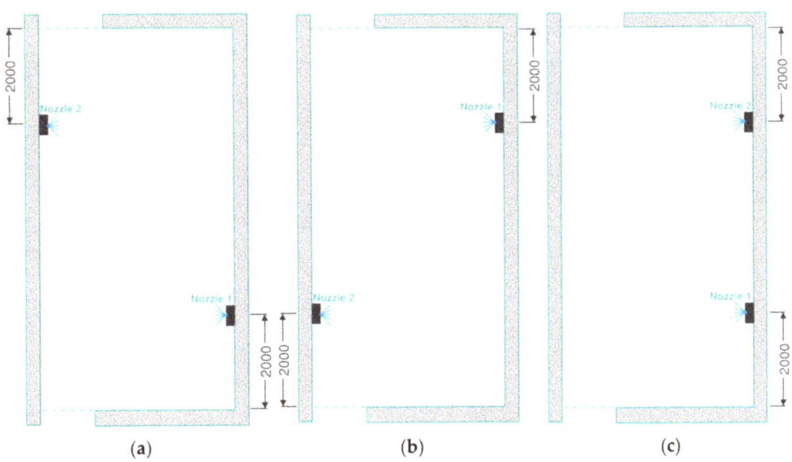

Figure 6. Nozzle arrangements for the tests: (**a**) arrangement 1 (A-01 to A-04); (**b**) arrangement 2 (A-05 to A-08); and (**c**) arrangement 3 (A-09 to A-11). All dimensions in mm.

2.5. Fire Test Results

A summary of the 11 tests, in relation to the fire location, room enclosure, nozzle arrangement, presence of a fan, detector activation time, and nozzle activation time, is shown in Table 2.

Table 2. A summary of the 11 test arrangements and nozzle activation times.

Test Number	Fire Location	Room Enclosure	Nozzle Arrangement	Fan Ventilated	Detector Activation Time [mm:ss]	Nozzle Activation Time [mm:ss]
A-01	Corner	Four walls	Arrangement 1	No	00:39	01:24
A-02	Centre 1	Four walls	Arrangement 1	No	00:31	01:08
A-03	Centre 2	Four walls	Arrangement 1	No	00:29	00:54
A-04	Centre 1	Four walls	Arrangement 1	Yes	00:28	01:06
A-05	Corner	Four walls	Arrangement 2	No	00:35	01:14
A-06	Centre 1	Four walls	Arrangement 2	No	00:27	00:58
A-07	Centre 2	Four walls	Arrangement 2	No	00:33	00:52
A-08	Corner	Four walls	Arrangement 2	Yes	00:37	01:06
A-09	Centre 1	Two walls	Arrangement 3	No	00:29	02:12
A-10	Centre 2	Two walls	Arrangement 3	No	00:26	02:16
A-11	Corner	Two walls	Arrangement 3	No	00:33	00:53

As noted previously, a focus of the BS 8458:2015 [26] Annex C tests is the thermocouple temperatures observed at 75 mm from the ceiling and 1.6 m above floor level. Figure 7 presents the thermocouple temperatures for each test, separating the thermocouples out depending on their height and proximity to the fuel package. The electronically controlled water mist system was shown to pass the Annex C acceptance criteria for the tested arrangements.

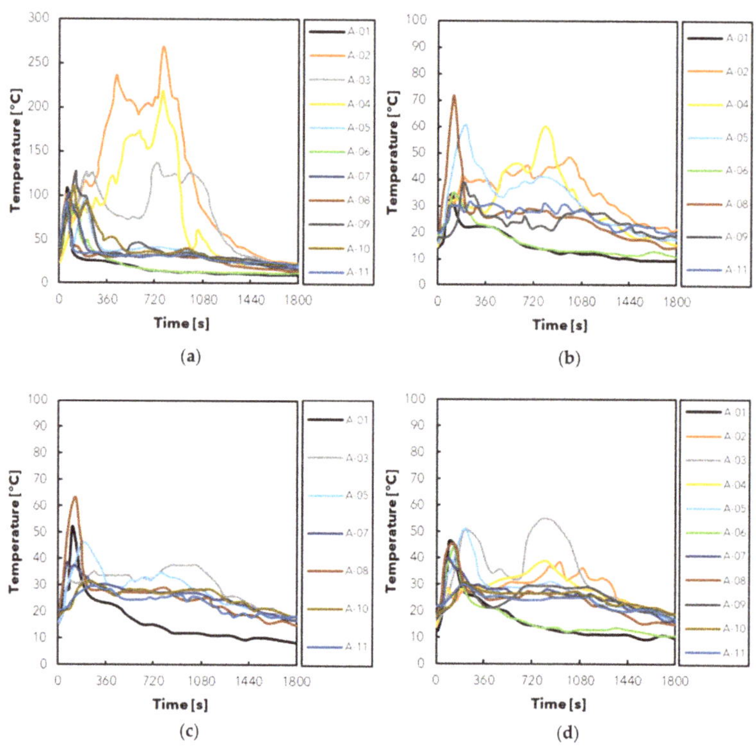

Figure 7. Thermocouple plots for each test: (**a**) 75 mm below ceiling; (**b**) 1.6 m above floor level [closest to fire]; (**c**) 1.6 m above floor level [centre of enclosure]; and (**d**) 1.6 m above floor level [furthest from the fire].

The thermocouple temperatures provide some indication of the variability of the fire growth of the fuel package, as it is unlikely such substantial variations are solely down to the observed system performance for each test. This view is supported from previous discussions detailed by Bill et al. [27], where it is suggested that the ignition and fuel package has the potential to produce a wide range of growth times and maximum HRR values.

There are three tests in the series where greater temperatures are observed for a prolonged period, particularly 75 mm below ceiling level, which are A-02, A-03, and A-04. Observing the time for the first nozzle to activate, these tests are shown to result in an initiation of the system between 54 to 68 s, which is similar to many of the other tests. Therefore, it does not appear that a delay in activation has resulted in the increase in temperature.

Figure 8 presents the specific arrangements of tests A-02 to A-04, with the room enclosure, fire location, and nozzle arrangement shown. For tests A-02 and A-04, which exhibit the highest temperatures presented in Figure 7a, the fire may have been shielded from the direct nozzle spray by the backing board and wooden frame (indicated by the two brown lines at 90° in Figure 8). It is particularly noted that in test A-02, the nozzle furthest from the fire activated rather than the closer nozzle, despite the fire being shielded from the furthest nozzle by the backing board and the closer nozzle having a direct line of sight of the ignition package. While this same shielding behaviour may also have been observed in tests A-09 and A-10, these two tests only incorporated two walls for the enclosure, potentially minimising the build-up of heat. This observation highlights that the direct application of water mist droplets to the fire is an important factor in its suppression, although other benefits such as cooling and dilution will still contribute to a certain extent.

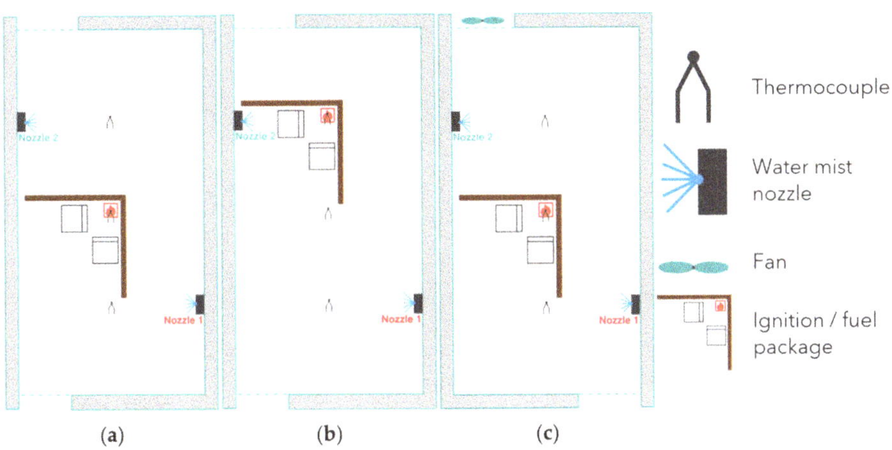

Figure 8. Indicative diagram of test arrangements: (**a**) A-02; (**b**) A-03; and (**c**) A-04. The first nozzle to activate is indicated in red text.

3. Representing the Fire Tests in a Zone Model

3.1. Modelling Tool and Methodology

To represent the fire tests computationally, fire modelling has been undertaken using the zone modelling software B-RISK [22], version 2021.2, developed by the Building Research Association of New Zealand (BRANZ) in Porirua, NZ. B-RISK incorporates a two-zone zone model to calculate fire dynamics, smoke dispersion, and temperature throughout rectilinear enclosures. Further discussion on the validation of B-RISK for application in representing suppression systems can be found in publications by others [28–30], including for electronically controlled water mist system nozzles [21].

The purpose of the modelling exercise described in this paper is to determine whether assumptions for suppression typically applied to sprinklers can potentially be extended to the performance of an electronically controlled water mist system. It is recognised that higher fidelity and more precise modelling tools could be used, such as the computational fluid dynamics (CFD) based tool, Fire Dynamics Simulator (FDS) [31]. However, the intent of the exercise is to provide practicing engineers with simple assumptions around the performance of water mist systems, and thus the adoption of complex modelling to explore the problem would introduce a potential layer of complication should engineers wish to extend outcomes to more simple tools. In addition, the value of an output of a model is directly tied to the specification of its inputs. Should these inputs carry with them a high degree of uncertainty, as is the case for both the information available in the BS 8458:2015 fire tests and potential residential fire scenarios [32] when the system is in use, then there is a reasonable argument that a more precise model provides limited additional value. Elms [33] referred to this a 'consistent crudeness', with Buchanan [34] applying the concept to structural fire engineering to suggest "*there is no point in obtaining highly accurate data for one part of the analysis, with a level of accuracy out of balance with the crudeness of at least the well defined part of the problem*".

3.2. Water Mist Suppression Performance

The electronically controlled water mist system initiates and begins to scan the room with a thermopile sensor following the activation of a combined smoke and heat detector. It is generally not feasible to represent this type of complex interaction in simple modelling tools and it would remain difficult to quantify in more complex tools, given that the scanning process is reliant on estimating the infrared radiation in some capacity, assessing the rate of change in temperature through the use of elaborate algorithms.

In a separate work, Spearpoint et al. [21] propose that the activation of the electronically controlled water mist nozzles can be represented by thermal sensitivity parameters for an 'equivalent' sprinkler system. That is, the system can be represented as a series of ceiling-mounted sprinklers with effective parameters for the RTI and C factor. To achieve this, a comparison was made between system activation times in experiments and activation times estimated in B-RISK zone models. This comparison was undertaken for the BS 8458:2015 tests detailed herein, as well as a separate series of experiments for slow-growing, shielded fires. Spearpoint et al. suggest that the system can be reasonably represented with effective values for the RTI and C factor of 20 $m^{\frac{1}{2}}s^{\frac{1}{2}}$ and 0.25 $m^{\frac{1}{2}}s^{-\frac{1}{2}}$, respectively, with an effective rated temperature of 68 °C and a ceiling offset of 20 mm. For the radial distance, Spearpoint et al. estimated this by using the distance between the centreline of the fire and the detection element (i.e., the nozzle). Therefore, the same approach to representing the system activation has been adopted in the modelling for this paper.

Once the system has activated and began introducing water droplets to the enclosure, consideration needs to be given to the impact suppression has on the HRR. For this, the modelling considers three possibilities, in line with those discussed in the opening of this paper and presented in Figure 1 previously, i.e., a sprinkler-controlled/capped HRR, the Nystedt [7] model, and the Evans [8] model. For the Evans model, a spray density of 0.07 mm/s (4.2 mm/min) has been applied, consistent with the minimum density identified for the majority of experimental data assessed in the original study. This spray density is not intended to be representative of the discharge rate of the water mist system (refer to discussion in Section 1.4) and is only used for the purposes of comparison. If the local discharge density of the system was to be applied to the Evans model, then the decay curve would closely align with the assumption of a sprinkler-controlled fire with a capped HRR (see Figure 1 previously).

3.3. Defining the Fire Parameters

The HRR for the BS 8458:2015 fire test is neither defined nor measured during the testing procedure. However, Spearpoint et al. [21] make reference to the work of Hostikka et al. [35] to determine a reasonable HRR assumption for representing the test fuel package. Hostikka et al. discuss previous full-scale laboratory measurements by Underwriters Laboratories (UL) in a UL 1626 [36] fire test, where the HRR for the corner fire was described as being initially around 100 kW, increasing in a t^2 manner to 300 to 500 kW after 60 s, and then reaching 1500 kW in 80 to 95 s. Spearpoint et al. note that the fuel package used in the UL 1626 fire test is the same as that used in the BS 8458:2015 test. It was proposed that a lower bound HRR relationship from the work of Hostikka et al. be used, as this was shown to provide closest agreement between estimated temperatures in the tests and models. The adopted HRR curve has been reproduced in Figure 9.

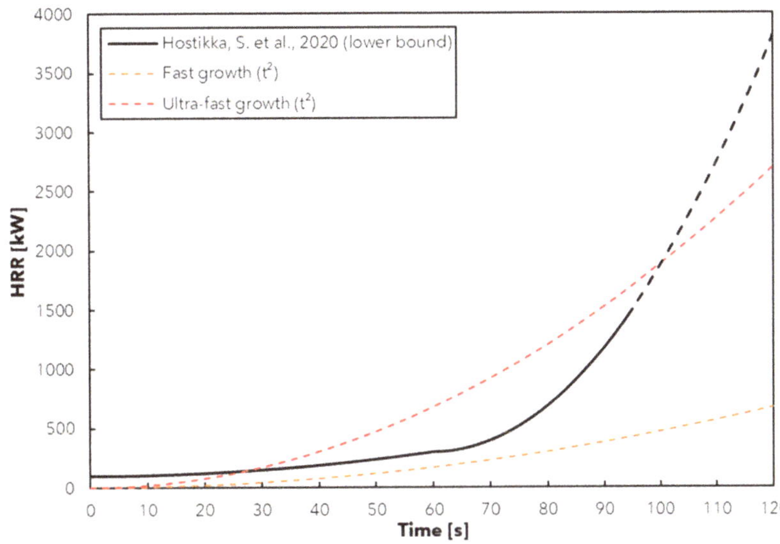

Figure 9. HRR relationship adopted for the zone modelling from Hostikka et al. [35]. Dashed black line indicates modelling assumption that the HRR grows beyond 1500 kW.

For other fire parameters, a value of 22,700 kJ/kg has been adopted for the effective heat of combustion of foam slabs [35], with a soot yield of 0.227 kg/kg and a radiative fraction of 0.46 for polyurethane foam [6]. A burner area of 1.075 m by 1.075 m has been adopted, and is representative of the combined dimensions of the wood crib and two foam slabs attached to the backing board. A 0.4 m elevation is adopted to align with the midpoint of the foam slabs and the top of the wood crib.

3.4. Enclosure Surface Properties

As noted in the test description, wall and ceiling surfaces to the enclosure were covered by 12.5 mm thick Type F fire-rated plasterboard. To represent this in the zone model, gypsum plasterboard properties have been approximated from Hopkin et al. [37], with a thickness of 12.5 mm, a density of 780 kg/m^3, a specific heat capacity of 0.095 kJ/kg/K, a thermal conductivity of 0.25 W/m/K, and an emissivity of 0.7.

The slab of the enclosure has been simulated with concrete properties estimated from BS EN 1992-1-2:2004 [38], with a thickness of 100 mm, a density of 2300 kg/m^3, a specific heat capacity of 0.9 kJ/kg/K, a thermal conductivity of 1.4 W/m/K, and an emissivity of 0.7.

3.5. Ventilation and Enclosure Openings

The test enclosure arrangements incorporate full height openings. For these openings, a coefficient of discharge of 1 has been applied, with no aerodynamic losses through the opening. This approach aligns with recommendations given in the B-RISK user guide [22] in situations where the top of the opening is flush with the ceiling.

Two of the tests include a mechanical fan blowing air into the enclosure. To represent this fan in the simulations, a volumetric supply flow rate of 0.2 m^3/s is assumed, for a velocity at the fan of 1 m/s and a fan area of 0.2 m^2. However, the mechanical ventilation is previously shown to have a negligible impact on the observed simulation results [21].

3.6. Estimating the Thermocouple Temperature

The B-RISK zone model does not estimate gas temperatures for a specific location of the domain but instead estimates the temperature uniformly for upper and lower gas layers. Therefore, the layer height estimated from the modelling has been used to determine whether the thermocouples sit within the upper layer or lower layer. For example, where a layer height of 2.3 m has been estimated, then the upper layer temperature (ULT) has been equated to the thermocouple which is 75 mm from the ceiling (i.e., 2.5 minus 0.075 is greater than 2.3). The thermocouples which are 1.6 m from floor level are then equated to the lower layer temperature (LLT).

4. Zone Model Simulation Results and Discussion

4.1. Estimated System Activation Times

As discussed previously, the modelling considers the activation of the system for the HRR presented in Figure 9 by applying the thermal sensitivity parameters and methodology recommended by Spearpoint et al. [21].

Table 3 presents the nozzle activation times in the tests compared to those estimated in the zone modelling, where these times are consistent with those presented in the work of Spearpoint et al. In observing these activation times, it was noted that five of the tests resulted in a comparable match between the measured and simulated activation times, with the corner fire tests (A-01, A-05, and A-08) producing much quicker estimated activation times. It was postulated by Spearpoint et al. that this underprediction was due to the short simulated radial distance from the fire to the nozzle head.

Table 3. Nozzle activation times and maximum HRRs estimated in the zone modelling.

Test Number	Nozzle Activation Time [mm:ss]	Zone Model Nozzle Activation Time [mm:ss]	Zone Model Maximum HRR [kW]
A-01	01:24	00:44	208
A-02	01:08	01:03	309
A-03	00:54	01:39	185
A-04	01:06	01:04	316
A-05	01:14	00:18	118
A-06	00:58	00:50	239
A-07	00:52	00:59	293
A-08	01:06	00:18	118
A-09	02:12	01:20	692
A-10	02:16	01:16	551
A-11	00:53	00:21	125

Spearpoint et al. did not include tests A-09 to A-11 in their assessment, as these were two-walled ('open') experiments. For tests A-09 and A-10 in particular, a delayed activation time of the system was observed, up to a time of 2 min 16 s. Given that these arrangements resulted in a similar initial detection time from the smoke and heat detector (shown in Table 2 previously), this delay is attributed to the time for the activated nozzle head and thermopile sensor to subsequently identify a fire. As with the corner fire tests, the adopted parameters are shown to produce a relative underestimation in the activation times for these tests. This could be a result of the simplification of the open boundaries and substantial variations in air, smoke, and heat flow in the well-ventilated conditions, where these conditions may have contributed to the delayed activation time in the tests. The general trend for quicker activation times estimated within the simulations may also be influenced by the HRR curve adopted from Hostikka et al. [35], where this curve could generate greater HRRs than those realised in the tests.

Table 3 also includes the maximum HRR for each test, noting that this correlates with the HRR at the time that the system is estimated to first active within the model. After this, the HRR is either capped or decays, depending on the suppression model which is applied. The implications of this assumption on the temperatures estimated within the enclosure are discussed later. The maximum HRRs upon system activation are estimated to range between 118 kW for tests A-05 and A-08 up to 692 kW for test A-09.

4.2. Estimated Layer Heights

Figure 10 presents the layer heights estimated by B-RISK for each of the 11 simulations for a sprinkler-controlled/capped HRR. As the tests did not provide a means for observing or estimating the smoke layer heights, no experimental comparisons can be made. However, the simulated layer heights have been used to estimate where the thermocouples are likely to sit within the smoke layer. In all but one simulation (Test A-11), the thermocouple positioned 75 mm from the ceiling is estimated to sit within the upper smoke layer, while the thermocouples positioned 1.6 m from floor sit within the lower clear layer. Thus it appears reasonable, for the purposes of this paper, to assume that the ULT is largely representative of temperatures observed 75 mm from the ceiling.

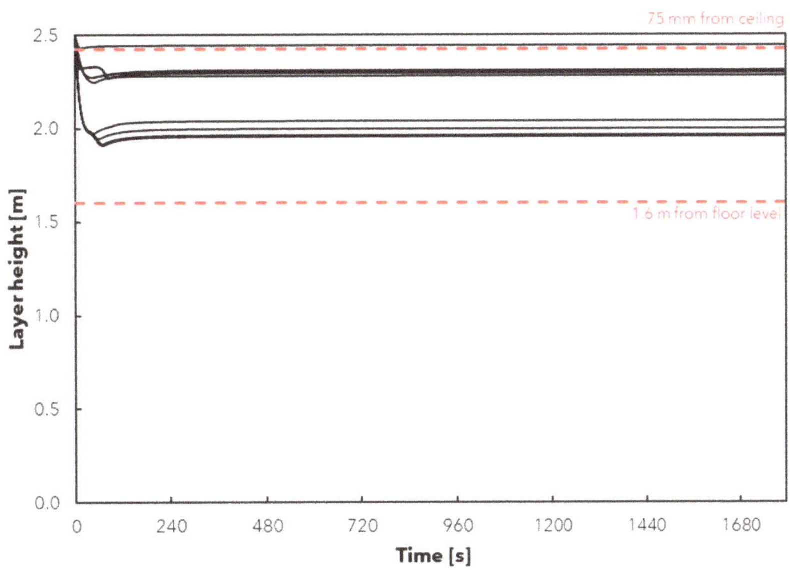

Figure 10. Layer heights estimated in the zone model simulations for a sprinkler-controlled HRR.

4.3. Estimated Temperatures and Suppression Performance Comparisons

Given the observation that the thermocouple position 75 mm from the ceiling sits within the upper smoke layer, the following section focusses on a comparison between the test data for this thermocouple location and the ULT estimated in the zone modelling.

Figure 11 presents this comparison for the three suppression assumptions, namely: (a) the HRR is capped upon system activation; (b) the HRR decays following the Nystedt [7] model; and (c) the HRR decays following the Evans [8] model. Tests A-02 to A-04 have been presented on separate plots (denoted with subscript 2) as these three tests produced higher temperatures and were not observed to result in a 'rapid' decay following system activation compared to the other eight tests (subscript 1).

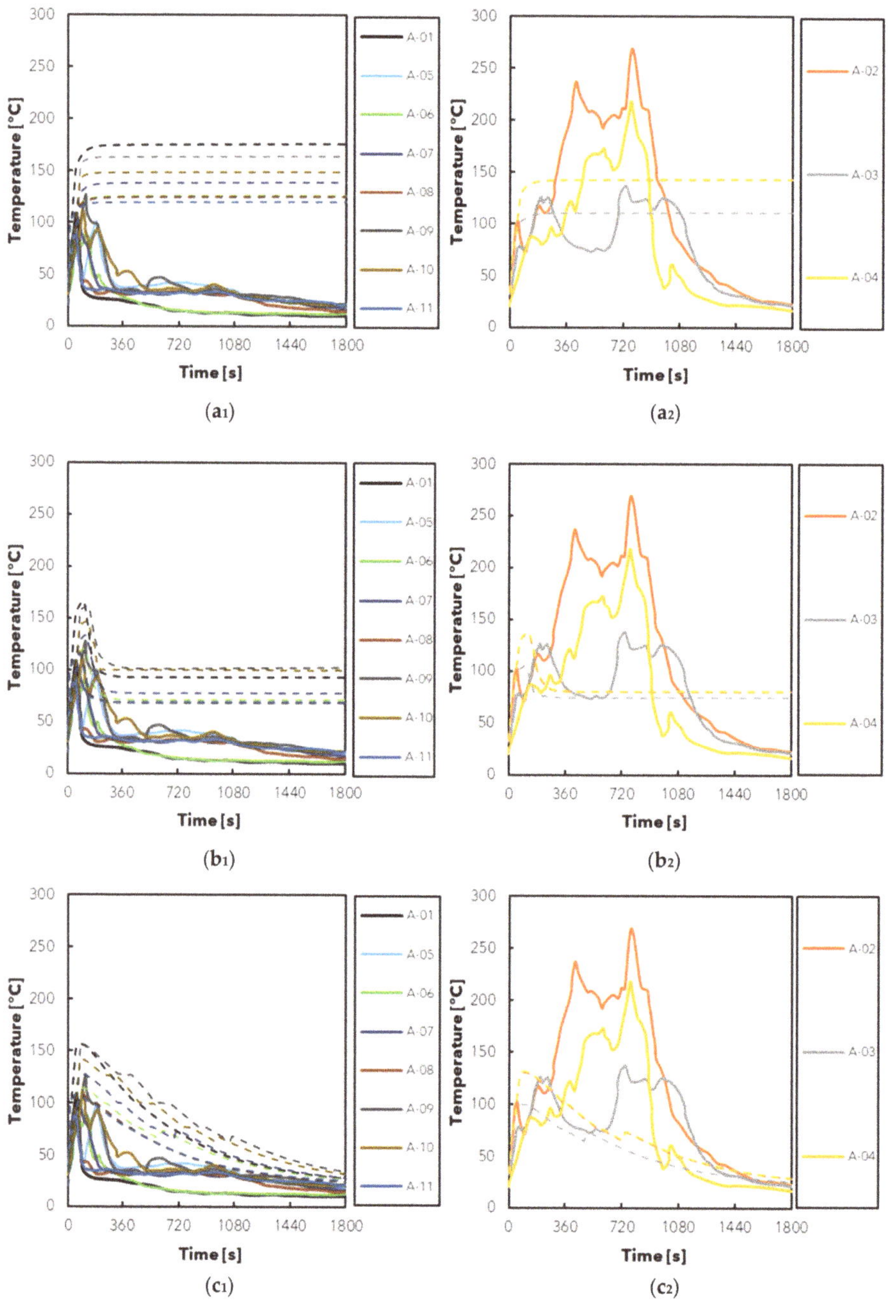

Figure 11. Comparison of thermocouple data (75 mm below ceiling) to ULT (estimated in B-RISK model) for three suppression assumptions: ($\mathbf{a_1}$,$\mathbf{a_2}$) sprinkler-controlled HRR; ($\mathbf{b_1}$,$\mathbf{b_2}$) Nystedt model; and ($\mathbf{c_1}$,$\mathbf{c_2}$) Evans model. Solid lines indicate thermocouple data and dashed lines indicate model estimations. Subscript 1 denotes tests where rapid temperature decay was observed following system activation and subscript 2 where rapid temperature decay was not observed after activation.

When considering tests in the A-01 to A-11 group where rapid temperature decay was observed, it appears that assuming the HRR is capped upon system activation is conservative, with the estimated ULT being much greater than observed in the fire test data. In contrast, the peak temperature in the Nystedt model more closely fits the test data although, following the initial decay, temperatures are still estimated to be higher than the test data. Of the three methods, the Evans exponential decay curve appears to most closely align with the test data.

Observing the three tests (A-02 to A-04) where rapid decay was not observed after system activation, the two decay models (Nystedt and Evans) appear quite favourable in comparison to the test data. In this instance, the closest assumption appears to be a sprinkler-controlled HRR, although the maximum temperature is still under-estimated.

Examining the comparisons collectively, it does not appear unreasonable to apply suppression assumptions, historically estimated from sprinkler performance, to the performance of the electronically controlled water mist system assessed in this paper.

4.4. Limitations

In undertaking the tests and modelling presented in this paper, it is acknowledged that there are a number of limitations. The key limitations are mentioned below, noting that the list is not intended to be exhaustive:

- The BS 8458:2015 fire tests represent a limited range of fire scenarios, albeit with a fire growth rate on the more 'severe' side of potential residential fires (in the range of a fast to ultra-fast growing fire). In comparison, Hopkin et al. [39] estimate that a medium growth rate sits close to the 95th percentile of residential fire incidents. Spearpoint et al. [21] noted that the harmonisation of test standards (such as BS 8458:2015, BS 9252:2011 [40], and BS EN 12559-14:2020 [41]) for 'legacy hazards' can lead to erroneous assumptions that a system is suitable for a broader range of hazards.
- The enclosure dimensions for the tests were 8 m long by 4 m wide by 2.5 m high, with the nozzles spaced at a distance of 4 m apart. The electronically controlled water mist system is currently designed to protect an area within 6 m of each nozzle for a 90° radius [25], and therefore a greater number of nozzles would need to be incorporated to achieve adequate coverage for larger enclosures.
- By representing the water mist nozzles as equivalent to sprinkler heads which are mounted at the ceiling, there are limitations in how these assumptions can then be applied in enclosures with different ceiling heights, e.g., tall and double height spaces. The electronically controlled water mist system incorporates nozzles which are positioned within the enclosure walls at a height of approximately 1.45 m from floor level, and it is designed to discharge water in the direction of the fire rather than in the upper smoke layer. It could be hypothesised that taller ceilings would not significantly slow the system's activation, or alter its performance, when compared to ceiling-mounted sprinkler heads. However, this would need to be verified through further experimentation.
- The fire tests and modelling methods do not consider the reliability of the system, i.e., it is assumed the system activates and operates as intended. A common criticism levied against 'novel' fire safety systems is the lack of knowledge or availability of data for their reliability and performance. However, it can be difficult to identify a system's reliability in a practical sense without their frequent inclusion in buildings, since reasonable quantification of reliability usually requires that a system is subject to a number of 'real' (i.e., non-experimental) incidents to build an adequate dataset of events. In an attempt to address this issue, preliminary work is underway which considers fault tree analyses and reliability targets for adequate performance of the system in specific applications, such as for open plan apartments and loft conversions.

5. Conclusions

Fire safety guidance documents are provided throughout the world to support performance-based calculations of suppression systems, such as PD 7974-1:2019 [5] in the UK. These documents make reference to different assumptions and practices which can be adopted and applied to the HRR of a fire following the activation of a suppression system. However, many of these recommended practices, such as the sprinkler-controlled fire (with a capped HRR), the Nystedt [7] model, or the Evans [8] model, are derived from experimental observations for sprinkler performance. Whether their application can be considered appropriate for other types of suppression system, such as water mist systems, remains largely unknown.

To explore the topic of suppression performance of water mist, this paper details a series of BS 8458:2015 [26] fire tests undertaken for an electronically controlled water mist system. Thermocouple test data for the fire tests has been compared to a series of B-RISK zone model simulations, where the activation performance of the water mist system has been represented in the models as an 'equivalent' sprinkler system, applying the recommended assumptions of Spearpoint et al. [21]. Three different recommended practices for representing suppression of the HRR following system activation have been applied, to observe whether their application can be reasonably extended to the electronically controlled water mist system.

When comparing thermocouple test data to the estimated layer temperatures of the zone modelling output, it has been found that suppression assumptions traditionally applied for sprinklers appear to remain appropriate for the electronically controlled water mist system detailed in this paper. In most cases, the Evans exponential decay model (adopting an equivalent sprinkler spray density of 0.07 mm/s rather than the local discharge density of the system) provides the closest agreement between the models and test data, while the assumption of a controlled fire is shown to be largely conservative. However, for three of the more 'severe' fire tests, the assumption of a decaying fire comes across as favourable and a controlled fire is likely more reasonable.

The fire tests considered in this paper represent a limited range of rapid-growth fire scenarios in specific enclosure arrangements. It would therefore be beneficial to undertake further experiments within different enclosure dimensions, which also involve the measurement of the fuel load, e.g., by use of a load cell, to verify that the suppression assumptions remain appropriate for other potential residential arrangements, fuel types, and different fire growth rates.

Finally, it would be useful to investigate whether B-RISK could be modified to include a water mist suppression model such as that proposed by Wighus and Brandt [20] and then compare the experimental results with predictions for the HRR and gas layer temperature.

Author Contributions: Conceptualisation, C.H. and M.S.; Formal analysis, C.H. and M.S.; Investigation, C.H., M.S., Y.M. and W.M.; Methodology, C.H. and M.S.; Visualisation, C.H.; Writing—Original draft, C.H.; Writing—review and editing, M.S., Y.M. and W.M.; Resources, Y.M. and W.M.; Project administration, Y.M. and W.M.; Funding acquisition, Y.M. and W.M. All authors have read and agreed to the published version of the manuscript.

Funding: The experimental work described in this paper was part funded by a Smart Grant through Innovate UK Smart Award, Project 66965.

Institutional Review Board Statement: Not applicable.

Informed Consent Statement: Not applicable.

Data Availability Statement: Data sharing not applicable.

Acknowledgments: The authors would like to acknowledge WarringtonFire for carrying out the BS 8458:2015 tests presented in this paper.

Conflicts of Interest: Authors Y.M. and W.M. are co-founders of Plumis Ltd. Plumis Ltd. are involved in the product development and sale of electronically controlled water mist systems. Ashton Fire Ltd.

and OFR Consultants Ltd. are the employers of authors C.H. and M.S., respectively. Ashton Fire Ltd. and OFR Consultants Ltd. have received partial funding from Plumis Ltd. to carry out research in relation to the performance of electronically controlled water mist systems.

Nomenclature

Symbols	
$\dot{Q}$	Heat release rate, kW
$\dot{Q}_{(t_{act})}$	Heat release rate at the time of system activation, s
$\dot{Q}_{(t-t_{act})}$	Decaying heat release rate following system activation, s
t	Time, s
t_{act}	Time of system activation, s
$\dot{w}''$	Water spray density, mm/s
α	Fire growth rate, kW/s^2
Abbreviations, acronyms, and initialisms	
BBRAD	Boverket's building regulations general recommendations on the analytical design of a building's fire protection
BRANZ	Building Research Association of New Zealand
BS	British Standard
C factor	Conductivity factor
CFD	Computational fluid dynamics
CO	Carbon monoxide
C/VM2	Verification method for New Zealand building code clauses C1–C6
FDS	Fire Dynamics Simulator
HRR	Heat release rate
IR	Infrared
LLT	Lower layer temperature
NFPA	National Fire Protection Association
PD	Published Document
RTI	Response time index
TC	Thermocouple
UL	Underwriters Laboratories
ULT	Upper layer temperature
VDI	Verein Deutscher Ingenieure

References

1. Nolan, D.P. Methods of fire suppression. In *Handbook of Fire and Explosion Protection Engineering Principles*; Nolan, D.P., Ed.; William Andrew Publishing: Oxford, UK, 2011; pp. 211–242. [CrossRef]
2. Fleming, R. Automatic sprinkler system calculations. In *SFPE Handbook of Fire Protection Engineering*, 5th ed.; Springer: New York, NY, USA, 2016; pp. 1423–1449.
3. Mawhinney, J.; Back, G. Water mist fire suppression systems. In *SFPE Handbook of Fire Protection Engineering*, 5th ed.; Springer: New York, NY, USA, 2016; pp. 1587–1645.
4. Ruff, G.A.; Urban, D.L.; Pedley, M.D.; Johnson, P.T. Fire Safety. In *Safety Design for Space Systems*; Musgrave, G.E., Larsen, A.M., Sgobba, T., Eds.; Butterworth-Heinemann: Burlington, VT, USA, 2009; pp. 829–883. [CrossRef]
5. *PD 7974-1:2019*; Application of Fire Safety Engineering Principles to the Design of Buildings. Initiation and Development of Fire within the Enclosure of Origin (Sub-System 1). BSI: London, UK, 2019.
6. Hopkin, C.; Spearpoint, M.; Bittern, A. Using experimental sprinkler actuation times to assess the performance of Fire Dynamics Simulator. *J. Fire Sci.* **2018**, *36*, 342–361. [CrossRef]
7. Nystedt, F. Verifying Fire Safety Design in Sprinklered Buildings. Department of Fire Safety Engineering and Systems Safety, Lund, 3150. 2011. Available online: https://portal.research.lu.se/ws/files/3912725/1832676.pdf (accessed on 7 February 2022).
8. Evans, D. *Sprinkler Fire Suppression Algorithm for HAZARD*; NISTIR 5254; National Institute of Standards and Technology: Gaithersburg, MD, USA, 1993.
9. Liu, Z.; Kim, A.K. A review of water mist fire suppression systems—Fundamental studies. *J. Fire Prot. Eng.* **1999**, *10*, 32–50. [CrossRef]
10. Ministry of Business, Innovation & Employment. *C/VM2, Verification Method: Framework for Fire Safety Design, for New Zealand Building Code Clauses C1–C6 Protection from Fire*; Amendment 6; New Zealand Government: Wellington, NZ, 2020.
11. *VDI 6019-1*; Engineering Methods for the Dimensioning of Systems for the Removal of Smoke from Buildings: Fire Curves, Verification of Effectiveness. Verein Deutscher Ingenieure: Düsseldorf, Germany, 2006.

12. The Swedish National Board of Housing, Building and Planning's General Recommendations on the Analytical Design of a Building's Fire Protection, BBRAD. Boverket. 2013. Available online: https://www.boverket.se/en/start/publications/publications/2013/the-swedish-national-board-of-housing-building-and-plannings-general-recommendations-on-the-analytical-design-of-a-buildings-fire-protection-bbrad/ (accessed on 19 July 2022).
13. *NFPA 92*; Standard for Smoke Control Systems. National Fire Protection Association: Quincy, MA, USA, 2021.
14. Arvidson, M. An Evaluation of Residential Sprinklers and Water Mist Nozzles in a Residential Area Fire Scenario; Research Institutions of Sweden, Borås, Sweden, 2017; RISE Report 2017:04.
15. Arvidson, M.; Larsson, I. *Residential Sprinkler and High-Pressure Water Mist Systems*; SP Swedish National Testing and Research Institute: Borås, Sweden, 2001; SP Report 2001:16.
16. Chow, W.K. Heat release rate of an open kitchen fire of small residential units in tall buildings. In Proceedings of the 3rd International Performance Buildings Conference, Purdue, IN, USA, 14–17 July 2014.
17. Qin, J.; Yao, B.; Chow, W.K. Experimental study of suppressing cooking oil fire with water mist using a cone calorimeter. *Int. J. Hosp. Manag.* **2004**, *23*, 545–556. [CrossRef]
18. Vaari, J. A study of total flooding water mist fire suppression system performance using a transient one-zone computer model. *Fire Technol.* **2001**, *37*, 327–342. [CrossRef]
19. Li, Y.F.; Chow, W.K. A zone model in simulating water mist suppression on obstructed fire. *Heat Transf. Eng.* **2006**, *27*, 99–115. [CrossRef]
20. Wighus, R.; Brandt, A. WATMIST—A one-zone model for water mist fire suppression. In Proceedings of the Halon Options Technical Working Conference, Albuquerque, NM, USA, 24–26 April 2001.
21. Spearpoint, M.; Hopkin, C.; Muhammad, Y.; Makant, W. Replicating the activation time of electronically controlled watermist system nozzles in B-RISK. *Fire Saf. J.* **2022**, *130*, 103592. [CrossRef]
22. Wade, C.; Baker, G.; Frank, K.; Harrison, R.; Spearpoint, M. *B-RISK 2016 User Guide and Technical Manual*; Building Research Association of New Zealand: Porirua, NZ, 2016; SR364.
23. Plumis Ltd. Plumis—Intelligent Fire Suppression. Available online: https://plumis.co.uk/ (accessed on 16 April 2022).
24. Plumis Ltd. Automist Smartscan Hydra BS8458 Fire Sprinkler Performance. 2020. Available online: https://plumis.co.uk/smartscan (accessed on 2 November 2020).
25. Plumis Ltd. *Automist Smartscan Hydra Design, Installation, Operation and Maintenance (DIOM) Manual, version 2.00.0*; Plumis Ltd.: London, UK, 2020.
26. *BS 8458:2015*; Fixed Fire Protection Systems. Residential and Domestic Watermist Systems. Code of Practice for Design and Installation. BSI: London, UK, 2015.
27. Bill, R.G.; Kung, H.-C.; Anderson, S.K.; Ferron, R. A new test to evaluate the fire performance of residential sprinklers. *Fire Technol.* **2002**, *38*, 101–124. [CrossRef]
28. Wade, C.; Spearpoint, M.; Bittern, A.; Tsai, K. Assessing the sprinkler activation predictive capability of the BRANZFIRE fire model. *Fire Technol.* **2007**, *43*, 175–193. [CrossRef]
29. Hopkin, C.; Spearpoint, M. Evaluation of sprinkler actuation times in FDS and B-RISK. *Int. Fire Prof. J.* **2019**, *28*, 22–27.
30. Hopkin, C.; Spearpoint, M. Numerical simulations of concealed residential sprinkler head activation time in a standard thermal response room test. *Build. Serv. Eng. Res. Technol.* **2020**, *42*, 98–111. [CrossRef]
31. McGrattan, K.; Hostikka, S.; McDermott, R.; Floyd, J.; Vanella, M. *Fire Dynamics Simulator User's Guide*; NIST SP 1019; National Institute of Standards and Technology: Gaithersburg, MD, USA, 2019. [CrossRef]
32. Hopkin, C.; Spearpoint, M.; Hopkin, D.; Wang, Y. Using probabilistic zone model simulations to investigate the deterministic assumptions of UK residential corridor smoke control design. *Fire Technol.* **2022**, *58*, 1711–1736. [CrossRef]
33. Elms, D.G. Consistent Crudeness in System Construction. In *Optimization and Artificial Intelligence in Civil and Structural Engineering: Volume I: Optimization in Civil and Structural Engineering, Topping, B.H.V., Ed.*; Springer: Dordrecht, The Netherlands, 1992; pp. 71–85. [CrossRef]
34. Buchanan, A. The Challenges of Predicting Structural Performance in Fires. *Fire Saf. Sci.* **2008**, *9*, 79–90. [CrossRef]
35. Hostikka, S.; Veikkanen, E.; Hakkarainen, T.; Kajolinna, T. Experimental investigation of human tenability and sprinkler protection in hospital room fires. *Fire Mater.* **2020**, *45*, 823–832. [CrossRef]
36. *UL 1626*; Standard for Safety for Residential Sprinklers for Fire-Protection Service. 4th ed, Underwriters Laboratories Inc.: Northbrook, IL, USA, 2008.
37. Hopkin, D.; Lennon, T.; El-Rimawi, J.; Silberschmidt, V. A numerical study of gypsum plasterboard behaviour under standard and natural fire conditions. *Fire Mater.* **2012**, *36*, 107–126. [CrossRef]
38. *BS EN 1992-1-2:2004+A1:2019*; Eurocode 2. Design of Concrete Structures. General Rules. Structural Fire Design. BSI: London, UK, 2005.
39. Hopkin, C.; Spearpoint, M.; Wang, Y.; Hopkin, D. Design fire characteristics for probabilistic assessments of dwellings in England. *Fire Technol.* **2019**, *56*, 1179–1196. [CrossRef]
40. *BS 9252:2011*; Components for Residential Sprinkler Systems. Specification and Test Methods for Residential Sprinklers. BSI: London, UK, 2011.
41. *BS EN 12259-14:2020*; Fixed Firefighting Systems. Components for Sprinkler and Water Spray Systems. Sprinklers for Residential Applications. BSI: London, UK, 2020.

MDPI
St. Alban-Anlage 66
4052 Basel
Switzerland
www.mdpi.com

Fire Editorial Office
E-mail: fire@mdpi.com
www.mdpi.com/journal/fire

www.ingramcontent.com/pod-product-compliance
Lightning Source LLC
Chambersburg PA
CBHW061846210926
51850CB00006B/981
* 9 7 8 3 7 2 5 8 0 3 1 4 9 *